Theory and Applications of Computability

In cooperation with the association Computability in Europe

Series Editors

Prof. P. Bonizzoni
Università degli Studi di Milano-Bicocca
Dipartimento di Informatica Sistemistica e Comunicazione (DISCo)
Milan
Italy
bonizzoni@disco.unimib.it

Prof. V. Brattka
Universität der Bundeswehr München
Fakultät für Informatik
Neubiberg
Germany
vasco.brattka@unibw.de

Prof. E. Mayordomo
Universidad de Zaragoza
Departamento de Informática e Ingeniería de Sistemas
Zaragoza
Spain
elvira@unizar.es

Prof. P. Panangaden
McGill University
School of Computer Science
Montréal
Canada
prakash@cs.mcgill.ca

Founding Editors: P. Bonizzoni, V. Brattka, S.B. Cooper, E. Mayordomo

More information about this series at http://www.springer.com/series/8819

Books published in this series will be of interest to the research community and graduate students, with a unique focus on issues of computability. The perspective of the series is multidisciplinary, recapturing the spirit of Turing by linking theoretical and real-world concerns from computer science, mathematics, biology, physics, and the philosophy of science.

The series includes research monographs, advanced and graduate texts, and books that offer an original and informative view of computability and computational paradigms.

Series Advisory Board

Robert I. Soare

Turing Computability

Theory and Applications

Robert I. Soare
Department of Mathematics
The University of Chicago
Chicago, Illinois, USA

03Dxx (Computability and recursion theory).

ISSN 2190-619X   ISSN 2190-6203   (electronic)
Theory and Applications of Computability
ISBN 978-3-662-56858-3   ISBN 978-3-642-31933-4   (eBook)
DOI 10.1007/978-3-642-31933-4

Cover illustration: Damir Dzhafarov designed the image of the Turing machine used in the book
cover.

Printed on acid-free paper

This Springer imprint is published by Springer Nature
The registered company is Springer-Verlag GmbH Berlin Heidelberg

I dedicate this book to my wife, Pegeen.

Contents

II   Trees and Π_1^0 Classes   163

III Minimal Degrees   195

IV Games in Computability Theory   209

V   History of Computability   225

17 History of Computability   227

References   251

Preface

The title of this book, *The Art of Turing Computability: Theory and Applications,* emphasizes three very important concepts: (1) *computability* (effective calculability); (2) *Turing* or classical computability in the sense of Turing and Post; and (3) the *art* of computability: as a skill to be practiced, but also emphasizing an esthetic sense of beauty and taste in mathematics.

The Art of Classical Computability

Mathematics is an art as well as a science. We use the word "art" in two senses. First "art" means a *skill* or craft which can be acquired and improved by practice. For example, Donald Knuth wrote *The Art of Computer Programming,* a comprehensive monograph in several volumes on programming algorithms and their analysis. Similarly, the present book is intended to be a comprehensive treatment of the *craft* of computability in the sense of knowledge, skill in solving problems, and presenting the solution in the most comprehensible, elegant form. The sections have been rewritten over and over in response to comments by hundreds of readers about what was clear and what was not, so as to achieve the most elegant and easily understood presentation.

However, in a larger sense this book is intended to develop the art of computability as an *artistic endeavor,* and with appreciation of its mathematical beauty. It is not enough to state a valid theorem with a correct proof. We must see a sense of beauty in how it relates to what came before, what will come after, the definitions, why it is the right theorem, with the right proof, in the right place.

One of the most famous art treasures is Michelangelo's statue of *David* displayed in the Accademia Gallery in Florence. The long aisle to approach the statue is flanked with the statues of Michelangelo's unfinished slaves struggling as if to emerge from the block of marble. There are practically no details, and yet they possess a weight and power beyond their physical proportions. Michelangelo thought of himself, not as carving a statue, but as seeing the figure within the marble and then chipping away the marble to release it. The unfinished slaves are perhaps a more revealing example of this talent than the finished statue of David.

Similarly, it was Alan Turing in 1936 and 1939 who saw the figure of computability in the marble more clearly than anyone else. Finding a formal definition for effectively calculable functions was the first step, but *demonstrating* that it captured computability was as much an artistic achievement as a purely mathematical one. Gödel himself had expressed doubt that it would be possible to do so. The other researchers thought in terms of mathematical formalisms like recursive functions, λ-definable functions, and arithmetization of syntax. It was Turing who saw the computer itself in the marble, a simple intuitive device equipped with only a finite program and using only a finite sequence of strokes at each stage in a finite computation, the vision closest to our modern computer. Even more remarkable, Turing saw how to explicitly *demonstrate* that this mechanical device captured all effectively calculable processes. Gödel immediately recognized this achievement in Turing and in no one else.

The first aim of this book is to present the *craft* of computability, but the second and more important goal is to teach the reader to see the figure inside the block of marble. It is to allow the reader to understand the nature of a computable process, of a set which can be computably enumerated, of the process by which one set B is computed relative to another set A, of a method by which we measure the information content of a set, an algebraic structure, or a model, and how we approximate these concepts at a finite stage in a computable process.

The Great Papers of Computability

During the 1930's, educators suggested that college students should read the great books of Western culture in the original. At the University of Chicago the principal proponents were President Robert Maynard Hutchins and his colleague Professor Mortimer Adler. The curriculum relied on *primary* sources as much as possible and a discussion under the supervision of a professor. For decades the Great Books Program became a hallmark of a University of Chicago education.

In the first two decades of Computability Theory from 1930 to 1950 the primary sources were *papers* not books. Most were reprinted in the book by Martin Davis [1965] *The Undecidable: Basic Papers on Undecidable Propositions, Unsolvable Problems, and Computable Functions.* Of course, all of these papers are important, shaped the subject, and should be read by the serious scholar. However, many of these papers are written in a complicated mathematical style which is difficult for a beginner to comprehend. Nevertheless, at least two of these papers are of fundamental importance and are easily accessible to a beginning student. My criteria for selecting these papers are the following.

1. The paper must have introduced and developed a topic of fundamental importance to computability.

2. The topic and its development must be as important today as then.

3. The paper must be written in a clear, informal style, so appealing that any beginning student will enjoy reading it.

There are two papers in computability which meet these criteria.

Turing's 1936 Paper, Especially §9

Turing's 1936 paper is probably the single most important paper in computability. It introduces the Turing machine, the universal machine, and demonstrates the existence of undecidable problems. It is most often used in mathematics and computer science to define computable functions. It is perhaps Turing's best known and most influential paper.

I am especially recommending Turing *The Extent of the Computable Numbers,* §9, pp. 249–254 in Turing's 1936 paper. Here Turing gives a demonstration that the numbers computable by a Turing machine "include all numbers which would naturally be regarded as computable." This is a brilliant demonstration and is necessary for the argument. Without it we do not know that we have diagonalized against *all* potential decidable procedures and therefore we have no undecidable problems. Books on computability rarely give this demonstration even though it is critical, perhaps because of its nonmathematical nature. Every student of computability should read this very short section.

Post's 1944 Paper, Especially §11

Turing's 1939 paper very briefly introduced the notion of an "oracle machine," a Turing machine which could consult an oracle tape (database), but he did not develop the idea. In his paper *Recursively Enumerable Sets of Positive Integers and Their Decision Problems,* Emil Post in 1944 developed two crucial ideas, the structure and information content of computably enumerable (c.e.) sets, and the idea of a set B being *reducible* to another set A.

Turing never thought of his oracle machine as a device for reducing one set to another. It was simply a local machine interacting with an external database as today a laptop might query the Internet. Post was the first to turn the oracle machines into a reducibility of a set B to a set A, written $B \leq_T A$, which Post generously called *Turing reducibility*. Post's entire paper is wonderfully written and easily accessible to a beginner. He begins with simpler reducibilities such as many-one reducibility and truth-table reducibility and works up to Turing reducibility which was not understood at the time.

The last section §11 *General (Turing) Reducibility*, is especially recommended. Here Post explored informally the idea of a c.e. set B being Turing reducible to another c.e. set A. For the next decade 1944–1954 Post continued to develop the notions of Turing reducibility and information content. In 1948 Post introduced the idea of *degrees of unsolvability*, now called *Turing degrees*, which are the key to measuring the information content of a set or algebraic structure. Post gave his notes to Kleene before his death in 1954. Kleene revised them and published the Kleene-Post 1954 paper, introducing a finite forcing argument as in Chapter 6 to define Turing incomparable sets. These two notions, *computability* by Turing's automatic machine (a-machine) in 1936, and *reducibility* of one set B to another set A in Turing's 1939 paper and Post's 1944 paper, are the two *most important* ideas in computability theory. Therefore, these papers should be read by anyone taking a course from this book.

Other excellent computability papers are reprinted in [Davis 1965], especially the Gödel Incompleteness Theorem in [Gödel 1931] with the improvement by Rosser. Some of these papers may be difficult for a beginner to read, but they will be more accessible after a first course in computability. Gödel's collected works can be found in the three volumes, [Gödel 1986], [Gödel 1990], and [Gödel 1995].

Introduction

Turing Machines

A Turing machine (a-machine) is a kind of idealized typewriter with an infinite tape and a reading head moving back and forth one cell at a time (§1.4) according to a finite state program. In 1936 Turing demonstrated convincingly that this mathematical model captured the informal notion of effectively calculable. Turing's model and analysis have been accepted ever since as the most convincing model. It is the one on which we base the results in this book.

Oracle Machines and Turing Reducibility

Immediately after this paper, Turing went to Princeton where he wrote a PhD dissertation with Alonzo Church. The dissertation was mainly about ordinal logics, a topic suggested by Church, but one page described an *oracle machine (o-machine)* which is of the greatest importance in computability theory. Turing's oracle machine consisted of a Turing machine connected to an "oracle" which it could query during the computation. This is analogous to the modern model of a local server, such as a laptop computer, connected to a large database, such as the Internet which contains too much information to be stored on the local server.

Turing's oracle machine concept lay dormant for five years until Emil Post's extraordinary 1944 paper revived it, greatly expanded it, and cast the subject in an informal, intuitive light. Post defined a set B to be *Turing reducible* to a set A, written $B \leq_T A$, if there is an oracle machine which computes B when the characteristic function of A is written on the oracle tape. The oracle machines include the ordinary machines because if a set B is computed by an ordinary Turing machine, then it is computed by an oracle machine with $A = \emptyset$ on the oracle tape. But the oracle machines do much more. Turing reducibility is a crucial concept because in computablility theory and applications we rarely prove results about computable functions on computable sets. We compare *noncomputable* (undecidable) sets B and A with respect to their relative information content. We say that sets A and B are *Turing equivalent*, written $A \equiv_T B$ if $A \leq_T B$ and $B \leq_T A$, in which case we view A and B as coding the same information. Turing reducibility gives us a precise measure of the information they encode relative to other sets and the *Turing degrees* (§3.4) are equivalence classes containing sets with the same information content.

Computable Enumerable Sets

In 1936 Church and Kleene introduced the concept of a *computably enumerable* (c.e) set, also called a recursively enumerable (r.e.) set, as one which can be effectively listed, such as the theorems in a formal system like Peano arithmetic. In 1944 Post realized the importance of these sets in many areas of mathematics, and Post devoted much attention to studying their information content. His work on the structure of these sets and their information content under stronger reducibilities has had a great influence on the topics in this book.

Of the effective listing of c.e. sets, $\{W_e\}_{e \in \omega}$ Post reminded us that the Gödel diagonal set $K = \{e : e \in W_e\}$ is c.e. and noncomputable. The famous *Post Problem* was to determine whether there is only one such set up to Turing degree.

Bounded Turing Reductions

At first, Post did not make much progress on the general case of Turing reducibility. To progress toward it, he considered various stronger reducibilities called bounded reducibilities. A Turing reduction $\Phi_e^A = B$ witnessing $B \leq_T A$ is a *bounded Turing reduction*, written $B \leq_{bT} A$, if there is a computable function $h(x)$ bounding the use function, namely $\varphi_e^A(x) \leq h(x)$, where the use function $\varphi_e^A(x)$ is the maximum element *used* (scanned) during the computation.

For example, every c.e. set B is many-one reducible to K, $B \leq_m K$ by a computable function f, i.e., $x \in B$ iff $f(x) \in K$. Post introduced several structural properties of a c.e. set B in an attempt to prove incompleteness. For example, he proved that a $K \not\leq_m B$ for a simple set B. The varieties of simple and nonsimple sets he introduced and his various bounded reducibilities had a profound effect on the subject for decades and led indirectly to most of the results in this book.

Finally Understanding Turing Reducibility

Post realized that the bounded reducibilities would not solve his problem. It required a deeper understanding of Turing reducibility. His understanding increased over the next decade from 1944 until his death in 1954. Post introduced the notion of *degree of unsolvability* to collect into one equivalence class sets coding the *same information content*. He wrote notes on his work. As he became terminally ill in 1954, Post gave them to Kleene who expanded them and published it as [Kleene and Post 1954]. This paper was a fundamental advance toward solving Post's problem and toward understanding Turing reducibility. The key idea was the continuity of Turing functionals that if $\Phi_e^A(x) = y$ then $\Phi_e^\sigma(x) = y$ for some finite initial segment $\sigma \prec A$, and that if $B \succ \sigma$ then $\Phi_e^B(x) = y$ also.

Using this, Kleene and Post constructed sets A and B computable in K such that $A \not\leq_T B$ and $B \not\leq_T A$. Hence, $\emptyset <_T A <_T K$. This did not explicitly solve Post's Problem because the sets were not c.e., but Kleene and Post divided the conditions into requirements of the form $\Phi_e^A \neq B$, which could be arranged in a priority list of order type ω and processed one at a time using the Use Principle. This became the model for most arguments in the subject. It became the key step in the later solution of Post's Problem by Friedberg in 1957 and Muchnik in 1956 since they combined this strategy with a computable approximation stage by stage. From these ideas emerged the understanding that a Turing functional Φ_e is continuous as a map on Cantor space 2^ω and is not only continuous but *effectively continuous* because the inverse image of a basic open set is the *computable* union of basic open sets.

Priority Arguments

The Kleene-Post construction had produced finite initial segments $\sigma \prec A$ and $\tau \prec B$ such that for some x, $\Phi_e^\sigma(x) \neq \tau(x)$. Hence $\Phi_e^A(x) \neq B(x)$. To make the sets A and B *computably enumerable*, Muchnik and Friedberg had to abandon the K-oracle and computably enumerate the sets, letting A_s be the finite set of elements enumerated in A by the end of stage s and likewise for B_s. They attempted to preserve strings $\sigma \prec A_s$ and $\tau \prec B_s$ when it seemed to give $\Phi_e^\sigma(x) \neq \tau(x)$. This action might later be *injured* because action by a higher priority requirement forces $\sigma \not\prec A_{s+1}$ causing this condition for e to begin all over again. These results led to much more complicated infinite injury arguments.

Other Parts of This Book

The introduction so far explains the background and motivation for most of Part I up to finite injury priority arguments in Chapter 7. For the summary

and motivation of the other parts see the next section about how to read
this book.

How to Read This Book

Part I: Foundations of Computability

The core of the subject is Part I, Chapters 1–7, from the definition of Turing machines in Chapter 1 up to finite injury priority arguments in Chapter 7. Traditionally, a beginning undergraduate or graduate course of ten or fifteen weeks would go through the sections here one by one. Part I has been streamlined, with more complicated chapters moved to later parts in order to make this schedule feasible. After finishing Chapter 7 on finite injury, the reader will have a firm grasp of the fundamental results and methods of computability theory. One can also cover Part I more quickly as an initial segment of an advanced computability course by concentrating on the starred sections in Part I and then moving to other advanced topics.

Part II: Trees and Π_1^0 Classes

A *tree* is a set of strings closed under initial segments and a Π_1^0 *class* is the set of paths through a computable binary tree. These classes play an important role in model theory, extensions of Peano arithmetic, algorithmic randomness, and other applications. We study open and closed computable classes of reals, and basis and nonbasis theorems for Π_1^0 classes. We give a proof of the Superlow Basis Theorem, proved but not published, by Jockusch and Soare about 1969. We also give a proof by Dzhafarov and Soare of the Low Antibasis Theorem by Kent and Lewis. We show how Π_1^0

classes and their basis theorems are related to models of Peano arithmetic, with results by Jockusch and Soare, Scott, Shoenfield, and Solovay. Finally, we relate Π_1^0 classes to Martin-Löf randomness, computably dominated (hyperimmune-free) degrees, and to computably traceable sets.

Part III: Minimal Degrees

A Turing degree $\mathbf{a}$ is *minimal* if $\mathbf{a} > \mathbf{0}$ but there is no degree $\mathbf{b}$ such that $\mathbf{0} < \mathbf{b} < \mathbf{a}$. In Chapter 12 we present Spector's proof of a minimal degree $\mathbf{a} < \mathbf{0}''$. The proof uses a forcing argument like those in Chapter 6 but with more complicated forcing conditions of perfect trees instead of finite strings. In Chapter 13 we present the Sacks construction of a minimal degree $\mathbf{a} < \mathbf{0}'$. This is an approximation to Spector's method and uses a finite injury priority argument. We also sketch a limit computable (full approximation) construction of a minimal degree below $\mathbf{0}'$ using a computable construction. It can also be done below any nonzero computably enumerable degree, thereby producing a low minimal degree. Chapters 6 and 7 are the only prerequisites for this material.

Part IV: Games in Computability Theory

Games are very important in understanding the nature of computability, how to prove theorems, and how to solve problems. In Chapter 14 we present the classical Banach-Mazur games which are closely related to the finite extension constructions of Chapter 6, Sections 1–3, and may be read simultaneously with them and with Chapter 8 on open and closed classes in Cantor space. Players I and II alternately construct strings σ_{2n} and σ_{2n+1}, jointly constructing a point $f = \cup_n \sigma_n$ in Cantor space 2^ω. There is a predetermined class $\mathcal{A} \subseteq 2^\omega$. Player I wins the game according to whether $f \in \mathcal{A}$ or not. Winning strategies are described in terms of properties of $\mathcal{A}$. We also discuss the Cantor-Bendixson rank of points in a closed subclass $\mathcal{A} \subseteq 2^\omega$.

In Chapter 15 we make a very brief excursion into Gale-Stewart games. We consider the complexity of the winning strategy for a very simple computable game.

We finish in Chapter 16 by returning to the topic of more Lachlan games, first introduced in §2.5. These games are the *principal tool* in proving theorems and solving problems in computability theory.

Symbols Marking Importance and Difficulty

In Part I we use the following notation for sections, theorems, and exercises.

⋆⋆   Most important.

⋆   Very important.

No Marking   Average importance.

⊘   Skim or defer on a first reading until needed in a later chapter.

◇   Difficult exercise, do not assign lightly.

◇◇   Very difficult exercise.

Notation

Notation will be defined when introduced. We now summarize the most common notation and definitions.

Notation for Sets

The universe is the set of nonnegative integers $\omega = \{0, 1, 2, 3, \dots\}$ which sometimes appears in the literature as $\mathbb{N}$. Most of the objects we study can be associated with some $n \in \omega$ called a "code number" or "Gödel number." We can think of operations on these objects as being presented by a corresponding function on these numbers and our functions will have domain and range contained in ω.

Uppercase Latin letters A, B, C, D and X, Y, Z normally represent subsets of $\omega = \{0, 1, 2, 3, \dots\}$ with the usual set operations $A \cup B$, $A \cap B$; $|A|$, or $\mathrm{card}(A)$ denotes the cardinality of A; $\max(A)$ denotes the maximum element $x \in A$ if A is finite; $A \subseteq B$ denotes that A is a subset of B, and $A \subset B$ that it is a *proper* subset; $A - B$ denotes the set of elements in A but not in B; $\overline{A} = \omega - A$, the complement of A; $A \sqcup B$ denotes the *disjoint union*, i.e., $A \cup B$ provided that $A \cap B = \emptyset$; the *symmetric difference* is $A \bigtriangleup B = (A - B) \cup (B - A)$; a, b, c, $\dots x$, y, z, $\dots$ represent integers in ω; $A \times B$ is the Cartesian product of A and B, the set of ordered pairs (x, y) such that $x \in A$ and $y \in B$; $\langle x, y \rangle$ is the integer that is the image of the pair (x, y) under the standard pairing function from $\omega \times \omega$ onto ω; $A \subseteq^* B$ denotes that $|A - B| < \infty$; $A =^* B$ denotes that $A \bigtriangleup B$ is finite; $A \subset_\infty B$

denotes that $|B - A| = \infty$. Given a simultaneous enumeration (see p. 46) of A and B let $A \setminus B$ denote the set of elements enumerated in A before B and $A \searrow B = (A \setminus B) \cap B$, the set of elements appearing in A and later in B.

Logical Notation

We form predicates with the usual notation of logic where $\&, \vee, \neg, \Longrightarrow, \exists$, $\forall$ denote respectively, and, or, not, implies, there exists, for all; $(\mu x)\, R(x)$ denotes the least x such that $R(x)$ if it exists, and is undefined otherwise; $(\exists^{\infty} x)$ denotes "there exist infinitely many x," and $(\forall^{\infty} x)$ denotes "for almost all x" as in Definition 3.5.1. These quantifiers are dual to each other. The latter is written $(\exists x_0)(\forall x \geq x_0)$. We use $x, y, z < w$ to abbreviate $x < w$, $y < w$, and $z < w$. In a partially ordered set we let $x \mid y$ denote that x and y are incomparable, i.e., $x \not\leq y$ and $y \not\leq x$. We often use the dot convention to abbreviate brackets before and after the principal connective of a logical expression. For example, if α and β are well-formed formulas, then α . $\Longrightarrow$. β abbreviates $[\,\alpha\,] \Longrightarrow [\,\beta\,]$. The algorithm is to insert a right bracket just before $\Longrightarrow$ and then a matching left bracket just before the first symbol in α. Do the corresponding algorithm for β. The dots increase readability of a long expression. TFAE abbreviates "The following are equivalent."

We use the usual Church *lambda notation* for defining partial functions. Suppose $[\ldots x \ldots]$ is an expression such that for any integer x the expression has at most one corresponding value y. Then $\lambda x\,[\ldots x \ldots]$ denotes the associated partial function $\theta(x) = y$, for example $\lambda x\,[x^2]$. The expression $\lambda x\,[\uparrow]$ denotes the partial function which is undefined for all arguments. We also use the lambda notation for partial functions of k variables, writing $\lambda x_1 x_2 \ldots x_k$ in place of λx. An expression such as $\lambda x\, y\,[x + y]$, denotes addition as a function of x and y. However, $\lambda x\,[x + y]$ indicates that the expression is viewed as a function of x with y as a parameter, such as $\lambda x\,[x + 2]$. One advantage is that with an expression of several arguments, such as in the *s-n-m* Theorem 1.5.5 (Parameter Theorem) we can make clear which arguments are variables and which are parameters, for example as explained in Remark 1.5.6. Define $f(x) = 1 \doteq x$ to be 1 if $x = 0$ and 0 if $x \geq 1$. We call this the *monus* function. It produces a 0-1 valued function $f(x) \neq x$.

Lattices and Boolean Algebras

A *lattice* $\mathcal{L} = (L; \leq, \vee, \wedge)$ is a partially ordered set (poset) in which any two elements a and b have a least upper bound (lub) $a \vee b$ and greatest lower bound (glb) $a \wedge b$. An *upper semi-lattice* has lub only. For example, the Turing degrees under Turing reducibility form an upper semi-lattice. If $\mathcal{L}$ contains a least element and greatest element these are called the *zero* element 0 and *unit* element 1, respectively. In such a lattice a is the *complement* of b if $a \vee b = 1$ and $a \wedge b = 0$, and $\mathcal{L}$ forms a *Boolean algebra* if every element has a complement. A nonempty subset $\mathcal{I} \subseteq \mathcal{L}$ forms an *ideal* $\mathcal{I} = (I; \leq, \wedge, \vee)$ of $\mathcal{L}$ if $\mathcal{I}$ satisfies the conditions:

(1) $[a \in L \ \& \ a \leq b \in I] \implies a \in I$, and

(2) $[a \in I \ \& \ b \in I] \implies a \vee b \in I$.

A *filter* $F \subset L$ satisfies the dual conditions. For example, the subsets of ω form a Boolean algebra with the finite sets as an ideal and the cofinite sets as a filter.

Notation for Strings and Functionals

We let $2^{<\omega}$ denote the set of all finite sequences of 0's and 1's called strings and denoted by σ, ρ, and τ. Let 2^ω denote the set of all functions f from ω to $2 = \{0,1\}$, and ω^ω the set of all functions f from ω to ω. The integers $n \in \omega$ are *type 0* objects, (partial) functions $f \in 2^\omega$ or subsets $A \subseteq \omega$ (which are identified with their characteristic function $\chi_A \in 2^\omega$) are *type 1* objects, a (partial) *functional* Ψ is a map from type 1 objects to type 1 objects, i.e., a map from 2^ω to 2^ω and is called a *type 2* object. Identifying a set A with its *characteristic function* χ_A we often write $A(x)$ for $\chi_A(x)$. Uppercase Latin letters A, B, C, $\ldots$, represent subsets of ω. Script letters $\mathcal{A}$, $\mathcal{B}$, $\mathcal{C}$ represent subsets of 2^ω and are called *classes* to distinguish them from sets.

Gödel Numbering of Finite Objects

In his Incompleteness Theorem [1931] Gödel introduced the method of assigning a *code number* or *Gödel number* to every formal (syntactical) object such as a formula, proof, and so on. We now present two ways to effectively code a sequence of n-tuples of integers $\{a_1, a_2, \ldots a_n\}$, define

(1) $$a = p_1^{a_1+1} p_2^{a_2+1} \ \ldots \ p_k^{a_k+1}$$

where p_i is the ith prime number. Given a we can effectively recover the prime power $(a)_i = a_i + 1$. This coding is injective but not surjective on ω.

The second method uses the following standard pairing function and has the added advantage that the n-tuple coding below is an injective and surjective map from ω onto ω^n.

Standard Pairing Function

(i) Let $\langle x, y \rangle$ denote the integer that is the image of the ordered pair (x, y) under the standard pairing function $\frac{1}{2}(x^2 + 2xy + y^2 + 3x + y)$ which is a 1:1 computable function from $\omega \times \omega$ onto ω. Let π_1 and π_2 denote the *inverse pairing functions* $\pi_1(\langle x, y \rangle) = x$, and $\pi_2(\langle x, y \rangle) = y$.

(ii) Let $\langle x_1, x_2, x_3 \rangle$ denote $\langle \langle x_1, x_2 \rangle, x_3 \rangle$. Let the *n-ary pairing function* be

$$(2) \qquad \langle x_1, x_2, \ldots, x_n \rangle \;=\; \langle \cdots \langle \langle x_1, x_2 \rangle, x_3 \rangle, \ldots, x_n \rangle.$$

(All these functions are clearly computable and even primitive recursive.)

   If the sequences are all of fixed length n we may use method 2, the n-ary pairing function of (2),

$$(3) \qquad f(a_1, a_2, \ldots a_n) \;=\; \langle a_1, a_2, \ldots a_n \rangle$$

Otherwise, we use the first method above of coding using prime powers. There are many other coding algorithms. The important point for coding is that the method be effective and invertible, but it is often useful to have it surjective as well.

   Note that both methods are effectively invertible. Let θ be any 1:1 computable partial function. Then θ is effectively invertible on its range. Just enumerate the pairs (u, v) with $\theta(u) = v$ until, if ever, a pair (x, y) is found and then define $\psi(y) = x$. See also the Definition 2.1.7 of graph(θ) and the Uniformization Theorem 2.1.8.

Effective Numbering of Finite Sets and Strings

Given a finite set $F = \{x_1, x_2, \ldots x_k\}$ where $x_1 < x_2 \ldots x_k$ we give F the (strong) index $y = 2^{x_1} + 2^{x_2} \ldots + 2^{x_k}$ and write that $D_y = F$. Let $D_0 = \emptyset$. Likewise, give every string $\sigma \in 2^{<\omega}$ an effective index from using either the strong index coding or Gödel numbering so that from the index we can recover the length $k = |\sigma|$ and every component $\sigma(i)$, for $i < k$. Such a numbering of strings σ_z is given in Definition 2.3.6.

Partial Computable (P.C.) Functions

Let $\{P_e\}_{e \in \omega}$ be an effective numbering of all Turing machine programs (as in Definition 1.5.1). We write $\varphi_e(x) = y$ if program P_e with input x halts and yields output y, in which case we say that $\varphi_e(x)$ *converges* (written $\varphi_e(x) \downarrow$), and otherwise $\varphi_e(x)$ *diverges* (written $\varphi_e(x) \uparrow$); $\{\varphi_e\}_{e \in \omega}$ is an effective listing of all *partial computable (p.c.)* functions; the domain and range of φ_e are denoted by $\operatorname{dom}(\varphi_e)$ and $\operatorname{rng}(\varphi_e)$. A set A is *computably enumerable (c.e.)* if it can be effectively listed, i.e., if $A = \operatorname{dom}(\varphi_e)$ for some e.

If $\text{dom}(\varphi_e) = \omega$ then φ_e is a *total computable function* (abbreviated *computable function*); we let f, g, h, ... denote total functions; $f \circ g$ or fg denotes the composition of functions, applying first g to an argument x and then applying f to $g(x)$. Let $f \upharpoonright x$ denote the restriction to elements $y < x$ and $f \upharpoonright\!\!\upharpoonright x$ the restriction to elements $y \leq x$.

Turing Functionals Φ_e^A

Let $\{\widetilde{P}_e\}_{e \in \omega}$ be an effective numbering of all Turing machine oracle programs, finite sets of sextuples defined in §3.2.1. Write $\Phi_e^A(x) = y$ if oracle program $\widetilde{P}_e$ with A on its oracle tape and input x halts and yields output y. Let the use function $\varphi_e^A(x)$ be the greatest element z for which the computation scanned the square $A(z)$ on the oracle tape. We regard Φ_e as a (partial) functional (type 2 object) from 2^ω to 2^ω mapping A to B if $\Phi_e^A = B$.

The use function $\varphi_e^A(x)$ has an exponent A to distinguish from the p.c. function $\varphi(x)$. They usually come in matched pairs, $\Psi^A(x)$ and $\psi^A(x)$, $\Theta^A(x)$ and $\theta^A(x)$, where the lowercase function denotes the use function corresponding to the uppercase functional. See Definition 3.2.2 (vi) for a function f as oracle in place of the set A.

Lachlan Notation

When $E(A_s, x_s, y_s, \ldots)$ is an expression with a number of arguments subscripted by s denoting their value at stage s, Lachlan introduced the notation $E(A, x, y, \ldots)[s]$ to denote the evaluation of E where all arguments are taken with their values at the end of stage s.

(4) $\Phi_e^A(x)[s]$ denotes $\Phi_{e,s}^{A_s}(x_s)$ and $\varphi_e^A(x)[s]$ denotes $\varphi_{e,s}^{A_s}(x_s)$.

This Lachlan notation has become very popular and is now used in most papers and books.

Acknowledgements

Among others I would like to thank colleagues and students for careful reading of preliminary versions of this book and for their suggestions and corrections.

This includes my University of Chicago colleagues, Denis Hirschfeldt, Joseph Mileti, Antonio Montalban, and Maryanthe Malliaris; and former students, Eric Astor, William Chan, Barbara Csima, Chris Conidis, David Diamondstone, Damir Dzhafarov, Rachel Epstein, Kenneth Harris, Karen Lange, Russell Miller, Jonathan Stephenson, and Matthew Wright.

It includes colleagues at other universities, Ted Slaman, Richard Shore, Carl Jockusch, Douglas Cenzer, Leo Harrington, Manuel Lerman; and colleagues at the University of Wisconsin, Steffen Lempp, Bart Kastermans, Arnie Miller, Joe Miller, and Wisconsin students, Asher Kach, Dan Turetsky, and Nathan Collins.

I am grateful to the students of Barbara Csima at the University of Waterloo, Vladimir Soukharev, Jui-Yi Kao, Atul Sivaswamy, David Belanger, and Carolyn Knoll; to Piet Rodenburg and Tom Sterkenbert in Amsterdam and their students, including Frank Nebel; to Notre Dame colleagues Julia Knight and Peter Cholak and their students, Joshua Cole, Yang Lu, Stephen Flood, Quinn Culver and John Pardo; to Valentina Harizanov at George Washington University and her students, Jennifer Chubb and Sarah Pingrey; to Aaron Sterling at Iowa State University; to Russell Miller's student at Queens College CUNY, Rebecca Steiner; to Iraj Kalantari at Northern Illinois University and his student, Abolfazi Karimi; to Rachel Epstein and her students at Harvard who covered Part I of the book in detail; to Damir Dzhafarov who drew the diagrams in tikz; to Linda Westrick at the University of Connecticut. Carl Jockusch and Damir Dzhafarov read some of the advanced chapters in detail and made a number of mathematical corrections and suggestions. Damir also designed the excellent cover diagram of a Turing machine.

I am grateful to Ronan Nugent, Senior Editor at Springer-Verlag, who read the manuscript in detail, made a number of corrections, and handled the editing and production of this book.

Part I

Foundations of Computability

1
Defining Computability

1.1 Algorithmically Computable Functions

In this chapter we define the notion of a computable function using Turing machines in §1.4. Then we develop its most important properties such as the Enumeration Theorem 1.5.3 and the s-m-n Theorem 1.5.5 (Parameter Theorem), which we shall use often. The historical development of other definitions of computable functions is discussed in the history chapter, Chapter 17.

1.1.1 Algorithms in Mathematics

Mathematicians have studied calculation and algorithms since earlier than the time of the Babylonians. Specifying algorithms goes back at least to Euclid's *Elements* (written about 330–320 B.C.), and his famous greatest common divisor algorithm. Another famous algorithm is the sieve of Eratosthenes for calculating whether an integer k is prime by crossing out all proper multiples of numbers less than k, beginning with 2, then 3, and so on, to determine whether k has been crossed out, in which case it is not prime.

These procedures have the properties we recognize in an algorithmic procedure. The procedure is specified by a finite set of instructions; the calculation proceeds in a finite sequence of steps, eventually halting with the answer; it proceeds in a deterministic, completely specified fashion without any recourse to human intelligence or random devices.

Prior to 1930 it was generally not considered necessary to define the informally computable functions, or *effectively calculable* functions, as they were called in the 1930s. Rather, Hilbert had called for the discovery of algorithmic solutions to specific problems, such as Hilbert's tenth problem on the solution of Diophantine equations, or the *Entscheidungsproblem (decision problem)*, the problem of finding an algorithm to decide which formulas of first-order logic are valid.

However, Gödel's *first Incompleteness Theorem* [Gödel 1931] (stated in modern terms and with an improvement by Rosser) showed that any effectively axiomatizable formal system T which includes elementary number theory is either *inconsistent* or *incomplete*. This dramatic blow to Hilbert's famous consistency program suggested to some mathematicians that Hilbert's *Entscheidungsproblem* might also have a negative solution. However, to prove that a given problem is algorithmically unsolvable they first had to formally characterize the informal class of effectively calculable functions and then prove that no function in that class constituted a solution to the specific problem. The program to precisely define the informal class of intuitively computable functions began in earnest after Gödel's results in 1931.

1.1.2   The Obstacle of Diagonalization and Partial Functions

The objective was first to formally define the effectively calculable functions and then give an effective list of them all. We begin with functions on ω, and therefore we can identify all finite objects with code numbers (Gödel numbers), as explained in the Notation Section. To characterize the effectively calculable functions we would like to produce a list $\mathcal{C} = \{f_n\}_{n \in \omega}$ of functions with three properties: (1) the list $\{f_n\}_{n \in \omega}$ includes all and only algorithmically computable functions; (2) there is a uniformly effective listing of them, namely an algorithmically computable function $g(n, x) = f_n(x)$; and (3) every f_n is a *total* function, i.e., is defined on *all* $x \in \omega$. However, these three conditions are contradictory because we can define the *diagonal function* $h(x) = g(x, x) + 1$, which is algorithmically computable because g is, but $h \notin \mathcal{C}$ because $h(x) \neq f_x(x)$.

Hence, we must abandon one of the three conditions, but we do not abandon either (1) or (2), both of which are crucial to the entire theory. Therefore, we must give up (3). If we are dealing with *partial* computable functions, then (3) is no longer an obstacle because in the definition of h we may no longer have that $g(x, x)$ is defined and hence cannot argue that $h(x) \neq g(x, x)$. It is more natural to consider *partial* computable functions anyway, because we shall effectively list all algorithms in some formalism, and certain algorithms may be naturally defined only on *some* but not all arguments. All the formal classes we consider produce algorithmic functions which are *partial* and are called *partial computable (p.c.)* functions. Those partial functions in the class which are total are called *total computable functions* or simply *computable functions*.

1.1.3 The Quest for a Characterization

To produce an undecidable problem one had to first formally define the class of effectively calculable functions. Kurt Gödel, realizing that the primitive recursive functions he had used in his 1931 Incompleteness Theorem were not sufficient to capture all effectively calculable functions, proposed in lectures at Princeton University in 1934 the more general class, Herbrand-Gödel general recursive functions, now called simply *recursive functions*. In 1936 Church proposed *Church's Thesis*, that the effectively calculable functions be identified with the recursive functions. By the beginning of 1936 Gödel was not convinced that his own recursive functions captured the effectively calculable functions. However, when Gödel saw the following analysis by Turing later in 1936, he was immediately convinced.

1.1.4 Turing's Breakthrough

Upon this scene came a twenty-three-year-old graduate student at Cambridge University, who was unaware of the work in recursive functions by Gödel, Kleene, and Church at Princeton University. In 1935 Turing had heard the lectures of Cambridge don M.H.A. Newman on Gödel's paper and on the Hilbert *Entscheidungsproblem*. Turing worked on the problem for the remainder of 1935 and submitted his solution to the incredulous Newman on April 15, 1936. Turing's monumental paper [Turing 1936] was distinguished because: (1) Turing analyzed an idealized *human* computing agent which brought together the intuitive concept of a "function produced by a mechanical procedure" which had been evolving for more than two millennia from Euclid to Leibniz to Babbage and Hilbert; (2) Turing specified a remarkably simple formal device (*Turing machine*) and informally but convincingly demonstrated the equivalence of (1) and (2), namely he demonstrated *Turing's Thesis (TT)*; (3) Turing proved the unsolvability of Hilbert's *Entscheidungsproblem,* which established mathematicians had been studying intently for some time; (4) Turing proposed a universal Turing machine, one which carried within it the capacity to duplicate any other, an idea of great theoretical importance which was also later to have considerable impact on the development of high-speed digital computers. Gödel enthusiastically accepted Turing's Thesis and his analysis and thereafter gave Turing credit for the definition of mechanical computability.

1.2   ** Turing Defines Effectively Calculable

We can divide Turing's 1936 paper into two parts: describing the informal notion of effectively calculable; and giving the precise mathematical model of a Turing machine which captures it. By 1936–1937 it had been shown that various formal definitions of computable functions were mathematically equivalent, including Turing computable functions, recursive functions, and others. Nevertheless, we have to argue that at least one of the formal definitions captures the informal notion of a function being effectively calculable by a human being. Turing did this more convincingly than anyone else in [Turing 1936] Sections 1 and 9.

Turing Sections 2–8 presented in [Turing 1936] first give the definition of a Turing machine, and then a universal machine. Only at the end, in Section 9, did he give the following informal argument which captures the notion of effectively calculable functions. From this follows almost at once the formal definition of a Turing machine in the next section. We present his results in the reverse order. The essence of Turing's definition in Section 9 is the following.

- Turing proposed a number of simple operations "so elementary that it is not easy to imagine them further subdivided."

- He divided the work space into squares, which he assumed to be one-dimensional.

- He assumed finitely many symbols. Each square contains one symbol.

- He assumed finitely many states (of mind).

- The action of the machine is determined by the present state and the squares observed.

- The squares are bounded by B, say $B = 1$.

- The reading head examines one symbol in one square.

- We may assume the machine moves to only squares within a radius C of the current square. We may assume $C = 1$.

- The machine may print a symbol in the current square, change state, and move to an adjacent square.

[Sieg 2006] gave an axiomatization of a more general form of these properties and showed that any procedure satisfying these axioms produces a Turing machine computable function.

1.3 ** Turing's Thesis, Turing's Theorem, TT

Turing believed that he had established in [Turing 1936] that a function is effectively calculable by a human being iff it is computable by a Turing machine. Turing wrote in [Turing 1954] that "its status is something between a theorem and a definition." Neither Turing nor Gödel ever described this as a "thesis," as Kleene named it in his book [Kleene 1952]. So that readers can recognize Kleene's terminology, we shall call it "Turing's Thesis," but also and more accurately "Turing's Theorem," letting the symbol TT label the statement with no ambiguity in content, only in name. (See the history chapter, Chapter 17.) The result is stated only for numbers but clearly carries over to finite structures by Gödel numbering.

Theorem 1.3.1 (Turing's Thesis, Turing's Theorem (TT)). *A function is effectively calculable by a human being iff it can be computed by a Turing machine.*

1.4 ** Turing Machines

Definition 1.4.1. (Turing) A *Turing machine (automatic machine, a-machine) M* includes a two-way infinite *tape* divided into *cells*, a *reading head* which scans one cell of the tape at a time, and a finite set of internal *states* $Q = \{q_0, q_1, \ldots, q_n\}$, $n \geq 1$. Each cell is either blank (B) or has written on it the symbol 1. In a single step the machine may simultaneously: (1) change from one state to another; (2) change the scanned symbol s to another symbol $s' \in S = \{1, B\}$; and (3) move the reading head one cell to the right (R) or left (L). The operation of M is controlled by a partial map $\delta : Q \times S \to Q \times S \times \{R, L\}$ (which may be undefined for some arguments).

The interpretation is that if $(q, s, q', s', X) \in \delta$ then the machine M in state q, scanning symbol s, changes to state q', replaces s by s', and moves to scan one square to the right if $X = R$ (or left if $X = L$). The map δ viewed as a finite set of quintuples is called a *Turing program*. The *input* integer x is represented by a string of $x + 1$ consecutive 1's (with all other cells blank). The Turing machine is pictured in Figure 1.1.

We begin with M in the *starting state* q_1 scanning the leftmost cell containing a 1, called the *starting cell*. If the machine ever reaches the *halting state* q_0, after say s steps, then we say M *halts* and the *output* y is the total number of 1's on the tape. (Note that $f(x) = \max\{x + 1, s\}$ bounds the maximum distance from the starting cell to any cell which is either scanned or contains an input symbol. Hence the determination of y is effective.)

We may assume that M never makes any further moves after reaching state q_0, i.e., that the domain of δ contains no element of the form (q_0, s). We say that M *computes* the partial function ψ provided that $\psi(x) = y$

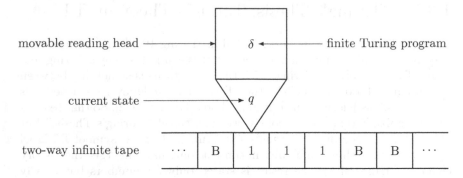

movable reading head ——————→ δ ←——————— finite Turing program

current state ——————————→ q

two-way infinite tape $\cdots$ | B | 1 | 1 | 1 | B | B | $\cdots$

Figure 1.1. Turing machine

if and only if M with input x eventually halts and yields output y. For example, the following machine computes the function $f(x) = x + 3$.

$$
\begin{array}{cccccc}
& q_1 & 1 & q_1 & 1 & R \\
(1.1) & q_1 & B & q_2 & 1 & R \\
& q_2 & B & q_0 & 1 & R
\end{array}
$$

The instantaneous condition of M during each step in a Turing calculation is completely determined by: (1) the current state q_i of the machine; (2) the symbol s_1 being scanned; (3) the symbols on the tape to the right of symbol s_1 up to the last 1, i.e., $s_2, s_3, \ldots, s_k$; and (4) the symbols to the left of s_1 up to the first 1, i.e., $t_1, t_2, \ldots t_m$. This is the (instantaneous) *configuration* of the machine at that step and is written

$$
(1.2) \qquad c \;\; = \;\; t_m t_{m-1} t_{m-2} \cdots t_2\, t_1\, q_i\, s_1\, s_2\, s_3 \cdots s_k.
$$

For example, the machine of (1.1) in calculating on input $x = 0$ passes through the configurations, $q_1 1, 1q_1 B, 11q_2 B$, and $111q_0 B$, and it yields output $y = 3$. (Recall that the input x is coded by $x + 1$ consecutive 1's while the output y is coded by the total number of 1's on the tape. Also notice that the tape contains only finitely many non-blank symbols at the beginning of any calculation and that this condition persists at all later stages, whether the machine halts or not, so that the integers k and m in (1.2) exist.)

If the machine M enters a state $q \neq q_0$ and reads a symbol s from which $\delta(q, s)$ gives no moves, then the machine *stalls*, i.e., makes no further moves, and gives no output. We do not refer to this as halting even though the machine stalls and stops forever. We use the term *halting* only if M enters the halting state q_0.

Definition 1.4.2. A *Turing computation* according to Turing program P with input x is a sequence of configurations, $c_0, c_1, \ldots, c_n$ such that c_0 represents the machine in the starting state q_1 reading the leftmost symbol of the input x, c_n represents the machine in the halting state q_0, and the transition $c_i \to c_{i+1}$, for all $i < n$, is given by the Turing program P.

Thus, from now on a *computation* will always refer to a halting, i.e., *convergent*, calculation. A partial function of n variables is associated with each Turing machine M by representing the input $(x_1, x_2, \ldots, x_n)$ by the following initial configuration of M: $q_1 \alpha_1 B \alpha_2 \ldots B \alpha_n$ where α_i consists of $x_i + 1$ consecutive 1's.

A Simple Example of a Computation

In (1.1) we gave a simple Turing program to compute $f(x) = x + 3$. Let us consider the computation as the machine proceeds on input 2 (represented by three 1's, 111, on the tape).

$q_1 \, 1 \, 11 \, B \, B$

$1 \, q_1 \, 11 \, B \, B$

$11 \, q_1 \, 1 \, B \, B$

$111 \, q_1 \, B \, B$

$1111 \, q_2 \, B$

$11111 \, q_0$

The input is 2 (denoted by 111) and the output is 5 (denoted by a total of five 1's on the final tape. All the other cells are blank (B) and not explicitly mentioned. The machine halts in state q_0 and never moves again on input 2.

Definition 1.4.3. Given n inputs $x_1, \ldots x_n$, we represent these as an input to a Turing machine by writing each as a block of $(x_k + 1)$ 1's and separating each block by a B.

1.4.1   Exercises on Turing Machines

Exercise 1.4.4. Write Turing machines which compute functions $f(x) = 0$, $f(x) = \lambda x[k]$, $f(x) = 2x$, and $f(x, y) = x + y$.

1.5 ** The Basic Results

1.5.1 Numbering Turing Programs P_e

Definition 1.5.1. (Indices of Turing Programs). Since each Turing program is a finite set of quintuples, we can list all Turing programs in such a way that for any program we can effectively find its number on the list, and conversely. Use the Gödel numbering of the Notation Section.

(i) Let P_e be the Turing program with code number e in this coding, also called the Gödel number or index e.

(ii) Let $\varphi_e^{(n)}$ be the partial function of n variables computed by P_e. Let φ_e denote $\varphi_e^{(1)}$. We call φ_e a *partial computable (p.c.) function*. From now on we *identify* the Turing program P_e and p.c. function φ_e with index e. (Some recent books use upper case Φ_e instead of φ_e, but this will cause no difficulty.)

This is called the *standard numbering* or *canonical numbering* of Turing programs and partial computable functions. By Exercise 1.7.8, any other effective numbering will be computably isomorphic to this one.

Lemma 1.5.2 (Padding Lemma). *Each partial computable (p.c.) function* φ_x *has infinitely many indices. Furthermore, for each x we can effectively find an infinite set A_x of indices for the same partial function (i.e., $\varphi_y = \varphi_x$ for all $y \in A_x$).*

Proof. Given φ_x, we consider the associated Turing machine P_x. If Turing program P_x mentions only internal states $\{q_0, \ldots, q_n\}$, add extraneous instructions $q_{n+1}B\ q_{n+1}B\,R,\ \ q_{n+2}B\ q_{n+2}B\,R, \ldots$ to get new programs for the same function. □

1.5.2 Numbering Turing Computations

In (1.2) we described an *(instantaneous) configuration*, now denoted by c below, of a Turing machine M at a given stage v during the computation:

$$(1.3) \qquad\qquad c \;=\; \bar{t}\, \underline{q_i}\, s_1\, \bar{s}$$

where s_1 is the symbol being scanned, $\bar{t} = t_w \ldots t_2\, t_1$ represents the sequence of symbols on the tape to the left of s_1, and $\bar{s} = s_2 \ldots s_v$ is the sequence of symbols on the tape to the right of s_1. Since the reading head moves at most one square at every stage, the sequences $\bar{t}$, $\bar{s}$ and s_1 include the symbols in any cell which has been scanned up to stage v. For example, M always starts in the *initial* configuration

$$(1.4) \qquad\qquad c_1 \;=\; \underline{q_1}\, s_1 s_2 \ldots s_{k+1}$$

i.e., in the starting state q_1, reading the first symbol s_1 of the input of $k+1$ consecutive 1's representing integer k. As soon as M (if ever) reaches at stage v a configuration c_i which contains the *halting state* q_0,

(1.5) $$t_w \ \ldots \ t_2 \ t_1 \ \underline{q_0} \ s_1 \ s_2 \ldots s_v,$$

M outputs the total number of 1's among the tape symbols in (1.5); it turns off; and it never enters another configuration. In this case M *converges* on its input. However, M may either go on forever, entering infinitely many different configurations, or loop through finitely many infinitely often, or else stall at some configuration c_i if its Turing program P_e (being only a *partial* set of instructions) gives it no option to enter another configuration c_{i+1}. In the latter three cases M *diverges* on its input. By choosing from the Turing program P_e the first possible move (if any), we can ensure that the passage of M through configurations c_1, c_2, ... is deterministic. (We asked that the program be a partial *function* so it is deterministic, but if it had been nondeterministic we could have made it deterministic in this fashion.) Hence, from every c_i there is at most one consequent configuration c_{i+1}.

This action of a Turing machine in passing through an effective sequence of configurations can now be coded by Gödel numbers as follows. A configuration c of the form (1.3) is a finite sequence of symbols. Once we have assigned numbers to each tape symbol s_i and state q_j, we can assign a number to every configuration using the prime power equation (1) of the Notation Section. If the sequence results in a configuration in the halting state q_0 then the resulting sequence of configurations is a *computation*.

1.5.3   The Enumeration Theorem and Universal Machine

In §1.1.2 we showed that there could be no effective enumeration of all *total* computable functions. However, the effective numbering in the preceding subsections for the syntax of Turing machines allows us to now give an effective enumeration of all *partial* computable functions. This result is crucial in virtually all our theorems.

Theorem 1.5.3 (Enumeration Theorem and Universal Machine).   *There is a partial computable (p.c.) function of two variables* $\psi(e, x)$ *such that* $\psi(e, x) = \varphi_e(x)$. *By TT there is some i such that* $\varphi_i(e, x) = \psi(e, x)$.

Proof. A Turing machine $M(e, x)$ which computes $\psi(e, x)$ is called a *universal* Turing machine. We give the instructions in ordinary mathematical notation but by TT they can be performed by a Turing machine.
Step 1.   Given the pair of inputs (e, x) convert e to the Turing program P_e by using the numbering of Turing programs in Definition 1.5.1 and unique factorization to recover the exponents in (1) of the prime power representation in the Notation section.
Step 2.   Simulate the action of P_e on input x. The simulation begins with state q_1 on the leftmost symbol of x in standard input form. If the simulation is in state q_j reading symbol s_k, then M searches through the

tuples in P_e until it finds the first of the form (q_j, s_k, q, t, X) and then performs the indicated action on the simulation tape.   □

Remark 1.5.4. The Enumeration Theorem 1.5.3 is crucial for all our work. For example, to show that a set A is undecidable we must show that $\chi_A \neq \varphi_e$ for all e. That requires effectively listing all algorithmic partial functions $\{\varphi_e\}_{e \in \omega}$. In 1936 Turing approached this in machine fashion by explicitly specifying a *universal machine,* "a single machine which can compute any computable sequences." Turing's universal machine took as input a number x and a Turing program, which in our notation would be a finite sequence of 5-tuples for the Turing program, and his machine performed as in step 2 above.

1.5.4   The Parameter Theorem or s-m-n Theorem

Theorem 1.5.5 (Parameter Theorem or s-m-n Theorem). *There is a 1:1 computable function $s(x, y)$ such that for all x, y, and z,*

$$\varphi_{s(x,y)}(z) = \varphi_x(y, z).$$

Proof. (informal). The algorithm for Turing program $P_{s(x,y)}$ on input z first obtains program P_x by inverting the coding[1] of Turing programs in §1.5.1, and then applies P_x to input (y, z). This procedure defines an algorithm with inputs x and y. By TT it has a Turing machine index $s(x, y)$. It is easy to check that $s(x, y)$ is already 1:1. However, instead of checking we may replace $s(x, y)$ by an obviously 1:1 computable function $s'(x, y)$ such that $\varphi_{s'(x,y)} = \varphi_{s(x,y)}$ by using the Padding Lemma 1.5.2 and by defining $s'(x, y)$ in increasing order of $\langle x, y \rangle$, where $\langle x, y \rangle$ is the image of (x, y) under the standard pairing function.   □

Remark 1.5.6 (Intuition for Parameter Theorem). The Parameter Theorem 1.5.5 asserts that y may be treated as a fixed *parameter* in the program $P_{s(x,y)}$ which operates on z, and furthermore that the index $s(x, y)$ of this program is computable in x and y. Suppose $\varphi_x(y, z) = y + z$. Fix the parameter as $y = 3$ and consider the resulting function $f(z) = \varphi_x(3, z) = 3 + z$. Now $f(x)$ is computable and must have some index $\varphi_i(z) = f(z)$. The Parameter Theorem shows that we can pass computably from x and the parameter y to such an index i for $f(z)$.

---

[1] The Parameter Theorem is often stated for an m-tuple $(y_1, y_2, \ldots y_m)$ and an n-tuple $(z_1, z_2, \ldots z_n)$. Then the function s is called $s_n^m(x, \overrightarrow{y})$, which gives the theorem its name. Here $m = n = 1$ and the function is $s(x, y) = s_1^1(x, y)$. By the pairing function (2) we can identify the m-tuple of $\overrightarrow{y}$ with a single y and likewise the n-tuple $\overrightarrow{z}$ with a single z. We achieve notational convenience without loss of generality.

Theorem 1.5.7 (Unbounded Search Theorem). *If $\theta(x, y)$ is a partial computable function, and*

$$(1.6) \quad \psi(x) = (\mu y) \lfloor \theta(x, y)\!\downarrow \ = \ 1 \quad \& \quad (\forall z < y) [\ \theta(x, z)\!\downarrow \ \neq 1\ ]\ ],$$

then $\psi(x) = y$ is a partial computable function.

Proof. For a fixed x compute $\theta(x, y)$ in order of y as follows. For fixed s compute s steps for each $y \leq s$ and then go to step $s + 1$. Continue until (if ever) the first y is found such that $\theta(x, y)\!\downarrow \ = 1$. Output $\psi(x) = y$. $\square$

Here $\psi(x)$ diverges if there is no such y. Note that this applies if θ is total, but we cannot effectively determine whether θ is total. Therefore, we state it more generally as above. The second conjunct is necessary for ψ to be partial computable as noted in Exercise 1.6.28.

1.6 ** Unsolvable Problems

Convention 1.6.1. If $R \subseteq \omega^n$, $n \geq 1$, then R has property P if the set $\{\langle x_1, \ldots, x_n \rangle : R(x_1, x_2, \ldots, x_n)\}$ has property P, such as being computable, c.e., Σ_1, etc. (Note that this agrees with the Definition 2.3 of R being computable iff the characteristic function χ_R is computable.)

1.6.1 Computably Enumerable Sets

Definition 1.6.2. (i) A set A is *computably enumerable (c.e.)* if A is the domain of some partial computable (p.c.) function.

(ii) Let the eth c.e. set be denoted by

$$(1.7) \qquad W_e := \operatorname{dom}(\varphi_e) = \{\ x \ : \ \varphi_e(x)\!\downarrow\ \}.$$

Note that any computable set is c.e. since if A is computable, then $A = \operatorname{dom}(\psi)$, where $\psi(x) = 1$ if the characteristic function $\chi_A(x) = 1$ and $\psi(x)\!\uparrow$ otherwise. We shall show in Theorem 2.1.10 that a nonempty set A is c.e. iff A is the range of a computable function (i.e., iff there is an algorithm for listing the members of A). The frequent occurrence of c.e. sets in other branches of mathematics and the existence of noncomputable c.e. sets such as K below have yielded numerous undecidability results, such as the Davis-Matijasevič-Putnam-Robinson resolution of Hilbert's tenth problem on the unsolvability of certain Diophantine equations, and the Boone-Novikov theorem on the unsolvability of the word problem for finitely presented groups.

1.6.2 Noncomputable C.E. Sets

Definition 1.6.3. Let $K := \{\ x : \varphi_x(x) \text{ converges}\ \} = \{\ x \ : \ x \in W_x\ \}.$

Theorem 1.6.4. K *is c.e.*

Proof. K is the domain of the following p.c. function:

$$\psi(x) = \begin{cases} x & \text{if } \varphi_x(x) \text{ converges} \\ \text{undefined} & \text{otherwise.} \end{cases}$$

Now ψ is p.c. because $\psi(x)$ can be computed by applying program P_x to input x and giving output x only if $\varphi_x(x)$ converges. Alternatively, and more formally, $K = \text{dom}(\theta)$, where $\theta(x) = \varphi_z^{(2)}(x, x)$ for $\varphi_z^{(2)}$ the p.c. function defined in the Enumeration Theorem 1.5.3. $\qquad\square$

Theorem 1.6.5. K *is not computable.*

Proof. If K had a computable characteristic function then the following function would be computable:

$$f(x) = \begin{cases} \varphi_x(x) + 1 & \text{if } x \in K, \\ 0 & \text{if } x \notin K. \end{cases}$$

However, f cannot be computable because $f \neq \varphi_x$ for every x. $\qquad\square$

Therefore, there is no algorithm for deciding on a given input x whether $x \in K$. This is our first example of an unsolvable problem.

Definition 1.6.6. $K_0 = \{ \langle x, y \rangle : x \in W_y \}$.

Note that K_0 is also c.e. Indeed, $K_0 = \text{dom}(\theta)$, where $\theta(\langle x, y \rangle) = \varphi_z^{(2)}(y, x)$ for $\varphi_z^{(2)}$ as in the Enumeration Theorem 1.5.3.

Corollary 1.6.7. K_0 *is not computable.*

Proof. Note that $x \in K$ iff $\langle x, x \rangle \in K_0$. Thus if K_0 has a computable characteristic function, so does K, contrary to Theorem 1.6.5. $\qquad\square$

The *halting problem* is to decide for arbitrary x and y whether $\varphi_y(x)$ converges, i.e., whether $\langle x, y \rangle \in K_0$. Corollary 1.6.7 asserts the unsolvability of the halting problem. The proof of the corollary suggests an indirect method for proving unsolvability of new problems by reducing K to them.

Definition 1.6.8. (i) A is *many-one reducible* (*m-reducible*) to B (written $A \leq_m B$) if there is a computable function f such that $f(A) \subseteq B$ and $f(\overline{A}) \subseteq \overline{B}$, i.e., $x \in A$ iff $f(x) \in B$.

(ii) A is *one-one reducible* (*1-reducible*) to B ($A \leq_1 B$) if $A \leq_m B$ by a 1:1 computable function.

For example, the proof of Corollary 1.6.7 established that $K \leq_1 K_0$ via the function $f(x) = \langle x, x \rangle$. Note that if $A \leq_m B$ via f then $\overline{A} \leq_m \overline{B}$ via f also. It is obvious that $\leq_m$ and $\leq_1$ are reflexive and transitive, and hence induce the following equivalence relations.

Definition 1.6.9. Define equivalence relations and degrees as follows.

(i) $A \equiv_m B$ if $A \leq_m B$ and $B \leq_m A$.

(ii) $A \equiv_1 B$ if $A \leq_1 B$ and $B \leq_1 A$.

(iii) $\deg_m(A) = \{B : A \equiv_m B\}$.

(iv) $\deg_1(A) = \{B : A \equiv_1 B\}$.

The equivalence classes under $\equiv_m$ and $\equiv_1$ are called the *m-degrees* and
1-*degrees* respectively.

Proposition 1.6.10. *If* $A \leq_m B$ *and* B *is computable then* A *is computable.*

Proof. If $A \leq_m B$ via f, then $\chi_A(x) = \chi_B(f(x))$, so χ_A is computable if B is computable. $\qquad\square$

We often write $A(x)$ for the characteristic function $\chi_A(x)$. Proposition 1.6.10 provides a technique for proving the unsolvability of numerous problems, such as those of deciding, given x, whether φ_x is a constant function, a total function, whether $\mathrm{dom}(\varphi_x) \neq \emptyset$, whether φ_x is extendible to a total computable function, and so on. If we can reduce one unsolvable problem A to another one B, then B is also unsolvable.

Theorem 1.6.11. $K \leq_1 \mathrm{Tot} := \{x : \varphi_x \text{ is a total function}\}$.

Proof. Define the function:

$$
\psi(x,y) = \begin{cases} 1 & \text{if } x \in K; \\ \text{undefined} & \text{otherwise.} \end{cases}
$$

Clearly, ψ is p.c. because the program to compute $\psi(x,y)$ says: first attempt to compute $\varphi_x(x)$; if this fails to converge then output nothing; if it converges then output 1 for every argument y. By the Parameter Theorem there is a 1:1 computable function f such that $\varphi_{f(x)}(y) = \psi(x,y)$. Namely, choose e such that $\varphi_e(x,y) = \psi(x,y)$, and define $f(x) = \lambda x[s_1^1(e,x)]$. Now f is 1:1 because s_1^1 is 1:1. Note that

$$x \in K \Longrightarrow \varphi_{f(x)} = \lambda y[1] \Longrightarrow \varphi_{f(x)} \text{ total} \Longrightarrow f(x) \in \mathrm{Tot}, \text{ and}$$

$$x \notin K \Longrightarrow \varphi_{f(x)} = \lambda y[\uparrow] \Longrightarrow \varphi_{f(x)} \text{ not total} \Longrightarrow f(x) \notin \mathrm{Tot}.$$

$\qquad\square$

Notice that this proof shows that the problem of deciding, given x, whether φ_x is constant, or even whether $\mathrm{dom}(\varphi_x) \neq \emptyset$, is an unsolvable problem. Notice also that in this proof K could be replaced by an arbitrary c.e. set A. Suppose $A = \mathrm{dom}\,(\theta)$ for θ p.c. Then the program begins by attempting to compute $\theta(x)$ in place of $\varphi_x(x)$. However, we could not replace K by an arbitrary *non*-c.e. set A because there would be no computable counterpart to this first step in the program for ψ, and ψ must be computable.

Applications of the Parameter Theorem 1.5.5 such as the one above will occur often. The reader should verify in each case that the instructions for computing $\psi(x, y)$ are effective. From now on we can simply write $\varphi_{f(x)}(y)$ for $\psi(x, y)$ without explicitly mentioning the Parameter Theorem 1.5.5. The method in the above proof applies to numerous other sets A in place of Tot so long as the property defining A is a property of functions and not of certain indices for them, for example, if A is an index set defined as follows.

Definition 1.6.12. A set $A \subseteq \omega$ is an *index set* if for all x and y

$$[\, x \in A \ \& \ \varphi_x = \varphi_y \,] \quad \Longrightarrow \quad y \in A.$$

1.6.3   The Index Set Theorem and Rice's Theorem

Theorem 1.6.13 (Index Set Theorem). *If A is a nontrivial index set, i.e., $A \neq \emptyset$, $A \neq \omega$, then either $K \leq_1 A$ or $K \leq_1 \overline{A}$. Furthermore, choose e_0 such that $\varphi_{e_0}(y)$ is undefined for all y. If $e_0 \in \overline{A}$, then $K \leq_1 A$.*

Proof. If $e_0 \in \overline{A}$, then $K \leq_1 A$ as follows. (If $e_0 \in A$, then $K \leq_1 \overline{A}$ similarly.) Since $A \neq \emptyset$ we can choose $e_1 \in A$. Now $\varphi_{e_1} \neq \varphi_{e_0}$ because A is an index set. By the Parameter Theorem 1.5.5 define a 1:1 computable function f such that

$$\varphi_{f(x)}(y) \ = \ \begin{cases} \varphi_{e_1}(y) & \text{if } x \in K \\ \\ \text{undefined} & \text{if } x \notin K. \end{cases}$$

Now

$$x \in K \quad \Longrightarrow \quad \varphi_{f(x)} = \varphi_{e_1} \quad \Longrightarrow \quad f(x) \in A,$$

$$x \in \overline{K} \quad \Longrightarrow \quad \varphi_{f(x)} = \varphi_{e_0} \quad \Longrightarrow \quad f(x) \in \overline{A}.$$

The last implication in each line follows because A is an index set.   □

Corollary 1.6.14 (Rice's Theorem). *Let $\mathcal{C}$ be any class of partial computable functions. Then $A = \{n : \varphi_n \in \mathcal{C}\}$ is computable iff $\mathcal{C} = \emptyset$ or $\mathcal{C}$ is the class of all partial computable functions.*

Proof. By definition, A is an index set. If $A = \emptyset$ or $\overline{A} = \emptyset$, then trivially A is computable. Otherwise, the Index Set Theorem 1.6.13 implies that $K \leq_1 A$ or $K \leq_1 \overline{A}$. Therefore, A is not computable.   □

(Rice's results on c.e. sets can be found in [Rice 1953] and [Rice 1956].) It is possible that both $K \leq_1 A$ and $K \leq_1 \overline{A}$ for an index set A, for example if $A = \text{Tot}$.

Definition 1.6.15. In addition to K and K_0, which are *not* index sets (see Exercise 2.2.8), we shall use the following index sets which correspond to the natural unsolvable problems mentioned above.

$$
\begin{aligned}
K_1 &= \{x : W_x \neq \emptyset\}; \\
\text{Fin} &= \{x : W_x \text{ is finite}\}; \\
\text{Inf} &= \{x : W_x \text{ is infinite}\} = \omega - \text{Fin}; \\
\text{Tot} &= \{x : \varphi_x \text{ is total}\} = \{x : W_x = \omega\}; \\
\text{Con} &= \{x : \varphi_x \text{ is total and constant}\}; \\
\text{Cof} &= \{x : W_x \text{ is cofinite}\}; \\
\text{Rec} &= \{x : W_x \text{ is computable (recursive)}\}; \\
\text{Ext} &= \{x : \varphi_x \text{ is extendible to a total computable function}\}.
\end{aligned}
$$

Of the above index sets, K_1, Fin, Inf, Tot, Cof, and Rec turn out to be the most important in later work. Each of the above is a nontrivial index set and hence noncomputable by the Index Set Theorem 1.6.13. Only K, K_0 and K_1 are c.e. sets, and these all have the same 1-degree (see Exercise 1.6.22). Hence, by Theorem 1.7.4 these three are all computably isomorphic and may be regarded as interchangeable for all our purposes.

Definition 1.6.16. A c.e. set A is *1-complete* if $W_e \leq_1 A$ for every c.e. set W_e.

Clearly, K_0 is 1-complete because $x \in W_e$ iff $\langle x, e \rangle \in K_0$. Thus, K and K_1 are 1-complete by Exercise 1.6.22. The remaining index sets above (and their complements) are not c.e. as we shall see in §2.1. However, in Chapter 4 we shall show that

$$\text{Inf} \equiv_1 \text{Tot} \equiv_1 \text{Con} \qquad \text{and} \qquad \text{Cof} \equiv_1 \text{Rec} \equiv_1 \text{Ext}.$$

A major goal of computability theory is to classify exactly how unsolvable a problem is by computing its "degree" of unsolvability relative to other problems. For example, if $A \equiv_r B$, where $r = 1$, m or the more general Turing (T) reducibility of Chapter 3, then A and B have the same r-degree and intuitively they "code the same information." If $A <_r B$, then B has strictly higher r-degree than A and codes more information.

In Chapter 4 we shall give another characterization of these sets in terms of the arithmetical hierarchy. For each $n \in \omega$, we shall have a level in the hierarchy consisting of sets of the same T-degree. For example, the computable sets lie at level 0; the sets K, K_0, and K_1 at level 1; Inf, Con and Tot at level 2; Cof, Rec, and Ext at level 3; and so on. The sets at higher levels have strictly higher T-degree than those at lower levels.

1.6.4 Computable Approximations to Computations

Definition 1.6.17. We write $\varphi_{e,s}(x) = y$ if $x, y, e < s$ and y is the output of $\varphi_e(x)$ in $< s$ steps of the Turing program P_e. If such a y exists we say

$\varphi_{e,s}(x)$ *converges*, which we write as $\varphi_{e,s}(x) \downarrow$, and *diverges* $(\varphi_{e,s}(x) \uparrow)$ otherwise. Similarly, we write $\varphi_e(x) \downarrow$ if $\varphi_{e,s}(x) \downarrow$ for some s and we write $\varphi_e(x) \downarrow = y$ if $\varphi_e(x) \downarrow$ and $\varphi_e(x) = y$, and similarly for $\varphi_{e,s}(x) \downarrow = y$. Define $W_{e,s} := \mathrm{dom}(\varphi_{e,s})$.

(If $x \in W_{e,s}$ then $x, e < s$ by Definition 1.6.17.) Note that $\varphi_e(x) = y$ iff $(\exists s) [\varphi_{e,s}(x) = y]$ and $x \in W_e$ iff $(\exists s) [x \in W_{e,s}]$.

Theorem 1.6.18. *(i) The set $\{ \langle e, x, s \rangle : \varphi_{e,s}(x) \downarrow \}$ is computable, as is the set $W_{e,s} = \mathrm{dom}(\varphi_{e,s})$.*

(ii) The set $\{\langle e, x, y, s \rangle : \varphi_{e,s}(x) = y\}$ is computable.

Proof. We get (i) and (ii) because we calculate until an output is found or until the first s steps have been completed.   □

We often use the following operation of join to construct a set which codes two given sets A and B and we use it in the exercises below.

Definition 1.6.19. Let A *join* B, written $A \oplus B$, be

$$\{2x : x \in A\} \cup \{2x + 1 : x \in B\}.$$

1.6.5   Exercises

Exercise 1.6.20. Let $B = A \oplus \overline{A}$ for some set $A \subset \omega$. Prove that $B \leq_1 \overline{B}$.

Exercise 1.6.21. Prove that the m-degree of $A \oplus B$ is the least upper bound (lub) of the m-degrees of A and B, i.e.,

(i) $A \leq_m A \oplus B$ and $B \leq_m A \oplus B$; and

(ii) if $A \leq_m C$ and $B \leq_m C$ then $A \oplus B \leq_m C$. (It is false that 1-degrees, even of c.e. sets, always have an lub or a greatest lower bound (glb).)

Exercise 1.6.22. Prove that $K \equiv_1 K_0 \equiv_1 K_1$. (Note that the proof of Theorem 1.6.11 automatically shows that $K \leq_1 A$ for $A = K_1$, Con, or Inf.)

Exercise 1.6.23. Prove that $K \leq_1$ Fin without using the Index Set Theorem 1.6.13. (In Chapter 4 we shall show that K is Σ_1-complete and Fin is Σ_2-complete and that any Σ_1 set is m-reducible to any complete Σ_2 set. This gives an alternate proof.)

Exercise 1.6.24. For any x show that $\overline{K} \leq_1 \{y : \varphi_x = \varphi_y\}$, and also prove that $\overline{K} \leq_1 \{y : W_x = W_y\}$. *Hint.* Consider separately the cases W_x finite and infinite, and use the method of Exercise 1.6.23. (Also see Exercise 2.4.13 (ii) to see that this reduction cannot be made uniform.)

Exercise 1.6.25. Give a direct construction that Ext $\neq \omega$. Hence, not every partial computable function is extendible to a total computable function.

Exercise 1.6.26. Disjoint sets A and B are *computably inseparable* if there is no computable set C such that $A \subseteq C$ and $C \cap B = \emptyset$, and *computably separable* otherwise. (The existence of computable inseparable sets is very useful, for example, in model theory.) Define $A := \{ x : \varphi_x(x) = 0 \}$ and $B := \{ x : \varphi_x(x) = 1 \}$.

(i) Prove that A and B are computably inseparable by showing that no φ_e is the characteristic function of a separating set C.

(ii) Give an alternative proof that $\mathrm{Ext} \neq \omega$.

(iii) For A and B as in part (i), prove that $K \equiv_1 A$ and $K \equiv_1 B$.

There is a stronger notion called *effectively inseparable* defined in Exercise 2.4.17. Effectively inseparable sets are computably inseparable but not vice versa. The sets A and B defined here are in fact effectively inseparable.

Exercise 1.6.27. A set A is a *cylinder* if $(\forall B)\,[\, B \leq_m A \implies B \leq_1 A \,]$.

(i) Show that any index set is a cylinder.

(ii) Show that any set of the form $A \times \omega$ is a cylinder. (Recall that $X \times Y$ is the Cartesian product introduced in the Notation section.)

(iii) Show that A is a cylinder iff $A \equiv_1 B \times \omega$ for some set B.

Exercise 1.6.28. The partial computable functions are not closed under μ. Prove that there is a p.c. function θ such that $\lambda x \,[\, \mu y \,[\, \psi(x,y) = 1 \,]\,]$ is not p.c.

Exercise 1.6.29. If A is computable and B and $\overline{B}$ are each $\neq \emptyset$, prove that $A \leq_m B$. (Hence, neglecting the trivial sets $\emptyset$ and ω, there is a least m-degree.)

1.7 * Computable Permutations and Isomorphisms

Definition 1.7.1. (i) A *computable permutation* is a 1:1 computable function from ω onto ω.

(ii) A property of sets is *computably invariant* if it is invariant under all computable permutations.

Examples of computably invariant properties are the following:
   (i) A is c.e.;
   (ii) A has cardinality n (written $|A| = n$);
   (iii) A is computable.

The following properties are *not* computably invariant:
   (i) $2 \in A$;
   (ii) A contains the even integers;

(iii) A is an index set.

Definition 1.7.2. A is *computably isomorphic* to B (written $A \equiv B$) if there is a computable permutation p such that $p(A) = B$.

Definition 1.7.3. The equivalence classes under $\equiv$ are called *computable isomorphism types*.

We shall attempt to classify sets up to computable isomorphism just as the algebraist classifies structures up to isomorphism. One reason for introducing $\leq_1$ in Definition 1.6.8 (where $\leq_m$ would have sufficed for undecidability results) is the following theorem, which is an effective analogue of the classical Schröder-Bernstein Theorem for cardinal numbers.

1.7.1 Myhill Isomorphism Theorem

Theorem 1.7.4 (Myhill Isomorphism Theorem). *$A \equiv B$ iff $A \equiv_1 B$. (Indeed, we prove more in Exercise 1.7.7.)*

Proof. ($\Longrightarrow$). Trivial.
($\Longleftarrow$). Let $A \leq_1 B$ via f and $B \leq_1 A$ via g. Therefore,

$$(1.8) \quad (\forall x)(\forall y)\,[\, x \in A \iff f(x) \in B \quad \& \quad y \in B \iff g(y) \in A \,].$$

We define a computable permutation h by stages $s \in \omega$ so that $h(A) = B$. Suppose that by the end of stage $2s$ we have finite sets $X = \{x_1, x_2, \ldots x_n\}$ (on the left-hand side) and $Y = \{y_1, y_2, \ldots y_n\}$ (on the right-hand side) and a 1:1 map h such that $h(x_i) = y_i$ for all $1 \leq i \leq n$ such that

$$(1.9) \qquad\qquad (\forall x \in X)\,[\, x \in A \quad \iff \quad h(x) \in B \,].$$

Stage $2s + 1$. We shall define $h(s)$ if it is not already defined. Compute $f(s) = t_1$. If $t_1 \notin Y$ define $h(s) = t_1$. If $t_1 \in Y$, say $t_1 = y_i$, then take $x_i = h^{-1}(t_1)$. Note that $x_i \in A$ iff $s \in A$ by (1.8) and (1.9). Compute $f(x_i) = t_2$. If $t_2 \notin Y$ define $h(s) = t_2$. Otherwise, $t_2 = y_j$ for some j. Take $x_j = h^{-1}(y_j)$ and note that $x_j \in A$ iff $s \in A$ as before. Compute $f(x_j) = t_3$. If $t_3 \notin Y$ define $h(s) = t_3$. Otherwise, $t_3 = y_i$ for some i. Take $x_i = h^{-1}(y_i)$ and note that $x_i \in A$ iff $s \in A$ as before. Compute $f(x_i) = t_4$. Continue in this fashion until a new element $z \notin Y$ is found and define $h(s) = z$. Note that $X \cup \{s\}$ has $n{+}1$ elements but Y has only n elements, so the procedure must terminate.

Stage $2s + 2$. Find the value of $h^{-1}(s)$ in similar fashion using h^{-1} and g in place of h and f. $\qquad\square$

Note that the sets A and B may be highly noncomputable. All we have are the maps f and g and (1.8).

1.7.2   Acceptable Numberings

Definition 1.7.5. (Acceptable Numbering Conditions). Let $\mathcal{P}$ be the class of partial computable functions of one variable.

(i) A *numbering* of the p.c. functions is a map π from ω onto $\mathcal{P}$.

(ii) The numbering $\{\varphi_e\}_{e\in\omega}$ of Definition 1.5.1 is called the *standard numbering* or *canonical numbering* of the partial computable functions.

(iii) Let $\widehat{\pi}$ be another numbering and let ψ_e denote $\widehat{\pi}(e)$. Then $\widehat{\pi}$ is an *acceptable numbering* if there are computable functions f and g such that (1) $\varphi_{f(x)} = \psi_x$, and (2) $\psi_{g(x)} = \varphi_x$.

The following result is important because most obvious numberings are acceptable and any two acceptable numberings differ merely by a computable permutation. Therefore, it will not matter exactly *which* acceptable numbering we chose originally.

Theorem 1.7.6 (Acceptable Numbering Theorem, Rogers). *For any acceptable numbering $\{\psi_e\}_{e\in\omega}$ of the partial computable functions, there is a computable permutation h of ω such that $\varphi_e = \psi_{h(e)}$ for all e.*

Proof. Assume that $\{\psi_e\}_{e\in\omega}$ is an acceptable numbering of the p.c. functions. Let f and g satisfy (1) and (2) of Definition 1.7.5 (iii). The proof will follow from several other results and exercises to be considered later. First prove the Generalized Isomorphism Theorem 1.7.7 so that h exists if both f and g can be replaced by *one-one* functions f_1 and g_1. Second, find f_1 as in Exercise 1.7.8 and g_1 as in Theorem 2.3.9. Complete the proof as in Corollary 2.3.10.   □

1.7.3   Exercises

Exercise 1.7.7. (Generalized Isomorphism Theorem). Let $\omega = \bigcup_n A_n = \bigcup_n B_n$ where the sequences $\{A_n\}_{n\in\omega}$ and $\{B_n\}_{n\in\omega}$ are each pairwise disjoint. Let f and g be 1:1 computable functions such that $f(A_n) \subseteq B_n$ and $g(B_n) \subseteq A_n$ for all n. Show that the construction of the Myhill Isomorphism Theorem 1.7.4 produces a computable permutation h such that $h(A_n) = B_n$ for all n.

Exercise 1.7.8. (Rogers). Fill in the steps of the proof of the Acceptable Numbering Theorem 1.7.6 as follows.

(i) First use the Padding Lemma 1.5.2 to show that we may replace f by a 1:1 function f_1 if necessary.

(ii) In Theorem 2.3.9 we shall prove that we may also replace g by a 1:1 computable function g_1 if necessary. Assuming g_1 for now, use Exercise 1.7.7 and the 1:1 functions f_1 and g_1 to obtain the computable permutation h.

Exercise 1.7.8 is important because it proves that *any* acceptable numbering of Turing programs is as good as the one we chose in Definition 1.7.5 because the two numberings are computably isomorphic. Exercise 1.7.9 gives an example where the function f of Exercise 1.7.8 (i) is computable but the function g of Exercise 1.7.8 (ii) is *not*. Although the classes of functions $\{\psi_e\}_{e \in \omega}$ and $\{\varphi_e\}_{e \in \omega}$ *both* constitute the class of p.c. functions, the former are "hidden" so that we cannot effectively find the p.c. function $\psi_{g(e)}$ which corresponds to the standard function φ_e.

Exercise 1.7.9. Obtain an effective numbering ψ which is *not* acceptable as follows. Define

$$
\begin{aligned}
\psi_{\langle 0,q \rangle}(0) &= \text{undefined} \\
\psi_{\langle p+1,\,q \rangle}(0) &= p \\
\psi_{\langle p,q \rangle}(x) &= \varphi_q(x), \quad \text{if } x > 0.
\end{aligned}
$$

Show that $\{\psi_n\}_{n \in \omega} = \mathcal{P}$ but that ψ is not acceptable. *Hint.* Show that if there is a computable function g such that $\psi_{g(x)} = \varphi_x$ then we can decide the halting problem. (For an alternative example define a computable function f such that $\{\varphi_{f(p)} : p \in \omega\}$ consists of exactly the partial functions with nonempty domain and let $\psi_{p+1} = \varphi_{f(p)}$, $\psi_0(y) = \lambda y\,[\uparrow]$.)

2
Computably Enumerable Sets

2.1 ⋆⋆ Characterizations of C.E. Sets

This chapter is intended to give the reader a more intuitive feeling for c.e. sets, their alternative characterizations, and their most useful static and dynamic properties. We also prove the Recursion Theorem 2.2.1 (Fixed Point Theorem), which will be a very useful tool, and we apply it to prove Myhill's Theorem 2.4.6 that all creative sets are computably isomorphic.

In Definition 1.6.2 we defined a set A to be *computably enumerable (c.e.)* if it is the domain of a partial computable function φ_e. This immediately gave an effective *numbering* of c.e. sets $W_e = \text{dom}(\varphi_e)$ in (1.7) of Chapter 1. In this chapter we present two other particularly useful characterizations of c.e. sets: the Σ_1^0 quantifier characterization in Theorem 2.1.3; and the Listing Theorem 2.1.10, which says that A is c.e. iff it can be effectively *listed* (enumerated). Recall our Convention 1.6.1 that a relation $R \subseteq \omega^n$ is said to have some property P (such as being computable, c.e., or Σ_1^0) iff the set $A = \{\overline{x} : R(\overline{x})\}$ has property P.

2.1.1 The Σ_1^0 Normal Form for C.E. Sets

Definition 2.1.1. (i) A set A is a *projection* of some relation $R \subseteq \omega \times \omega$ if $A = \{x : (\exists y)\, R(x, y)\}$ (i.e., geometrically A is the projection of the two-dimensional relation R onto the x-axis).

(ii)   A set A is in Σ_1^0-*form*, abbreviated A is Σ_1^0, if A is the projection of some computable relation $R \subseteq \omega \times \omega$.

Convention 2.1.2.   The reason for the notation Σ_1^0 is that the corresponding predicate for A has the form of one $\exists$ quantifier followed by a computable predicate. The forms Σ_n^0, Π_n^0, and Δ_n^0, $n \geq 0$, will be formally defined in Chapter 4 on the arithmetical hierarchy. The superscript 0 indicates that the quantifiers range over integers, but later we shall usually drop it according to Convention 4.1.1.

Theorem 2.1.3 (Normal Form Theorem for C.E. Sets).   *A set A is c.e. iff A is Σ_1^0.*

Proof. ($\Longrightarrow$). If A is c.e. then $A = W_e := \mathrm{dom}(\varphi_e)$ for some e. Hence,

$$x \in W_c \quad \Longleftrightarrow \quad (\exists s)\,[\,x \in W_{c,s}\,].$$

By Theorem 1.6.18 the relation $\{ \langle e, x, s \rangle \,:\, x \in W_{e,s} \}$ is computable.

($\Longleftarrow$).   Let $A = \{ x \,:\, (\exists y)\, R(x, y) \}$, where R is computable. Then $A = \mathrm{dom}(\psi)$ where $\psi(x) = (\mu y)\, R(x, y)$. $\qquad\square$

Definition 2.1.4. If $A = W_e$ then e is a Σ_1^0-*index* or *c.e.-index* for A.

Theorem 2.1.5 (Quantifier Contraction Theorem).   *A is Σ_1^0 iff there is an $n \in \omega$ and a computable relation $R \subseteq \omega^{nS+1}$ such that*

$$A = \{x : (\exists y_1) \cdots (\exists y_n)\, R(x, y_1, \ldots, y_n)\}.$$

Proof. Define the computable relation $S \subseteq \omega^2$ by

$$S(x, z) \quad \Longleftrightarrow_{\mathrm{dfn}}\quad R(x, (z)_1, (z)_2, \ldots, (z)_n),$$

where $z = p_0^{(z)_0} p_1^{(z)_1} \cdots p_n^{(z)_n}$ is the prime decomposition.

$$(\exists z)\, S(x, z) \quad \Longleftrightarrow \quad (\exists z)\, R(x, (z)_1, (z)_2, \ldots, (z)_n)$$
$$\Longleftrightarrow \quad (\exists y_1)(\exists y_2) \cdots (\exists y_n)\, R(x, y_1, y_2, \ldots, y_n).$$

(Choose any computable function which maps an n-tuple $(x_1, \ldots, x_n)$ to a unique y.) $\qquad\square$

Corollary 2.1.6.   *The projection of a c.e. relation is c.e.* $\qquad\square$

Definition 2.1.7. (i)   The *graph* of a (partial) function ψ is the relation

(2.1) $$\mathrm{graph}\,(\psi) = \{\, (x, y) \,:\, \psi(x)\!\downarrow = y \,\}.$$

(ii)   Partial functions ψ and θ are *equal*, $\psi = \theta$, if $\mathrm{graph}\,(\psi) = \mathrm{graph}\,(\theta)$.

It follows from Theorem 2.1.5 or Corollary 2.1.6 that a set A is c.e. if A is of the form

$$\{\, x \,:\, (\exists y_1) \ldots (\exists y_n)\, R(x, y_1, \ldots, y_n) \,\},$$

where $n \in \omega$ and $R \subseteq \omega^{n+1}$ is computable. This is very useful for showing various sets to be c.e. For example, using this to handle the ($\exists$) quantifiers and using Theorem 1.6.18 to see that the matrices are computable, it is easy to check that the following sets and relations are c.e.

$$(2.2) \qquad K = \{e : e \in W_e\} = \{e : (\exists s)\,(\exists y)\,[\,\varphi_{e,s}(e) = y\,]\},$$

$$(2.3) \qquad K_0 = \{\langle x, e\rangle : x \in W_e\} = \{\langle x, e\rangle : (\exists s)\,(\exists y)\,[\,\varphi_{e,s}(x) = y\,]\},$$

$$(2.4) \qquad K_1 = \{e : W_e \neq \emptyset\} = \{e : (\exists s)\,(\exists x)\,[\,x \in W_{e,s}\,]\},$$

$$(2.5) \qquad \mathrm{rng}\,(\varphi_e) = \{y : (\exists s)\,(\exists x)\,[\varphi_{e,s}(x) = y]\},$$

$$(2.6) \qquad \mathrm{graph}\,(\varphi_e) = \{(x, y) : (\exists s)\,[\varphi_{e,s}(x) = y]\}.$$

2.1.2   The Uniformization Theorem

Theorem 2.1.8 (Uniformization Theorem). *If $R \subseteq \omega^2$ is a c.e. relation, then there is a p.c. function ψ (called a* selector function *for R) such that*

$$\psi(x) \text{ is defined} \iff (\exists y)\,R(x, y),$$

and in this case $(x, \psi(x)) \in R$. (Furthermore, an index for ψ can be found computably from a c.e. index for R.)

Proof. Since R is c.e. and hence Σ_1^0, there is a computable relation S such that $R(x, y)$ holds iff $(\exists z)\,S(x, y, z)$. Define the p.c. function

$$\theta(x) = (\mu u)\,S(x, (u)_1, (u)_2)$$

and define $\psi(x) = (\theta(x))_1$. $\qquad\qquad\qquad\qquad\qquad\qquad\qquad\qquad\square$

The important thing to grasp about the Uniformization Theorem is not that ψ *exists* (which is obvious) but that ψ is partial *computable*.

Theorem 2.1.9 (Graph Theorem). *A partial function ψ is partial computable iff its graph is computably enumerable.*

Proof. ($\Longrightarrow$). The graph of ψ is c.e. by Theorem 2.1.3 and (2.6).
($\Longleftarrow$). If the graph of ψ is c.e., then ψ is its own p.c. selector function in Definition 2.1.7 because $R = \mathrm{graph}\,(\psi)$ can have only ψ as its selector function. $\qquad\qquad\qquad\qquad\qquad\qquad\qquad\qquad\qquad\qquad\qquad\qquad\square$

The following is the basic theorem on c.e. sets, and justifies the earlier intuitive description of a c.e. set A as one whose members can be effectively listed, $A = \{\,a_0, a_1, a_2, \dots\,\}$. (Note that the listing may have repetitions and need not be in increasing order.)

2.1.3   The Listing Theorem for C.E. Sets

Theorem 2.1.10 (Listing Theorem).   *A set A is c.e. iff $A = \emptyset$ or A is the range of a total computable function f. Furthermore, f can be found uniformly in an index for a nonempty A as explained in Exercise 2.1.24.*

Proof. ($\Longleftarrow$). If $A = \emptyset$, then A is c.e. Now suppose $A = \text{rng } (f)$, where f is a total computable function. Then A is c.e. by (2.5).

($\Longrightarrow$). Let $A = W_e \neq \emptyset$. Choose any $a \in W_e$. Define the computable function f by

$$f(\langle s, x \rangle) \;=\; \begin{cases} x & \text{if } x \in W_{e,s+1} - W_{e,s}; \\ a & \text{otherwise.} \end{cases}$$

Note that each $x \in W_e$, $x \neq a$, is listed exactly once. Clearly, $A = \text{rng } (f)$, because if $x \in W_e$, we choose the least s such that $x \in W_{e,s+1}$. Then $f(\langle s, x \rangle) = x$ and hence $x \in \text{rng } (f)$. $\square$

2.1.4   The C.E. and Computable Sets as Lattices

Theorem 2.1.11 (Union Theorem).   *The c.e. sets are closed under union and intersection uniformly, i.e., there are computable functions f and g such that $W_{f(x,y)} = W_x \cup W_y$ and $W_{g(x,y)} = W_x \cap W_y$.*

Proof. Use the Parameter Theorem 1.5.5 to define $f(x,y)$ by enumerating $z \in W_{f(x,y)}$ if $(\exists s)[z \in W_{x,s} \cup W_{y,s}]$ and similarly for g with $\cap$ in place of $\cup$. $\square$

Corollary 2.1.12 (Reduction Principle for c.e. sets).   *Given any two c.e. sets A and B, there exist c.e. sets $A_1 \subseteq A$ and $B_1 \subseteq B$ such that $A_1 \cap B_1 = \emptyset$ and $A_1 \cup B_1 = A \cup B$.*

Proof. Define the relation $R := A \times \{0\} \cup B \times \{1\}$ which is c.e. by Theorem 2.1.11. By the Uniformization Theorem 2.1.8, let ψ be the p.c. selector function for R. Let $A_1 = \{x : \psi(x) = 0\}$, $B_1 = \{x : \psi(x) = 1\}$. (See also Exercise 2.6.7.) $\square$

Definition 2.1.13. A set A is in Δ_1-*form*, abbreviated A is Δ_1, if both A and $\overline{A}$ are Σ_1^0.

The following basic theorem of Post relates c.e. sets to computable sets.

Theorem 2.1.14 (Complementation Theorem).   *A set A is computable iff both A and $\overline{A}$ are c.e., i.e., iff $A \in \Delta_1$.*

Proof. ($\implies$). If A is computable, then $\overline{A}$ is computable and A and $\overline{A}$ are both c.e.

($\impliedby$). Let $A = W_e$, $\overline{A} = W_i$. Define the computable function

$$f(x) = (\mu s)\,[\,x \in W_{e,s} \quad \text{or} \quad x \in W_{i,s}\,].$$

Then $x \in A$ iff $x \in W_{e,\,f(x)}$ and hence A is computable.   □

Corollary 2.1.15. $\overline{K}$ *is not c.e.*

Proof. By Theorems 1.6.4 and 1.6.5 K is c.e. and not computable.   □

It is easily shown that if $\overline{K} \leq_m A$ then A is not c.e. Therefore, we can conclude that many sets such as Tot, Fin, and Cof are not c.e. (See Exercise 2.1.17.)

Definition 2.1.16. (i) By the Union Theorem 2.1.11 the c.e. sets form a distributive lattice $\mathcal{E}$ under inclusion with greatest element ω and least element $\emptyset$,

$$\mathcal{E} = (\{W_e\}_{e\in\omega}, \cup, \cap, \emptyset, \omega).$$

(ii) By the Complementation Theorem 2.1.14 a c.e. set $A \in \mathcal{E}$ is computable iff $\overline{A} \in \mathcal{E}$. Hence, the computable sets form a Boolean algebra $\mathcal{C} \subset \mathcal{E}$ consisting of the complemented members of $\mathcal{E}$. (By Corollary 2.1.15 $\mathcal{C} \neq \mathcal{E}$.)

2.1.5 Exercises

Exercise 2.1.17. (i) Prove that if $A \leq_m B$ and B is c.e. then A is c.e.

(ii) Show that Fin and Tot are not c.e.

(iii) Show that Cof is not c.e.

Exercise 2.1.18. Prove that if A is c.e. and ψ is p.c. then $\psi(A)$ is c.e. and $\psi^{-1}(A)$ is c.e.

Exercise 2.1.19. Let f be a total function. Prove that f is computable iff its graph is computable.

Exercise 2.1.20. Prove that every infinite c.e. set contains an infinite computable subset.

Exercise 2.1.21. A set A is *co-c.e.* (or equivalently Π_1^0) if $\overline{A}$ is c.e. Use Exercise 1.6.26 on computably inseparable sets to prove that the Reduction Principle 2.1.12 fails for Π_1^0 sets.

Exercise 2.1.22. The *separation principle* holds for a class $\mathcal{C}$ of sets if for every $A, B \in \mathcal{C}$ such that $A \cap B = \emptyset$ there exists C such that $C, \overline{C} \in \mathcal{C}$, $A \subseteq C$, and $B \subseteq \overline{C}$. By Exercise 1.6.26 the separation principle fails for c.e. sets. Use Corollary 2.1.12 to show that the separation principle holds for co-c.e. sets.

Exercise 2.1.23. Prove that if $A \leq_1 B$ and A and B are c.e. and A is infinite then $A \leq_1 B$ via some f such that $f(A) = B$.

Exercise 2.1.24. (i) Show that the proof of the Listing Theorem 2.1.10 is uniform in e in the sense that there is a p.c. function $\psi(e, y)$ such that if $W_e \neq \emptyset$ then $\lambda y \, [\psi(e, y)]$ is total and $W_e = \{\psi(e, y) : y \in \omega\}$. This uniformity is needed, for example, in Exercise 2.6.8.

(ii) If W_e is infinite, show that there is a total computable *one-one* function f such that $\mathrm{rng}(f) = W_e$.

2.2 ⋆ Recursion Theorem (Fixed Point Theorem)

2.2.1 Fixed Points in Mathematics

Fixed point theorems play an important role in mathematics. For example, the Brouwer Fixed Point Theorem asserts that every continuous function f from the closed unit sphere $S = \{x : ||x|| \leq 1\}$ to itself has a *fixed point* x_0 in the sense that $f(x_0) = x_0$. Not every computable function f can have a fixed point x_0 in this sense because the successor function $f(x) = x + 1$ cannot have such a fixed point.

2.2.2 Operating on Indices

However, in the spirit of the s-m-n Theorem 1.5.5 let us consider the computable function operating entirely on *indices* x of p.c. functions φ_x. It would be too much to expect f to have a fixed point n in the sense that on Turing programs themselves $P_n = P_{f(n)}$ (i.e., $f(n) = n$), but that is more than we required in the Padding Lemma 1.5.2. Define Turing programs P_i and P_j to be *equivalent*, written $P_i \sim P_j$, if $\varphi_i = \varphi_j$, i.e., the programs are *extensionally* equivalent because they compute the same p.c. function. The Fixed Point Theorem here asserts that every computable function f has a fixed point n such that $P_n \sim P_{f(n)}$, i.e., $\varphi_n = \varphi_{f(n)}$.

The following theorem of Kleene, known as the *Recursion Theorem* or *Fixed Point Theorem*, is one of the most elegant and important results in the field. It shows that certain implicitly defined functions are actually computable. The proof, which uses only the *s-m-n* Theorem 1.5.5, is very short but mysterious. To ensure understanding we present first the standard short proof in §2.2.3 and then explain the proof again in §2.2.4 as a diagonal argument which fails.

2.2.3   ** A Direct Proof of the Recursion Theorem

Theorem 2.2.1 (Recursion Theorem, Kleene).   *For every computable function f there exists an n, called a* fixed point *of f, such that $\varphi_n = \varphi_{f(n)}$.*

Proof. Define the computable *diagonal* function $d(u)$ by

$$(2.7) \qquad \varphi_{d(u)}(z) = \begin{cases} \varphi_{\varphi_u(u)}(z) & \text{if } \varphi_u(u) \text{ converges}; \\ \text{undefined} & \text{otherwise}. \end{cases}$$

By the *s-m-n* Theorem 1.5.5 d is 1:1, total, and d is independent of f. Given f, choose an index v such that

$$(2.8) \qquad \varphi_v = f \circ d.$$

We claim that $n = d(v)$ is a fixed point for f. First, note that f total implies $f \circ d = \varphi_v$ is total, so $\varphi_v(v)$ converges and $\varphi_{d(v)} = \varphi_{\varphi_v(v)}$. Now

$$(2.9) \qquad \varphi_n = \varphi_{d(v)} = \varphi_{\varphi_v(v)} = \varphi_{fd(v)} = \varphi_{f(n)}.$$

The second equality follows from (2.7) and the third follows from (2.8). $\square$

Corollary 2.2.2. *For every computable function f, there exists n such that $W_n = W_{f(n)}$.*

2.2.4   A Diagonal Argument Which Fails

Kleene first discovered the Recursion Theorem 2.2.1 by trying to refute Church's Thesis. The failure of attempts to diagonalize out of the class of partial computable functions helped convince him and Church of the robustness of the class and hence of the thesis. The above proof is best visualized as a "diagonal argument that fails" as in [Owings 1973].

Now consider the *matrix* whose rows are $R_y = \{\varphi_{\varphi_y(x)}\}_{x \in \omega}$ for $y \in \omega$. (Here $\varphi_{\varphi_y(x)}$ is the undefined function if $\varphi_y(x)$ diverges.) The entry $R_y(x)$ will be denoted in the diagram by $\varphi_y(x)$.

$$\underline{\varphi_0(0)} \quad \varphi_0(1) \quad \varphi_0(2) \quad \varphi_0(x) \quad \varphi_0(y) \qquad \varphi_0(v)$$

$$\varphi_1(0) \quad \underline{\varphi_1(1)} \quad \varphi_1(2) \quad \varphi_1(x) \quad \varphi_1(y) \quad \cdots \quad \varphi_1(v)$$

$$\varphi_2(0) \quad \varphi_2(1) \quad \underline{\varphi_2(2)} \quad \varphi_2(x) \quad \varphi_2(y) \quad \cdots \quad \varphi_2(v)$$

$$\vdots$$

$$\varphi_y(0) \quad \varphi_y(1) \quad \varphi_y(2) \quad \varphi_y(x) \quad \underline{\varphi_y(y)} \quad \cdots \quad \varphi_y(v)$$

$$\vdots$$

$$\varphi_e(0) \quad \varphi_e(1) \quad \varphi_e(2) \quad \varphi_e(x) \quad \varphi_e(y) \quad \cdots \qquad d(v) = R_e(v)^* = \varphi_e(v)$$

$$\vdots$$

$$\varphi_v(0) \quad \varphi_v(1) \quad \varphi_v(2) \quad \varphi_v(x) \quad \cdots \qquad \cdots \quad f \circ d(v) = R_v(v)^{**} = \varphi_v(v)$$

$$\vdots$$

Here the closure properties of the partial computable functions under the Parameter Theorem 1.5.5 guarantee that the diagonal sequence shown as underlined, $\underline{D} = \{\varphi_{\varphi_y(y)}\}_{y \in \omega}$, is *indeed* one of the rows, i.e., the eth row R_e, where $\varphi_e(x) = d(x)$. That is,

$$(2.10) \qquad \underline{D} = \{\varphi_{\varphi_y(y)}\}_{y \in \omega} = R_e := \{\varphi_{\varphi_e(x)}\}_{x \in \omega} = \{\varphi_{d(x)}\}_{x \in \omega}.$$

Any computable function f induces a transformation on the rows $R_u = \{\varphi_{\varphi_u(x)}\}_{x \in \omega}$ of this matrix, mapping R_u to the row $\{\varphi_{f \circ \varphi_u(x)}\}_{x \in \omega}$. In particular, f maps the *diagonal row* $R_e = \{\varphi_{d(x)}\}_{x \in \omega}$ to $R_v = \{\varphi_{f \circ d(x)}\}_{x \in \omega}$, where $\varphi_v(x) = f \circ d(x)$.

The key is to look at the vth *column* of the matrix. Consider its intersection with row R_e and row R_v. Now $R_e = \underline{D}$ is *indeed* the diagonal sequence. Hence, $R_e(v)$, the vth element of row R_e, must be $\varphi_{d(v)} = \varphi_{\varphi_v(v)}$, by the choice of e and the definition of the diagonal $d(v)$. Now $R_v(v)$, the vth element of row R_v, must be $\varphi_{\varphi_v(v)}$ by the definition of row R_v. Since these matrix entries, $R_e(v)^*$ and $R_v(v)^{**}$, are equal we conclude that $\varphi_{d(v)} = \varphi_{fd(v)}$. (These two matrix entries have been marked with * and ** to highlight them in the diagram.) Hence, $n = d(v)$ is a fixed point for f.

An Initial Application of the Recursion Theorem

A typical application of the Recursion Theorem 2.2.1 is the following.

Corollary 2.2.3. *There exists* $n \in \omega$ *such that* $W_n = \{n\}$.

Proof. By the Parameter Theorem 1.5.5 define $W_{f(x)} = \{x\}$. By the Recursion Theorem 2.2.1 there exists n such that $W_n = W_{f(n)} = \{n\}$. □

2.2.5 Informal Applications of the Recursion Theorem

The preceding example suggests a shortcut in applying the Recursion Theorem because the final line has three terms connected by equality signs. We often omit the middle term containing f and simply write, "define a p.c. function φ_n using in advance its own index n as follows:"

$$(2.11) \qquad \varphi_n \quad := \quad \{ \dots \quad n \quad \dots \}.$$

This apparent circularity is removed by the Recursion Theorem 2.2.1 because we are really using the Parameter Theorem 1.5.5 to define a computable function $f(x)$:

$$(2.12) \qquad \varphi_{f(x)}(z) \quad := \quad \{ \dots \quad x, z \quad \dots \}$$

where the portion $\{ \dots \quad x, z \quad \dots \}$ is given by a procedure uniformly computable in x and z. Finally, we apply the Recursion Theorem 2.2.1 to get a fixed point n for $f(x)$ and combine (2.11) and (2.12) to get

$$(2.13) \qquad \varphi_n(z) \quad = \quad \varphi_{f(n)}(z) \quad := \quad \{ \dots \quad n, z \quad \dots \}.$$

The only restriction here is that the algorithm $\{ \dots \quad x, z \quad \dots \}$ in (2.12) cannot use any special properties of φ_n, such as φ_n being total or convergent on a particular z, but rather must apply to *any* φ_x even if at the end we can show that φ_n is total.

Even though this seems technically simple, it leads to a big conceptual advance for us in applying the Recursion Theorem 2.2.1 since from now on we may assume we have the index before we begin to define φ_n.

2.2.6 Other Properties of the Recursion Theorem

We stated the Recursion Theorem 2.2.1 above in its simplest form although the proof actually yields considerably more information which will be useful and which we now state explicitly.

Proposition 2.2.4. *In the Recursion Theorem 2.2.1 the fixed point n can be computed from an index for f by a 1:1 computable function g.*

Proof. Let $v(x)$ be a computable function such that $\varphi_{v(x)} = \varphi_x \circ d$. Let $g(x) = d(v(x))$. Both d and v are 1:1 by the *s-m-n* Theorem 1.5.5.   □

Proposition 2.2.5. *In the Recursion Theorem 2.2.1 there is an infinite computable set of fixed points for f.*

Proof. By the Padding Lemma 1.5.2 there is an infinite c.e. set V of indices v such that $\varphi_v = f \circ d$, but d is 1:1 so $\{d(v)\}_{v \in V}$ is infinite and c.e. Choose an infinite increasing and therefore computable subset.   □

For more sophisticated applications such as Myhill's Theorem 2.4.6 we need the following form of the Recursion Theorem with parameters. (By coding finite sequences it suffices to consider a single parameter.) The proof exploits the uniformity of the earlier proof and is otherwise identical.

Theorem 2.2.6 (Recursion Theorem with Parameters, Kleene).   *If $f(x, y)$ is a computable function, then there is a 1:1 computable function $n(y)$ such that $\varphi_{n(y)} = \varphi_{f(n(y), y)}$.*

Proof. Define a 1:1 computable function d by

$$(2.14) \qquad \varphi_{d(x,y)}(z) := \begin{cases} \varphi_{\varphi_x(x,y)}(z) & \text{if } \varphi_x(x, y) \text{ converges;} \\ \text{undefined} & \text{otherwise.} \end{cases}$$

Choose v such that $\varphi_v(x, y) = f(d(x, y), y)$. Then $n(y) = d(v, y)$ is a fixed point, because

$$\varphi_{d(v,y)} = \varphi_{\varphi_v(v,y)} = \varphi_{f(d(v,y),y)}.$$

The first equality follows from the definition of d in (2.14), and the second from the definition of v.   □

The following theorem strengthens Theorem 2.2.6 by replacing the total function $f(x, y)$ by a partial function $\psi(x, y)$ using the same proof. This form will be useful later.

Theorem 2.2.7. *If $\psi(x, y)$ is a partial computable function, then there is a computable function $n(y)$ such that for all y*

$$\psi(n(y), y)\downarrow \qquad \Longrightarrow \qquad \varphi_{n(y)} = \varphi_{\psi(n(y), y)}.$$

Proof. Use the same proof as for Theorem 2.2.6.   □

2.2.7   Exercises

Exercise 2.2.8. Prove that K is not an index set.

Exercise 2.2.9. Define the following set of *minimal indices* of p.c. functions $M := \{ x : \neg(\exists y < x)\,[\, \varphi_x = \varphi_y \,] \}$.

(i) Prove that M is infinite but contains no infinite c.e. set. *Hint.* Use the Recursion Theorem.

(ii) (Jockusch). Let h be a computable function which is finite-one, i.e., for each y there are only finitely many e with $h(e) = y$. (Note that if $h(e)$ is the "size" of the program with Gödel number e, measured in any reasonable way, then h is finite-one.) Let

$$M_h = \{ e : (\forall i)\,[\, h(i) < h(e) \quad . \implies . \quad \varphi_i \neq \varphi_e \,] \}.$$

Show that M_h is an infinite set which does not contain any infinite c.e. subset.

Exercise 2.2.10. A set A is *self-dual* if $A \leq_m \overline{A}$. For example, if $A = B \oplus \overline{B}$ then A is self-dual.

(a) Use the Recursion Theorem to prove that no index set A can be self-dual.

(b) Use the Recursion Theorem to give a short proof of Rice's Theorem 1.6.14.

Exercise 2.2.11. Show that Corollary 2.2.2 is equivalent to: For every c.e. set A, $(\exists n) [W_n = \{x : \langle x, n \rangle \in A\}]$. *Hint.* Define the c.e. set

$$A_n = \{x : \langle x, n \rangle \in A\}$$

and compare the sequence $\{A_n\}_{n \in \omega}$ with $\{W_n\}_{n \in \omega}$.

2.3   Indexing Finite and Computable Sets

If $A = W_e$ then e is a Σ_1^0-index for A by Definition 2.1.4 and e gives a way to enumerate A. However, if A is computable, then we might want a stronger Δ_0^0-*index* i such that φ_i is the characteristic function $A(x)$ for A. Furthermore, if A is finite, we might want a still stronger *canonical index* which specifies all the members of A, its cardinality, and its maximum element. Here we explore these stronger indices for computable or finite sets, and use the Recursion Theorem 2.2.1 to prove that one cannot pass effectively from a weaker to a stronger index.

2.3.1   Computable Sets and Δ_0 and Δ_1 Indices

Definition 2.3.1. (i) We say that e is a Σ_1^0-index (Σ_1-index, c.e. index) for a set A if $A = W_e$.

(ii) e is a Δ_0^0-*index* (Δ_0-*index, characteristic index*) for a set A if φ_e is the characteristic function χ_A for A.

(iii) $\langle e, i \rangle$ is a Δ_1^0-*index* (Δ_1-*index*) for a computable set A if $A = W_e$ and $\overline{A} = W_i$.

We normally drop the superscript 0 following convention 2.1.2.

Remark 2.3.2. Clearly, from a Δ_0-index for a computable set A we can effectively obtain a Δ_1-index for A and therefore a Σ_1-index for A. The proof of the Complementation Theorem 2.1.14 allows us to pass effectively from a Δ_1-index for A to a Δ_0-index for A. From now on we consider Δ_0 and Δ_1 indices to be completely synonymous.

However, we cannot, in general, pass effectively from a Σ_1-index for a computable set A to a Δ_1-index for A, as the following theorem establishes.

Theorem 2.3.3 (Nonconversion Theorem). *There is no partial computable function ψ such that*

$$(\forall x)\,[\;W_x \text{ computable} \quad \Longrightarrow \quad (\exists y)\,[\;\psi(x)\!\downarrow\; = \; y \;\; \& \;\; \varphi_y = W_x\;].$$

(This asserts that there is no partial computable function which effectively converts any Σ_1 index x for a computable (even finite) set A, namely $W_x = A$, and outputs a Δ_0 index y such that φ_y is the characteristic function of W_x, written $\varphi_y = W_x$.)

Proof. Define a computable function f as follows and use the Recursion Theorem 2.2.1 in the manner described in §2.2.5 to obtain a fixed point n for f. Note that the following c.e. set W_n is computable because it is either empty or a singleton.

$$(2.15) \quad W_n = W_{f(n)} = \begin{cases} \{0\} & \text{if } \psi(n)\!\downarrow \text{ and } \varphi_{\psi(n)}(0)\!\downarrow = 0; \\ \emptyset & \text{otherwise.} \end{cases}$$

Now $\varphi_{\psi(n)}$ cannot be χ_{W_n}, the characteristic function of W_n, because $0 \in W_n$ iff $\varphi_{\psi(n)}(0) = 0$ iff $\chi_{W_n}(0) = 0$ iff $0 \notin W_n$. $\quad\square$

Corollary 2.3.4. *There is no partial computable function θ such that*

$$(\forall x)\,[\;W_x \text{ computable} \quad \Longrightarrow \quad (\exists z)\,[\;\theta(x)\!\downarrow\; = \; z \;\; \& \;\; W_z = \overline{W}_x\;].$$

Proof. This follows by Theorem 2.3.3 and the effective transfer between Δ_0-indices and Δ_1-indices in Remark 2.3.2. $\quad\square$

Corollary 2.3.5. *The computable sets are closed under $\cup$, $\cap$, and complementation. The closure under $\cup$ and $\cap$ is uniformly effective with respect to both Σ_1^0-indices and Δ_1^0-indices. The closure under complementation is uniformly effective with respect to Δ_1^0-indices only.*

2.3.2   Canonical Index y for Finite Set D_y and String σ_y

A finite set, being computable, has both a Σ_1-index and a Δ_0-index. Yet the latter does not effectively specify the maximum element of A or even the cardinality of A. Thus, for each finite set we introduce a third index which explicitly specifies all the elements of A.

Definition 2.3.6. (i) Given a finite set $A = \{x_1, x_2, \ldots, x_k\}$ we define $y = 2^{x_1} + 2^{x_2} + \cdots + 2^{x_k}$ to be the *canonical index* of A. Let D_y denote the finite set with canonical index y and D_0 denote $\emptyset$. (We usually imagine that $x_1 < x_2 < \cdots < x_k$ although the order does not matter.)

(ii) A sequence $\{D_{f(x)}\}_{x \in \omega}$ for some computable function f is called a *strong array* of finite sets.

(iii) Recall that $2^{<\omega}$ denotes the set of finite strings of 0's and 1's. For σ, τ in $2^{<\omega}$ let $\sigma \prec \tau$ denote that σ is an initial segment of τ. Let D_y be in as (i) above. Given $z = \langle y, m \rangle$, define the *string* σ_z *with canonical index* z:

$$(2.16) \qquad \sigma_z(x) = \begin{cases} 1 & \text{if } x \in D_y \quad \& \quad x < m \\ 0 & \text{if } x \notin D_y \quad \& \quad x < m \end{cases}$$

Note that the length $|\sigma_z| = m$. If $z = 0$, then define σ_0 to be the empty string $\emptyset$ sometimes denoted by ϵ.

If y is written in binary expansion then the elements of D_y are the positions where the digit 1 occurs. For example, the binary expansion of 5 is 101 and $D_5 = \{0, 2\}$. Clearly, there are computable functions f and g such that $f(y) = \max\{z : z \in D_y\}$ and $g(y) = |D_y|$. However, there is no computable way to go from a Δ_0-index for a finite set to its cardinality or to a maximum element. (See Exercise 2.3.13.)

Theorem 2.3.7. *There is no p.c. function φ_i such that if φ_x is the characteristic function of a finite set F then $\varphi_i(x)\!\downarrow = y$ where $D_y = F$.*

Proof. Assume φ_i exists. We build φ_n to be the characteristic function of a finite set. By the Recursion Theorem 2.2.1 we are given n in advance. Wait for the least stage s (if ever) such that $\varphi_{i,s}(n)\!\downarrow = y$. Meanwhile let $\varphi_n(x) = 0$ for all $x < s$. If stage s arrives, choose any $z \notin D_y$, $z > s$, and define $\varphi_n(z) = 1$. If s does not exist, then φ_n is the characteristic function of $\emptyset$ which is finite, so φ_i fails to have the necessary properties. $\square$

2.3.3 Acceptable Numberings of Partial Computable Functions

In Definition 1.7.5 we defined a sequence $\{\psi_e\}_{e \in \omega}$ to be an *acceptable numbering* of the p.c. functions if there exist computable functions f and g such that (1) $\varphi_{f(x)} = \psi_x$; and (2) $\psi_{g(x)} = \varphi_x$. Theorem 1.7.6 showed that for an acceptable numbering there is a computable permutation h such that $\psi_x = \varphi_{h(x)}$ provided that we can replace f and g by one-one functions f_1 and g_1 with the same properties.

It is easy to do this for f by the Padding Lemma 1.5.2 for the standard numbering $\{\varphi_e\}_{e \in \omega}$, but there was no obvious reason why g_1 existed. We now complete the proof by showing that g_1 exists. Therefore, the acceptable numbering $\{\psi_e\}_{e \in \omega}$ also satisfies the Padding Lemma. Note that in our proof we need only g (not f) to produce g_1.

Definition 2.3.8. For any p.c. function φ_x define the set of *indices* for φ_x,

$$(2.17) \qquad \text{Ind}_x = \{ y : \varphi_y = \varphi_x \}.$$

Theorem 2.3.9. *Suppose that $\{\psi_x\}_{x\in\omega}$ is any sequence of p.c. functions (not necessarily c.e.) and g is a computable function such that $\varphi_x = \psi_{g(x)}$. Then there is a one-one computable function $g_1(x)$ such that $\varphi_x = \psi_{g_1(x)}$.*

In the following theorem $g(\mathrm{Ind}_e)$ must be infinite or else Ind_e is computable, contradicting Rice's Theorem 1.6.14. However, given e and $g(e) = i$ at stage s it may be that i is already in the range of g_1. The objective is to find another index j not already in the range of g_1 with $\varphi_e = \psi_j$ and to do so computably and uniformly in e. The Recursion Theorem 2.2.1 supplies this index in a surprising fashion.

Corollary 2.3.10 (Acceptable Numbering Theorem). *If $\{\psi_x\}_{x\in\omega}$ is an acceptable numbering of the partial computable functions as in Definition 1.7.5, then there is a computable permutation h of ω such that $\varphi_e = \psi_{h(e)}$ for all e.*

Proof. (of Corollary 2.3.10). The one-one function g_1 of Theorem 2.3.9 completes the proof of the Acceptable Numbering Theorem 1.7.6.   □

Proof. (of Theorem 2.3.9). To obtain g_1 first define a function h such that

$$(2.18) \qquad (\forall x)\,[\,\varphi_{h(e,x)} = \varphi_e\,] \quad \text{and}$$

$$(2.19) \qquad (\forall x)\,(\forall y)\,[\,x \neq y \implies gh(e,x) \neq gh(e,y)\,].$$

Fixing e, define $h(e,k)$ by induction on k. Define $h(e,0) = e$. Define $B_k = \{gh(e,0),\ldots,gh(e,k)\}$. Use the Recursion Theorem 2.2.1 to define

$$(2.20) \qquad \varphi_n(z) = \begin{cases} \varphi_e(z) & \text{if } g(n) \notin B_k; \\ \text{undefined} & \text{if } g(n) \in B_k. \end{cases}$$

Case 1. $g(n) \notin B_k$. Then $\varphi_n = \varphi_e$ by the first clause of (2.20). Therefore, $n \in \mathrm{Ind}_e - g^{-1}(B_k)$. We define $h(e,k+1) = n$.

Case 2. $g(n) \in B_k$. Then $\varphi_n = \lambda z[\uparrow]$ by the second clause of (2.20). By inductive hypothesis, if $g(j) \in B_k$ then $\varphi_j = \varphi_e$ according to (2.18). Hence, $\varphi_n = \varphi_e = \lambda z[\uparrow]$. We now use the following Lemma 2.3.11 to find a new index $m \in \mathrm{Ind}_e - g^{-1}(B_k)$ for the undefined function φ_e and we define $h(e,k+1) = m$.

Note that the condition $g(n) \in B_k$ is computable because B_k is finite and g is computable by hypothesis. Therefore, we can tell which of the two cases in (2.20) will hold in defining φ_n.

Lemma 2.3.11. *Let $\varphi_e = \lambda z[\uparrow]$. There is a computable function $m(x)$ such that if $W_x \subseteq \mathrm{Ind}_e$, then $\varphi_{m(x)} = \varphi_e$ and $m(x) \notin W_x$.*

Proof. Fix $\varphi_i = \lambda z\,[1]$. Using the Recursion Theorem with Parameters 2.2.6 define the computable function $m(x)$ as follows:

$$(2.21) \qquad \varphi_{m(x)}(z) = \begin{cases} \varphi_i(z) & \text{if } m(x) \in W_x; \\ \\ \uparrow & \text{if } m(x) \notin W_x. \end{cases}$$

Suppose the first clause holds in (2.21). Since $m(x) \in W_x$, we know $\varphi_{m(x)} = \varphi_i \neq \varphi_e$ by the choice of φ_i. However, $\varphi_{m(x)} = \varphi_e$ because $m(x) \in W_x \subseteq \mathrm{Ind}_e$. This is a contradiction. Therefore, $\varphi_{m(x)}$ was defined by the *second* clause of (2.21) and $m(x) \notin W_x$. By the definition (2.21) we have $\varphi_{m(x)} = \varphi_e = \lambda z\,[\uparrow]$. This proves Lemma 2.3.11.   □

Given the finite set B_k above, choose an x such that $W_x = B_k$ and define $m = m(x)$. Then $m \in \mathrm{Ind}_e - g^{-1}(B_k)$, and we define $h(e, k+1) = m$. This completes the definition of h. Now define $g_1(x)$ by induction on x. Assume $g_1(y)$ has been defined for exactly those $y < x$. Given x, define

$$j = (\mu k)(\forall y < x)[\, g(h(x, k)) > g_1(y)\,].$$

Define $g_1(x) = gh(x, j)$. This proves Theorem 2.3.9.   □

Remark 2.3.12. Lemma 2.3.11 is a special case of the more general theorem that the index set Ind_x is *productive* as defined in Definition 2.4.3. However, the only fact we are using here is the productive property of Ind_e for $\varphi_e = \lambda z\,[\uparrow]$. In Exercise 2.4.13 we prove that the index set Ind_x is productive for *every* x but not *uniformly* in x.

2.3.4   Exercises

Exercise 2.3.13. Suppose we are given a Δ_0-index x for a finite set F. Prove that from any one of the following three items one can compute the remaining two: (1) a canonical index y for F; (2) $z = \max F$; (3) $v = |F|$. Conclude by Theorem 2.3.7 that there is no effective way to pass from x to any of these three.

Exercise 2.3.14. $^\diamond$ (Double Recursion Theorem, Smullyan). (i) Prove that for any computable functions $f(x, y)$ and $g(x, y)$ there exist a and b such that $\varphi_a = \varphi_{f(a,b)}$ and $\varphi_b = \varphi_{g(a,b)}$. *Hint.* Apply the Recursion Theorem with Parameters 2.2.6 to get a computable function $\widehat{a}(y)$ such that $\varphi_{\widehat{a}(y)} = \varphi_{f(\widehat{a}(y),y)}$. Let b be a fixed point for the computable function $h(y)$, where $\varphi_{h(y)} = \varphi_{g(\widehat{a}(y),y)}$. Let $a = \widehat{a}(b)$.

(ii) Use exactly the same method to prove the Double Recursion Theorem with Parameters, i.e., for any computable functions $f(x, y, z)$ and $g(x, y, z)$ there exist computable functions $a(z)$ and $b(z)$ such that

$$\varphi_{a(z)} = \varphi_{f(a(z), b(z), z)} \qquad \text{and} \qquad \varphi_{b(z)} = \varphi_{g(a(z), b(z), z)}.$$

2.4 * Complete Sets and Creative Sets

In §1.6 we defined the reducibilities $\leq_m$ and $\leq_1$. In Definition 3.2.2 we shall define the more general Turing reducibility $\leq_T$.

Definition 2.4.1. Let $r = 1$, m or T reducibility. A set A is r-*complete* if A is c.e. and $W \leq_r A$ for every c.e. set W.

The following Theorem 2.4.2 was given as Exercise 1.6.22. We now give the complete but short proof of this key result, because we need to pass back and forth between these three computably isomorphic sets.

Theorem 2.4.2. *The sets K, K_0, and K_1 are all 1-complete. Therefore, $K_0 \equiv K_1 \equiv K$ by Myhill's Isomorphism Theorem 1.7.4.*

Proof. Each is c.e. by §2.1 equations (2.2), (2.3), and (2.4). Furthermore, $K_0 = \{\langle x, y \rangle : x \in W_y\}$ is clearly 1-complete, since $x \in W_y$ iff $\langle x, y \rangle \in K_0$. Let W be any c.e. set. Now, as in Theorem 1.6.11, use the s-m-n-Theorem to define a 1:1 computable function f as follows:

$$W_{f(x)} = \begin{cases} \omega & \text{if } x \in W; \\ \emptyset & \text{otherwise.} \end{cases}$$

Now if $x \in W$, then $W_{f(x)} = \omega$ and $f(x) \in K \cap K_1$. If $x \in \overline{W}$ then $W_{f(x)} = \emptyset$, so $f(x) \notin (K \cup K_1)$. Thus, f witnesses that $W \leq_1 K$ and $W \leq_1 K_1$. □

2.4.1 Productive Sets

Definition 2.4.3. (i) A set P is *productive* if there is a p.c. function $\psi(x)$, called a *productive function* for P, such that

$$(\forall x)\,[\, W_x \subseteq P \quad\Longrightarrow\quad [\, \psi(x)\!\downarrow \ \& \ \psi(x) \in P - W_x \,]\,].$$

(ii) A c.e. set C is *creative* if $\overline{C}$ is productive.

For example, the set K is creative because $\overline{K}$ is productive via the identity function $\psi(x) = x$. Since $K \equiv K_0 \equiv K_1$ by Theorem 2.4.2, we know that K_0 and K_1 are also creative. (Note that if $W_x \subseteq \overline{K}$ then $x \notin W_x$ and $x \notin K$.) A creative set C is *effectively noncomputable* in the sense that for any candidate W_x for $\overline{C}$, $\psi(x)$ is an effective counterexample, because $\psi(x) \in \overline{C} - W_x$. These sets were called *creative* in [Post 1944] because their existence justifies "the generalization that every symbolic logic is incomplete and extendible relative to the class of propositions" necessary to code the relation $\{(x, y) : x \in W_y\}$. Post remarked: "The conclusion is inescapable that even for such a fixed, well-defined body of mathematical propositions, *mathematical thinking is, and must remain, essentially creative*."

Theorem 2.4.4. *Any productive set P has a total 1:1 computable productive function p.*

Proof. See Exercise 2.4.11.   □

Theorem 2.4.5. *(i) If P is productive then P is not c.e.*

(ii) If P is productive with productive function $p(x)$, then P contains an infinite c.e. set W.

(iii) If P is productive and $P \leq_m A$ then A is productive.

Proof. (i) Immediate.

(ii) Let $W_n = \emptyset$, and $W_{h(x)} = W_x \cup \{\, p(x)\, \}$. Define

$$(2.22) \qquad W \;=\; \{\, p(n),\, ph(n),\, ph^2(n),\, \dots\, \}.$$

(iii) Let $P \leq_m A$ via f, and let p be a productive function for P. Let $W_{g(x)} = f^{-1}(W_x)$. Then fpg is a productive function for A.   □

Using the fact that $\overline{K}$ is productive, Theorem 2.4.5 (iii) gives us a method for exhibiting new productive sets P by showing that $\overline{K} \leq_m P$. We now use the Recursion Theorem 2.2.1 to prove that *every* productive set has this property. An immediate corollary is that all creative sets are 1-complete and hence computably isomorphic to K.

2.4.2   ** Creative Sets Are Complete

Theorem 2.4.6 (Creative Set Theorem, Myhill, 1955).
(i) *If P is productive then $\overline{K} \leq_1 P$.*

(ii) *If C is creative, then C is 1-complete and $C \equiv K$.*

Proof. (i)   Let p be a total 1:1 productive function for P. Define the computable function f by

$$W_{f(x,y)} \;=\; \begin{cases} \{\, p(x)\, \} & \text{if } y \in K; \\ \emptyset & \text{otherwise.} \end{cases}$$

By the Recursion Theorem with Parameters 2.2.6 there is a 1:1 computable function $n(y)$ such that

$$(2.23) \qquad W_{n(y)} \;=\; W_{f(n(y),y)} \;=\; \begin{cases} \{\, p(\, n(y)\,)\, \} & \text{if } y \in K; \\ \emptyset & \text{otherwise.} \end{cases}$$

$$(2.24) \quad y \in K \implies W_{n(y)} = \{pn(y)\} \implies W_{n(y)} \not\subseteq P \implies pn(y) \in \overline{P},$$

$$(2.25) \quad y \in \overline{K} \implies W_{n(y)} = \emptyset \implies W_{n(y)} \subseteq P \implies pn(y) \in P.$$

In both lines the first implication follows from the definition of $n(y)$. The fact that p is a productive function for P yields the second implication in (2.24) and the third implication in (2.25). But (2.24) and (2.25) yield $\overline{K} \leq_1 P$ via the function $g(y) = p(n(y))$. Part (ii) follows by (i) and the Myhill Isomorphism Theorem 1.7.4. □

Corollary 2.4.7. *The following are equivalent:*
(i) *P is productive;*
(ii) *$\overline{K} \leq_1 P$;*
(iii) *$\overline{K} <_m P$.* □

Corollary 2.4.8. *The following are equivalent:*
(i) *C is creative;*
(ii) *C is 1-complete;*
(iii) *C is m-complete.* □

The next definition will be useful in Exercise 2.4.18 below and in Chapter 4.

Definition 2.4.9. Let (A_1, A_2) and (B_1, B_2) be two pairs of sets such that $A_1 \cap A_2 = \emptyset = B_1 \cap B_2$. Then

$$(2.26) \qquad\qquad (B_1, B_2) \leq_m (A_1, A_2)$$

if there is a computable function f such that $f(B_1) \subseteq A_1$, $f(B_2) \subseteq A_2$, and $f(\overline{B_1 \cup B_2}) \subseteq \overline{A_1 \cup A_2}$. We write "$\leq_1$" if f is 1:1.

2.4.3   Exercises

Exercise 2.4.10. Extend the proof of Theorem 2.4.5 (ii) to show that if P is productive, then there is a computable function f such that

$$(\forall x)\,[W_x \subseteq P \implies [[W_{f(x)} \subseteq (P \cap \overline{W_x})] \quad \& \quad |W_{f(x)}| = \infty]].$$

Apply the procedure to an arbitrary set W_n in place of $W_n = \emptyset$ in the proof of Theorem 2.4.5 (ii). Prove that if $W_n \subseteq P$ then $W \subseteq P$ for the set W in (2.22). Prove that if a repetition occurs in equation (2.22) then $W_n \not\subseteq P$. The algorithm needs to succeed only if $W_n \subset P$ but it needs to be defined for all indices because we cannot computably determine whether $W_n \subset P$.

Exercise 2.4.11. Prove Theorem 2.4.4 that every productive set has a total 1:1 computable productive function. *Hint.* (i) *Total.* Let P be productive via ψ. First define a computable function g such that

$$W_{g(x)} = \begin{cases} W_x & \text{if } \psi(x)\downarrow; \\ \emptyset & \text{otherwise.} \end{cases}$$

Show that $g(x)$ is computable. Define $q(x)$ to be either $\psi(x)$ or $\psi(g(x))$, whichever converges first.

(ii) *One-One.* Convert q to a 1:1 productive function p. Let $W_{h(x)} = W_x \cup \{q(x)\}$. Apply the method of proof of Theorem 2.4.5 (ii) and Exercise 2.4.10. Define $p(0) = q(0)$. To compute $p(x+1)$, enumerate the set

$$\{q(x+1),\, qh(x+1),\, qh^2(x+1),\, \ldots\}$$

until either some y not in $\{p(0), \ldots, p(x)\}$ is found, or a repetition occurs.

Exercise 2.4.12. If A is c.e. and $A \subset \overline{K}$ prove that $A \cup K \equiv K$. *Hint.* Use the Recursion Theorem as in (2.23) to prove that $K \leq_m A \cup K$.

Exercise 2.4.13. As in Definition 2.3.8, let $\text{Ind}_x = \{y : \varphi_y = \varphi_x\}$.

(i) Prove that for each x, Ind_x is productive. Note that Lemma 2.3.11 proved this for $\varphi_e(y)$ undefined for all y. That proof works if $\text{dom}(\varphi_x) \neq \omega$. See also Remark 2.3.12. *Hint.* If $\text{dom}(\varphi_x) = \omega$, then combine Corollary 2.4.7 and Exercise 1.6.24.

(ii) (Fejer). Show that this reduction $\overline{K} \leq_m \text{Ind}_x$ is not uniform in x, i.e., there is no computable function $f(x, y)$ in the sense that for all x, $y \in \overline{K}$ iff $f(x, y) \in \text{Ind}_x$. *Hint.* Choose $y_0 \in K$, consider $\lambda x \, [\, f(x, y_0)\,]$, and use the Recursion Theorem on f.

Exercise 2.4.14. Prove that if $K \leq_m A$ via f and A is not c.e. then there are $2^{\aleph_0}$ sets B such that $K \leq_m B$ via f. Hence there are $2^{\aleph_0}$ productive sets. *Hint.* If A is not c.e. then $S = A - f(K)$ is infinite.

Remark 2.4.15. In Exercise 1.6.26 we defined a disjoint pair (A, B) of c.e. sets to be *computably inseparable* if there is no computable set C such that $A \subseteq C$ and $C \cap B = \emptyset$. Now we define the stronger notion of *effectively inseparable*, which implies that any c.e. set C which separates must be creative and therefore noncomputable. Many pairs of c.e. sets in computability theory, such as the provable and refutable sentences in the language of Peano arithmetic, form pairs of effectively inseparable sets.

Definition 2.4.16. A disjoint pair (A, B) of c.e. sets is *effectively inseparable* if there is a partial computable function ψ, called a *productive* function for (A, B), such that for all x and y,

$$[A \subseteq W_x \; \& \; B \subseteq W_y \; \& \; W_x \cap W_y = \emptyset]$$
$$\implies \quad [\psi(\langle x, y \rangle)\downarrow \; \& \; \psi(\langle x, y \rangle) \notin (W_x \cup W_y)].$$

Note that if the pair (A, B) is effectively inseparable then it is computably inseparable. However, the converse is false.

Exercise 2.4.17. Assume the disjoint pair (A, B) of c.e. sets is effectively inseparable via ψ.

(i)   Use the method of Exercise 2.4.11 to prove that we may assume ψ is total and 1:1.

(ii)   (Let $W_x \setminus W_y$ be as in Definition 2.6.4.) Prove that the following c.e. sets are effectively inseparable: $A = \{x : \varphi_x(x) = 0\}$ and $B = \{x : \varphi_x(x) = 1\}$. *Hint.* Define ψ by

$$\varphi_{\psi(\langle x,y \rangle)}(z) = \begin{cases} 1 & \text{if } z \in W_x \setminus W_y \\ 0 & \text{if } z \in W_y \setminus W_x \\ \uparrow & \text{otherwise.} \end{cases}$$

(iii)   Show that if (A, B) is a pair of effectively inseparable c.e. sets, then each of A and B is creative.

Exercise 2.4.18. (Smullyan). Assume that (A_1, A_2) is a pair of effectively inseparable c.e. sets with productive function p. Prove that if (B_1, B_2) is a pair of disjoint c.e. sets then $(B_1, B_2) \leq_1 (A_1, A_2)$, as in Definition 2.4.9. *Hint.* By the Double Recursion Theorem with one parameter Exercise 2.3.14, define 1:1 computable functions $g(z)$, $h(z)$ such that

$$W_{g(z)} = \begin{cases} A_1 \cup \{p(\langle g(z), h(z) \rangle)\} & \text{if } z \in B_2 \\ A_1 & \text{otherwise,} \end{cases}$$

and

$$W_{h(z)} = \begin{cases} A_2 \cup \{p(\langle g(z), h(z) \rangle)\} & \text{if } z \in B_1, \\ A_2 & \text{otherwise.} \end{cases}$$

Let $f(z) = p(\langle g(z), h(z) \rangle)$ and show that $(B_1, B_2) \leq_1 (A_1, A_2)$ via f.

Exercise 2.4.19. (Jockusch-Mohrherr). Let A be any c.e. set except ω. Prove that A is creative iff

$$(\forall \text{ c.e. } W)\, [\, A \cap W = \emptyset \quad \Longrightarrow \quad A \equiv A \cup W\,].$$

Hint. Show that $\overline{A}$ contains an infinite c.e. subset W. Take an infinite computable subset $R \subset W$ and a creative subset $C \subset R$. Now $A \equiv A \cup C$ by hypothesis. Prove that A is creative.

Exercise 2.4.20.$^\diamond$ Let $\mathcal{C}$ be a class of c.e. sets. Show that the following two statements are equivalent.

(i) $\{e : W_e \in \mathcal{C}\}$ is c.e.

(ii)   There is a c.e. set I such that for all c.e. sets W,

$$W \in \mathcal{C} \quad \Longleftrightarrow \quad (\exists i)\,[\, i \in I \ \& \ D_i \subseteq W\,].$$

(Direction (ii) implies (i) is obvious, but (i) implies (ii) is harder. For this direction first prove from (i) that $U \in \mathcal{C}$ and $U \subseteq V$ implies $V \in \mathcal{C}$. Next prove that if $W \in \mathcal{C}$ then there is some finite set $F \subseteq W$ such that $F \in \mathcal{C}$.)

2.5 ** Elementary Lachlan Games

Lachlan proposed a new kind of game to analyze problems and prove theorems on c.e. sets. We introduce it briefly now and develop it more fully in Chapter 16.

2.5.1 The Definition of a Lachlan Game

Definition 2.5.1. A *Lachlan game* is a game between two players.

- Player I (RED) constructs a finite or infinite sequence of c.e. sets, $\{U_n\}_{n \in \omega}$ called the *red* sets.

- Player II (BLUE) constructs a finite or infinite sequence $\{V_n\}_{n \in \omega}$ of c.e. sets, called the *blue* sets.

- At every *move* (*stage*) of the game RED may enumerate at most finitely many integers into finitely many red sets and then BLUE may enumerate at most finitely many integers into finitely many blue sets. Let Z_s denote the finite set of elements enumerated in set Z by the end of stage s.[1]

- Each *play* of the game ends after ω many moves.

- At the start of the game we fix a computable sequence $\{R_e\}_{e \in \omega}$ of finite conditions on the red and blue sets called *requirements* which will be specified precisely in advance for each particular game.

- Player II *wins* the play of the game if the red and blue sets constructed during that play satisfy all the requirements, and Player I wins otherwise.

- A *game situation* at the end of stage s is the result $\{U_{n,s}\}_{n<s}$ and $\{V_{n,s}\}_{n<s}$ and can be effectively coded by an integer, say using the canonical indices for these finite sets.

- A *winning strategy* for a given player is a function from game situations to moves (and can be coded as a function f on ω) such that if the player follows this strategy he will win against any sequence of moves by his opponent. To prove a theorem we wish to prove that one player or the other has a winning strategy for that particular game.

---

[1]Without loss of generality we may assume that $[x \in U_{n,s} \vee x \in V_{n,s}] \implies x, n < s$.

- We begin by thinking of the strategies as computable functions and therefore of the sets U_n and V_n as c.e., but the games can be relativized to an oracle X, in which case the strategies are X-computable and the sets are X-c.e.

2.5.2 Playing Partial Computable (P.C.) Functions

This definition may seem too restrictive because in many theorems in computability we may allow one or both players to construct other objects such as partial computable (p.c.) functions or Turing reductions $\Phi_e^A(x)$, e.g. in the Friedberg-Muchnik theorem to meet a requirement such as $R_e : B \neq \Phi_e^A$. However, no generality is lost because these objects can be constructed as c.e. sets. For example, to define φ_e we enumerate the c.e. set

$$\mathrm{graph}(\varphi_e) := \{ \langle x, y \rangle \ : \ \varphi_e(x) = y \}$$

which we defined in Definition 2.1.7. Similarly, the Turing functionals Φ_e defined later can be viewed as c.e. sets. To play a p.c. function or Turing functional we simply enumerate axioms of the graph on moves during the game. Therefore, these Lachlan games are as general as we need to prove most theorems about c.e. sets.

2.5.3 Some Easy Examples of Lachlan Games

We begin by analyzing some known previous theorems in terms of Lachlan games. Since BLUE has a winning strategy for these games we can let RED play $V_n = W_n$ without loss of generality. We give a reference to the earlier result and sketch the winning strategy for BLUE as a game.

The construction of a noncomputable c.e. set K in Definition 1.6.3 is a game. RED plays W_n for all n and BLUE plays K. BLUE withholds n from K until, if ever, RED puts n into W_n, and then BLUE puts n into K.

The construction of a pair of effectively inseparable c.e. sets A and B in Exercise 2.4.17 is similar. RED plays φ_e for all e and BLUE plays A and B. BLUE withholds e from either A or B until $\varphi_e(e)\downarrow$, and then enumerates e into whichever of A and B witnesses that $\varphi_e(e)$ has the wrong value to separate them.

Theorem 2.3.3 on the lack of conversion of a Σ_1 index to a Δ_0 index for a computable set is a Lachlan game. RED plays a potential p.c. conversion function ψ and BLUE plays W_n. BLUE withholds 0 from W_n until there exists y such that $\psi(n) \downarrow = y$ and $\varphi_y(0) = 0$. If this never happens then $0 \notin W_n$ but $\varphi_y(0) \neq 0$. If y appears, then BLUE puts 0 into W_n so that $\varphi_y(0) \neq W_n(0) = 1$. (Notice that BLUE wins this game because he plays a Σ_1 object and RED plays a Δ_0 object, which gives BLUE one more move.)

The Myhill Creative Set Theorem 2.4.6 is also a Lachlan game. RED claims that a fixed set P is productive and plays a 1:1 total computable productive function $p(x)$ to witness productivity. BLUE plays a 1:1 computable function $n(y)$ and claims that $\overline{K} \leq_1 P$ via $p(n(y))$. For a fixed y, BLUE withholds $p(n(y))$ from $W_{n(y)}$ until $y \in K$ and then enumerates $p(n(y))$ in $W_{n(y)}$. If $y \in \overline{K}$, then $p(n(y)) \in P$ by (2.25). If $y \in K$ then the action by BLUE forces $p(n(y))$ to move "down" from P to $\overline{P}$ by (2.24). (This action is remarkable because we have no information about the productive set P, but only the property of its productive functon $p(x)$.)

2.5.4   Practicing Lachlan Games

In §2.5.3 we gave several examples of easy Lachlan games. It is not enough to simply *read* these examples. You must *practice* them as with any game. Go back over the original exercises or theorems and study them as a game. One of the new things to determine is which of the two players is playing which items. In general, for a known theorem, the given items are played by RED, and the items we construct are played by BLUE. We shall often use the games in future theorems and examples.

2.5.5   The Significance of Lachlan Games

The first advantage of a Lachlan game is that it helps to reveal, in dynamic terms, the rationale behind a given proof. The second advantage is that it helps to solve a new problem when we do not know which side will win. We assign BLUE and RED to opposite sides of an open question and let them struggle to find a winning strategy. When we sense a weakness for BLUE, we change sides and attempt to exploit it for RED. We keep changing sides back and forth with no loyalty to either side as we seek a solution. This dynamic tension is at the heart of many proofs in the subject, but authors rarely make any explicit reference to the game.

2.5.6   Exercises on Lachlan Games

Exercise 2.5.2. Let f be a computable function. Show that there is an n such that: (i) W_n is computable; and (ii) $(\mu y)\,[\,W_y = \overline{W}_n\,] > f(n)$. *Hint.* Let RED play f. Let BLUE play W_n by first computing $f(n)$ using the Recursion Theorem, and then defining a finite set W_n which is a counterexample.

Exercise 2.5.3. Prove that Inf $\leq_1$ Cof by the following function f. Let RED play W_x and BLUE play $W_{f(x)}$. Whenever RED adds a new element to A, BLUE puts the least element b_0^s of $\overline{B}^s$ into B.

Exercise 2.5.4. ° ** (Lachlan).

(i) Prove that if $K \leq_m A \times B$ and B is c.e., then $K \leq_m A$ or $K \leq_m B$.
(ii) Conclude that there is no pair of incomparable c.e. m-degrees whose l.u.b. is the m-degree of K. *Hint.* Note that $A \oplus B \leq_m A \times B$.

Hint for (i). Let RED play B c.e. and play a Turing reduction $K_0 \leq_m A \times B$ via computable functions $h(z)$ and $j(z)$, i.e.,

$$z \in K_0 \iff [\, h(z) \in A \ \& \ j(z) \in B \,].$$

BLUE constructs a c.e. set $D = W_e$ whose index e is known in advance by the Recursion Theorem 2.2.1. Therefore, we have

(2.27) $x \in W_e \iff \langle x, e \rangle \in K_0 \iff [\, f(x) \in A \ \& \ g(x) \in B \,]$

for $f(x) = h(\langle x, e \rangle)$ and $g(x) = j(\langle x, e \rangle)$. BLUE plays two strategies $\mathcal{S}_1$ and $\mathcal{S}_2$. Call x *good at stage* s if $x \notin D_s$ and $g(x) \in B_s$. Let $G = \{x_0, x_1, \ldots\}$ be a computable enumeration of the elements good at some stage. BLUE plays a strategy $\mathcal{S}_1$ which shows that if $|G| = \infty$, then

$$n \in K \iff x_n \in D \iff f(x_n) \in A.$$

Simultaneously, BLUE's backup strategy $\mathcal{S}_2$ shows that if G is finite, then $D =^* K$ and $x \in D$ iff $g(x) \in B$.

2.6 ⊘ The Order of Enumeration of C.E. Sets

2.6.1 Uniform Sequences and Simultaneous Enumerations

Definition 2.6.1. (i) A sequence of c.e. sets $\mathbb{V} = \{V_e\}_{e \in \omega}$ is *uniformly c.e. (u.c.e.)* if there is a computable function f such that $V_e = W_{f(e)}$. (In §3.5.2 we discuss the general concept of *uniform* computability and give examples.)

(ii) If $\mathbb{V} = \{V_e\}_{e \in \omega}$ is a u.c.e. sequence, and h is a 1:1 computable function on ω such that $h(\omega) = \{\langle x, e \rangle : x \in V_e\}$, then h is a *simultaneous computable enumeration (s.c.e.)* of $\mathbb{V}$. Intuitively, if $h(s) = \langle x, e \rangle$, then h *causes* x to be enumerated in V_e at stage s. Using h we define the array:

(2.28) $V_{e,s} = \{\, x : (\exists t \leq s)\,[\, h(t) = \langle x, e \rangle \,] \,\}.$

(iii) When dealing with an s.c.e. h we write $\mathbb{V}_h$ to indicate the dependence on h of the order of enumeration in (2.28). Define the *infinite join*:

(2.29) $\oplus_{y \in \omega} A_y := \{\, \langle x, y \rangle : x \in A_y \,\}_{y \in \omega}.$

If $\oplus_{e \in \omega} V_e$ is finite then we allow h to be *partial* provided that $h(s)$ defined implies $h(t)$ defined for all $t < s$ so that the domain of h is an initial segment of ω. This allows us to pass uniformly from any enumeration of $\oplus_e V_e$ to an s.c.e. h of it, because we do not have to ask whether

$\oplus_e V_e$ is finite. That is, we just continue to define h as in the Listing Theorem 2.1.10 (ii) so long as new elements enter $\oplus_e V_e$.

Convention 2.6.2. Definition 1.6.17 of the stage s approximation $W_{e,s}$ to W_e was convenient at the time, but it just bounded the search by stage s and may have allowed several elements to be enumerated at the same stage s in the same or different c.e. sets. From now on we fix some particular *simultaneous computable enumeration (s.c.e.)* h_0 of the standard array $\{W_e\}_{e \in \omega}$ of all c.e. sets and define $W_{e,s}$ to be the finite set derived from h_0 as in (2.28). We may assume that h_0 also satisfies:

$$(2.30) \qquad x \in W_{e,s} \quad \Longrightarrow \quad x, e < s$$

as in Definition 1.6.2. We now have

$$(2.31) \qquad (\forall s)\,(\exists \text{ at most one } \langle e, x \rangle)\,[\, x \in W_{e,s+1} - W_{e,s}\,].$$

This is convenient for many later constructions.

2.6.2 Static and Dynamic Properties of C.E. Sets

The properties of a c.e. set W_e presented so far have largely been *static* properties, but *dynamic* properties will play an important role.

Definition 2.6.3. (i) A property of a c.e. set W_e is *static* if it is described in terms of W_e as a completed object, such as a lattice property or the Σ_1^0 characterization of W_e in §2.1.1.

(ii) A property of a c.e. set W_e is *dynamic* if it is described in terms of a computable enumeration $\{W_{e,s}\}_{s \in \omega}$ of W_e and measures time-dependent properties, such as those which determine which elements have entered a certain set W_e before entering another given set W_i under a given simultaneous enumeration.

Theorems 2.6.5 and 2.7.1 in the following sections represent excellent examples of a combination of static and dynamic properties used to prove an easy but important theorem. We shall usually be doing a computable construction where the hypotheses and conclusions are c.e. sets and p.c. functions. Usually the only information we have is a Σ_1 index e for $A = W_e$, and such an e gives only a dynamic procedure for enumerating A as it is revealed stage by stage, and not any of its static properties as a completed object.

Definition 2.6.4. (Dynamic Definitions for C.E. Sets).   Fix a simultaneous computable enumeration (s.c.e.) function h of an array of c.e. sets $\{V_e\}_{e \in \omega}$ as in (2.28). Let X_s denote $V_{i,s}$ and Y_s denote $V_{j,s}$ for some i and j. Then the following sets are c.e.

(i) Define $X \setminus Y = \{z : (\exists s)\,[\, z \in X_s - Y_s \,]\}$, the elements enumerated in X before (if ever) being enumerated in Y.

(ii) Define $X \searrow Y = (X \setminus Y) \cap Y$, the elements enumerated in X and later in Y.

(Do not confuse $X \backslash Y$ with $X - Y$, which denotes $X \cap \overline{Y}$. Notice that the definitions $X \setminus Y$ and $X \searrow Y$ depend not only upon the *sets* X and Y but also upon the particular s.c.e. function h.)

Theorem 2.6.5 (Dynamic Flow Theorem).   *Fix a simultaneous computable enumeration (s.c.e.) of the c.e. sets* $\{W_e\}_{e \in \omega}$. *Suppose B is a noncomputable c.e. set. Fix some b with $W_b = B$ and define $B_s = W_{b,s}$ so that the simultaneous enumeration includes B. Then for every e,*

$$(2.32) \qquad W_e \supseteq \overline{B} \qquad \Longrightarrow \qquad W_e \searrow B \;\; \text{infinite},$$

and indeed $W_e \searrow B$ is noncomputable (see Exercise 2.6.7).

Proof. Note that $W_e \setminus B = (W_e \searrow B) \sqcup \overline{B}$. If $|W_e \searrow B| < \infty$, then $W_e \setminus B =^* \overline{B}$. Hence, $\overline{B}$ is c.e. and B is computable, contrary to hypothesis. □

2.6.3   Exercises

Exercise 2.6.6. If h is a simultaneous computable enumeration of a u.c.e. array $\{V_e\}_{e \in \omega}$ prove that there is a computable function $g(e, s)$ such that for $V_{e,s}$, defined in (2.28), we have $D_{g(e,s)} = V_{e,s}$ and hence that $V_e = \cup_s D_{g(e,s)}$.

Exercise 2.6.7. (i)   Given computable enumerations $\{X_s\}_{s \in \omega}$ and $\{Y_s\}_{s \in \omega}$ of c.e. sets X and Y prove that both $X \setminus Y$ and $X \searrow Y$ are c.e. sets.

(ii) Prove that $X \setminus Y = (X - Y) \cup (X \searrow Y)$.

(iii) Prove that if $X - Y$ is not c.e. then $X \searrow Y$ is infinite and indeed noncomputable.

(iv) Give an alternative proof of the Reduction Principle Corollary 2.1.12 by letting $A_1 = W_x \setminus W_y$ and $B_1 = W_y \setminus W_x$ where $W_x = A$ and $W_y = B$.

Exercise 2.6.8. Prove that there is a computable function f such that $\{W_{f(n)}\}_{n \in \omega}$ consists precisely of the computable sets. (Hence we can give an effective list of Σ_1^0-indices for the computable sets but not of Δ_1-indices for them.) *Hint.* Obtain $W_{f(n)} \subseteq W_n$ by enumerating W_n, placing in $W_{f(n)}$ only those elements enumerated in increasing order. (Note that we are using the uniformity shown in Exercise 2.1.24.)

2.7 ⊘ The Friedberg Splitting Theorem

The following theorem neatly combines the static and dynamic properties of c.e. sets. Its statement is static and traditional, but its proof relies on a key dynamic property.

2.7.1 The Priority Ordering of Requirements

The framework of the construction and proof is one we shall use many times in more complicated situations. We have a list of conditions to be satisfied called *requirements*, $\{R_e\}_{e \in \omega}$, which are given their natural ω ordering of priority: R_0, R_1, R_2, ... We say that requirement R_i has *higher priority* than R_j if $i < j$. The point is that every requirement R_i has at most finitely many predecessors of higher priority.

We construct a c.e. set A by stages during a computable construction, letting A_s denote the set of elements enumerated in A by the end of stage s, and we define $A = \cup_s A_s$. At stage $s + 1$ we examine every requirement R_i, $i < s$, which *requires attention* at stage $s+1$ in that: (1) R_i is not currently satisfied, and (2) some associated condition such as equation (2.34) below holds which gives us the opportunity to satisfy R_i now. From these R_i we select the highest priority requirement R_e (i.e., the least index e) and we allow R_e to *receive attention*, that is to *act* at stage $s + 1$.

In the following construction, if a requirement R_e acts at stage $s + 1$ it immediately becomes satisfied and remains satisfied at all stages $t > s + 1$. However, in more complicated constructions, such as the finite injury constructions of Chapter 7, a given requirement R_e, after being satisfied at stage s, may later be *injured* at some stage $t > s$ when some higher priority requirement R_i, $i < e$, acts, and R_e must start all over.

Theorem 2.7.1 (Friedberg Splitting Theorem). *If B is any noncomputable c.e. set there exist disjoint c.e. sets A_0 and A_1 such that:*

(i) $B = A_0 \sqcup A_1$ *(denoting disjoint union), and*

(ii) A_0 *and* A_1 *are noncomputable.*

Proof. Let f be a 1:1 computable function with range B, and define

$$B_s = \{f(0), \ldots, f(s)\}.$$

For (ii) we try to meet for all $e \in \omega$ and $i < 2$ the requirement

(2.33) $R_{\langle e,i \rangle}$: $W_e \neq \overline{A}_i$.

Stage $s = 0$. Put $f(0)$ in A_0. Let $A_{i,s}$ denote the set of elements in A_i by the end of stage s.

Stage $s + 1$. Choose the least $\langle e, i \rangle < s$ if it exists such that

(2.34) $f(s+1) \in W_{e,s}$ & $W_{e,s} \cap A_{i,s} = \emptyset$.

Enumerate $f(s+1)$ in A_i (so $W_{e,s} \cap A_{i,s+1} \neq \emptyset$ and $R_{\langle e,i \rangle}$ is satisfied forever after). We say that requirement $R_{\langle e,i \rangle}$ *acts* at stage $s+1$. If $\langle e,i \rangle$ fails to exist, then enumerate $f(s+1)$ in A_0.

Let $A_i = \cup_s A_{i,s}$ for $i = 0, 1$. Now $A_0 \cap A_1 = \emptyset$ and $A_0 \cup A_1 = B$. To verify $R_{\langle e,i \rangle}$ choose the least $\langle e,i \rangle$ such that requirement $R_{\langle e,i \rangle}$ is not satisfied. Hence, $W_e = \overline{A_i}$ so $W_e \supseteq \overline{B}$ because $A_i \subseteq B$. Therefore, $|W_e \setminus B| = \infty$ by the Dynamic Flow Theorem 2.6.5. But each $R_{\langle j,k \rangle}$, $\langle j,k \rangle < \langle e,i \rangle$, acts at most once. Therefore, $R_{\langle e,i \rangle}$ eventually corresponds to the minimal $\langle e,i \rangle$ satisfying (2.34), at which stage $R_{\langle e,i \rangle}$ acts and becomes satisfied forever thereafter, which is a contradiction.   □

2.7.2   Exercises

Exercise 2.7.2. Generalize Theorem 2.7.1 by using the same proof to show that if W is any c.e. set, then

$$(2.35) \qquad W - B \text{ non-c.e.} \quad \Longrightarrow \quad W - A_i \text{ non-c.e. for } i = 0, 1.$$

Furthermore, c.e. indices for A_0 and A_1 can be found uniformly from a c.e. index for B.

3
Turing Reducibility

3.1 The Concept of Relative Computability

3.1.1 Turing Suggests Oracle Machines (o-Machines)

After his epochal paper in 1936 on Turing machines, Turing wrote a Ph.D. dissertation with Alonzo Church which was published in 1939. A tiny part of this paper in section 4 contained the germ of one of the most important ideas in all of modern computability theory. In half a page Turing suggested augmenting his former a-machines by adding some kind of "oracle" which could supply the answers to specific questions during the computation.

> Let us suppose we are supplied with some unspecified means of solving number-theoretic problems; a kind of oracle as it were. We shall not go any further into the nature of this oracle apart from saying that it cannot be a machine. With the help of the oracle we could form a new kind of machine (call them o-machines), having as one of its fundamental processes that of solving a given number-theoretic problem.
>
> These machines may be described by tables of the same kind as those used for the description of a-machines.

3.1.2 Post Develops Relative Computability

The subject of oracle machines and relative computability lay dormant for five years from 1939 until Post's beautiful 1944 paper on computably enumerable sets, which heavily influenced the subject for decades. Post explored the concept of *reducing* a set B to another set A. He introduced first some stronger reducibilities, such as many-one reducibility $B \leq_m A$, previously studied in Definition 1.6.8, and truth-table reducibility $B \leq_{tt} A$, which we shall study in Definition 3.8.2. General Turing reducibility was not well understood in 1944. Post was the principal developer of it in a series of papers over the next decade in 1944, in 1948, and in 1954, the key paper by Kleene and Post.

In the crucial last section of Post's 1944 paper, *General (Turing) Reducibility*, Post named it "Turing reducibility," written $B \leq_T A$. Post described $B \leq_T A$ in intuitive terms as the most general form of the reducibilities he had explored. It is similar to the ordinary computation process to determine whether $x \in B$, except that at certain times the machine process chooses an element y and asks whether $y \in A$? If a correct answer is supplied, then the machine continues in this fashion until after a finite number of steps it yields the answer to the question, is $x \in B$? In a finite time at most a finite number of questions can be asked. Post asserted that a corresponding formulation of "Turing reducibility" should capture the intuitive notion of effective reducibility just as Turing computability captures effective calculability.

Consider a laptop computer connected to the Internet. This interaction between the local computer and a very large database is called *interactive computing* in modern computer terminology. Now remove the Internet connection. What remains is the finite oracle program on the laptop, which is independent of the database. This emphasizes the finite nature of the local *oracle machine (o-machine)* and its oracle program independent of the database.

3.2 ⋆⋆ Turing Computability

Neither Post nor Turing gave a formal definition of an o-machine. We present one such formulation of a Turing o-machine in §3.2.1.

3.2.1 An o-Machine Model for Relative Computability

Definition 3.2.1. (i) An *oracle Turing machine (o-machine)* is a Turing machine with the usual work tape and an extra "read only" tape, called the *oracle tape*, upon which is written the characteristic function of some set A, called the *oracle*, whose symbols $\{0, 1\}$ cannot be printed over, and are preceded by $B's$. Two reading heads move along these two tapes simultaneously.

As before, Q is a finite set of states, $S_1 = \{B, 0, 1\}$ is the oracle tape alphabet, $S_2 = \{B, 1\}$ is the work tape alphabet, and $\{R, L\}$ is the set of head-moving operations right and left. In a given state with the heads reading symbols s_1 and t_1 the machine can overprint t_1 on the work tape, change to a new state, and move each reading head one square right or left independently.

(ii) An *oracle Turing program* is a finite sequence of program lines as above. Fix an effective coding of all oracle Turing programs for o-machines. Let $\widetilde{P}_e$ denote the eth such oracle program under this effective coding.

(iii) If the oracle machine halts, let u be the maximum cell on the oracle tape scanned during the computation, i.e., the maximum integer tested for membership in A. We define u to be the *use of the computation*. We say that the elements $z \leq u$ are *used* in the computation. If no element is scanned we let $u = 0$.

3.2.2 Turing Computable Functionals Φ_e

Definition 3.2.2. (i) If the oracle program $\widetilde{P}_e$ with A on the oracle tape and input x halts with output y and if u is the maximum element *used* (scanned) on the oracle tape during the computation, then we write,

(3.1) $$\Phi_e^A(x) = y \quad \text{and} \quad \varphi_e^A(x) = u.$$

We refer to Φ_e^A as a *Turing functional* (*Turing reduction*), and we call φ_e^A the corresponding *use function*. Note that $\Phi_e^A(x)$ may be undefined for some x because $\widetilde{P}_e$ may be any oracle program. Such a functional is a *partial* functional but we often drop the word *partial*, unlike in the case of partial *functions*. The functional is simply determined by the program $\widetilde{P}_e$ and may be partial or total. However, for every x we have $\Phi_e^A(x)$ defined iff $\varphi_e^A(x)$ is defined.

(ii) If this happens in $< s$ steps and if e, x, y, and $u < s$, then we write,

(3.2) $$\Phi_{e,s}^A(x) = y \quad \text{and} \quad \varphi_{e,s}^A(x) = u$$

and we assume from now on that $\varphi_{e,s}^A(x) < s$.

(iii) If $\sigma \in 2^{<\omega}$ and this happens with σ on the oracle tape and $u < |\sigma|$ (therefore only σ is scanned) then we write,

(3.3) $\Phi_e^\sigma(x) = y$, $\varphi_e^\sigma(x) = u$, $\Phi_{e,s}^\sigma(x) = y$,   and $\varphi_{e,s}^\sigma(x) = u$.

(iv) If $\Phi_{e,s}^A(x)$ diverges, we write $\Phi_{e,s}^A(x) \uparrow$ and $\varphi_{e,s}^A(x) \uparrow$. Let $W_e^A = \mathrm{dom}(\Phi_e^A)$, and likewise for $W_{e,s}^A$, W_e^σ, and $W_{e,s}^\sigma$.

(v) A partial function θ is *Turing computable in A (A-Turing computable)*, written $\theta \leq_T A$, if there is an e such that $\Phi_e^A(x)\!\downarrow\, = y$ iff $\theta(x) = y$. A set B is *Turing reducible to A ($B \leq_T A$)* if the characteristic function $\chi_B \leq_T A$.

(vi) We also allow (total) functions f as oracles by defining Φ_e^f to be Φ_e^A where $A = \{\langle x, y \rangle : f(x) = y\}$, and allow partial functions ψ as outputs.

Definition 3.2.3. A functional Ψ on Cantor space 2^ω is *total* if $\Psi^A(x)$ is defined for *every* $A \subseteq \omega$ and $x \in \omega$, i.e., if $(\forall A)(\exists B)(\forall x)[\; \Psi^A(x) = B(x)\;]$.

Following Church and Turing, we say that a set B is *effectively reducible* to another set A if it is reducible in the intuitive sense.

Thesis 3.2.4 (The Post-Turing Thesis). *The set B is effectively reducible to another set A iff B is Turing reducible to A ($B \leq_T A$).*

This is implicit in Turing's 1939 paper and Post's 1944 paper. To justify the Post-Turing Thesis, note that if $B \leq_T A$ then clearly B is effectively reducible to A. For the converse assume that B is effectively reducible to A via some procedure P. Now either argue that P's being effective corresponds to some oracle program $\widetilde{P}_e$ or argue that the graph of P is effectively enumerable and therefore c.e.

Remark 3.2.5. We regard Φ_e as a partial function from 2^ω to 2^ω, which takes A to B if $\Phi_e^A = B$. We call this map a *functional* (type 2 object). The lower-case symbol φ_e with *no exponent* is the usual p.c. function in Definition 1.5.1. The symbol $\varphi_e^A(x)$ *with an exponent A* is the use function for $\Phi_e^A(x)$. The use function is defined only if there is an exponent A and a computation whose use is being measured. The notation for the use function $\varphi_e^A(x)$ always has an exponent such as A and does not conflict with the partial computable function $\varphi_e(x)$, which has no exponent.[1]

3.3   * Oracle Graphs of Turing Functional Φ_e

3.3.1   The Prefix-Free Graph F_e of Functional Φ_e

The oracle machine described in §3.2 is *self-delimiting* in the sense that if $\Phi_e^\sigma(x)$ on input x halts after reading exactly string σ on its oracle tape, i.e.,

---

[1]Some writers use upper-case Φ_e for the p.c. function φ_e, but we avoid this because uppercase Φ_e with no exponent represents a *type 2* partial functional on 2^ω. This oracle algorithm is identified with the oracle graph G_e and oracle program $\widetilde{P}_e$. However, φ_e is a *type 1* p.c. function on ω given by the Turing program P_e. We wish to keep the types distinct and consistent.

with use $\varphi_e^\sigma(x) = |\sigma| - 1$, then it must turn off and not read any more of the oracle tape. Therefore, the machine on input x cannot read any $\rho \succ \sigma$ and has not halted on any $\rho \prec \sigma$. The oracle machine is deterministic and on a fixed input x and oracle σ goes through the computation in a sequence of stages $\Phi_{e,s}^\sigma$ for $s \in \omega$.

Definition 3.3.1. For all e and x we define the set of *minimal halting strings,*

$$(3.4) \qquad H_{e,x} = \{ \sigma : (\exists s)[\ \Phi_{e,s}^\sigma(x)\downarrow \quad \text{and} \quad \varphi_{e,s}^\sigma(x) = |\sigma| - 1\ ].$$

These are the strings σ for which the computation of $\Phi_e^\sigma(x)$ enters the halting state q_0 after reading exactly σ.

Definition 3.3.2. A set of strings $S \subseteq 2^{<\omega}$ is *prefix-free* if S is an antichain with respect to the standard partial ordering $\prec$ on strings, i.e.,

$$(3.5) \qquad (\forall \sigma)(\forall \tau)[\ [\ \sigma \in S\ \&\ \sigma \prec \tau\ ] \implies \tau \notin S\ ].$$

Theorem 3.3.3. *For every e and x, $H_{e,x}$ of (3.4) is prefix-free.*

Proof. Suppose that $\sigma \in H_{e,x}$. This means that oracle program $\widetilde{P}_e$ on input x and oracle σ entered the halting state q_0 at some stage s. At this point the machine halts forever and never makes another move. □

In (2.1) we defined $\mathrm{graph}(\psi) = \{\langle x, y \rangle : \psi(x) = y\}$ for partial function ψ. This graph exactly characterizes ψ and is c.e. iff ψ is partial computable. From a p.c. funtion ψ we can compute the c.e. set of its graph, and from a c.e. set which is a graph we can compute the p.c. function it defines. In specifying a graph for a Turing *functional* in place of a function, we need to add the oracle string for the computation. We first do this in graph F_e for the minimal strings and then in G_e for all strings.

Definition 3.3.4. Given Definition 3.2.2 (iii) of $\Phi_e^\sigma(x)$, we define the *prefix-free oracle graph* of Φ_e as the following c.e. set of *axioms:*

$$(3.6) \qquad F_e = \{\ \langle \sigma, x, y \rangle\ :\ \Phi_e^\sigma(x) = y\ \&\ \sigma \in H_{e,x}\ \}.$$

Therefore, F_e includes only those σ which are minimal strings for e and x. Later, the oracle graph G_e of (3.7) will include *all* strings which give the computation.

Theorem 3.3.5. (Unique Use Property). *In (3.6), for every x, y, e, and oracle A there is at most one string $\sigma \prec A$ such that $\langle \sigma, x, y \rangle \in F_e$.*

Proof. Suppose $\Phi_e^A(x) \downarrow = y$. Then by Theorem 3.3.3 there is a unique $\sigma \prec A$ such that $\Phi_e^\sigma(x)$ converges and $\sigma \in H_{e,x}$. □

Remark 3.3.6. This unique use property will be applied in infinite injury, where we shall replace Φ_e^A by the hat-trick functional $\widehat{\Phi}_e^A$, which becomes undefined whenever an element less than the use $\varphi_e^A(x)$ enters A. For this to succeed we need to know that the current use $\varphi_{e,s}^A(x)$ is unique.

3.3.2   The Oracle Graph G_e of Functional Φ_e

Definition 3.3.7. Given the Definition 3.2.2 (iii) and (3.3) of $\Phi_e^\sigma(x)$, define the *oracle graph* of Φ_e as the following c.e. set of *axioms:*

$$(3.7) \qquad G_e = \{\, \langle \sigma, x, y \rangle \,:\, \Phi_e^\sigma(x) = y \,\}.$$

Of course, the oracle graph G_e cannot be prefix-free because monotonicity (3.9) contradicts the prefix-free property (3.5). However, every functional Φ_e determines a prefix-free graph $F_e \subset G_e$ which we may regard as a *basis* for G_e similar to a basis for a vector space.

Theorem 3.3.8 (Oracle Graph Theorem).   *Let the relation $\Phi_e^\sigma(x) = y$ be as in Definition 3.2.2, let F_e be as in (3.6), and let G_e be as in (3.7).*

(i) G_e *is single-valued in the sense that*

$$(3.8) \qquad [\, \langle \sigma, x, y \rangle \in G_e \quad \& \quad \langle \sigma, x, z \rangle \in G_e \,] \qquad \Longrightarrow \qquad y = z.$$

(ii) G_e *is c.e.*

(iii) G_e *is monotonic in the sense that*

$$(3.9) \qquad \langle \sigma, x, y \rangle \in G_e \quad \Longrightarrow \quad (\forall \tau \succ \sigma)[\, \langle \tau, x, y \rangle \in G_e \,].$$

(iv) $\langle \sigma, x, y \rangle \in G_e \qquad \Longleftrightarrow \qquad (\exists \tau \preceq \sigma)[\, \langle \tau, x, y \rangle \in F_e \,].$

Proof. (i)   This follows by the convention in Definition 3.2.1 (iii) of an o-machine that the machine can give an output only when it enters the halting state q_0, and thereafter it can make no moves and give no further output on that string and input. Therefore, for every pair (σ, x) there can be at most one y such that $\Phi_e^\sigma(x) = y$.

(ii)   Clearly, G_e is c.e. because $\langle \sigma, x, y \rangle \in G_e$ iff $(\exists s)[\Phi_{e,s}^\sigma(x) = y]$. This is Σ_1 and therefore c.e.

(iii)   For property (3.9), apply the definition of an o-machine computation. If $\Phi_{e,s}^\sigma(x) = y$ then when the machine produces the output y it enters the halting state q_0, and will never make any more moves on input x. Therefore, for $\tau \succeq \sigma$ the machine with τ on the oracle tape must eventually enter q_0 and give output y by exactly the same computation, and will never make any more moves.

(iv)   Suppose $\Phi_e^\tau(x)\!\downarrow = y$. Then $\langle \sigma, x, y \rangle \in F_e$ for the minimal such $\sigma \preceq \tau$. For every $\rho \succeq \sigma$, we have $\langle \rho, x, y \rangle \in G_e$ by (iii).   □

3.3.3   The Use Principle for Turing Functionals

Theorem 3.3.9 (Use Principle).   *Definition 3.2.2 guarantees:*

(i) $\Phi_e^A(x) = y$ $\Longrightarrow$ $(\exists s)\,(\exists \sigma \prec A)\,[\,\Phi_{e,s}^\sigma(x) = y\,];$

(ii) $\Phi_{e,s}^\sigma(x) = y$ $\Longrightarrow$ $(\forall t \geq s)\,(\forall \tau \succeq \sigma)\,[\,\Phi_{e,t}^\tau(x) = y\,];$

(iii) $\Phi_e^\sigma(x) = y$ $\Longrightarrow$ $(\forall A \succ \sigma)\,[\,\Phi_e^A(x) = y\,].$

Proof. For (i), any computation which converges does so at some finite stage, having used only finitely many elements. We may take $\sigma = A \restriction$ $\restriction \varphi_{e,s}^A(x)$. Now (ii) was proved in the monotonic property (3.9), and (iii) follows in the same manner from the definition of an oracle computation. □

Use Principle 3.3.9 is crucial for most of our subsequent theorems, where we often combine (i) and (iii). If $\Phi_e^A(x) = y$, we choose σ as in (i). Now for all $B \succ \sigma$ we must have $\Phi_e^B(x) = y$. We very often use the following fact.

(3.10) $[\,\Phi_{e,s}^A(x) = y\ \ \&\ \ A \restriction\!\restriction \varphi = B \restriction\!\restriction \varphi\,]$ $\Longrightarrow$ $\Phi_{e,s}^B(x) = y,$

where $\varphi = \varphi_{e,s}^A(x)$. This is because the B-computation $\Phi_{e,s}^B(x)$ will read only $\sigma = A \restriction\!\restriction \varphi$ on the oracle tape before it enters the halting state q_0.

3.3.4   Permitting Constructions

The oracle graph theorem plays a crucial role in permitting constructions in §5.2.4 and in building a Turing functional $\Theta^C = A$ in §5.5.3. When we add an axiom $\langle \sigma, x, y \rangle$ to the oracle graph G for Θ with $\sigma \prec C_s$, we must respect this commitment on $\Theta^C(x)$ until some element $v \leq |\sigma|$ enters C_t, in which case C *permits* us to change the value as explained in §5.2.

3.3.5   Lachlan Notation for Approximation by Stages

The following notation by Lachlan has become very popular and is now used in most papers and books. The idea is that a single object $\Phi_e^A(x)$ is being approximated in stages where the value of each parameter may depend on s. Therefore, we write $[s]$ *once* to apply to all occurrences.

Definition 3.3.10. *(Lachlan notation).* When $E(A_s, x_s, y_s, \ldots)$ is an expression with a number of arguments subscripted by s denoting their values at stage s, the notation $E(A, x, y, \ldots)[\,s\,]$ denotes the evaluation of E with all arguments taken with their values at s.

(3.11) $\Phi_e^A(x)\,[\,s\,]$ denotes $\Phi_{e,s}^{A_s}(x_s)$ and $\varphi_e^A(x)\,[\,s\,]$ denotes $\varphi_{e,s}^{A_s}(x_s)$.

3.3.6   Standard Theorems Relativized to A

Theorem 3.3.11 (Relativized Enumeration Theorem).   *There exists $z \in \omega$ such that for all sets $A \subseteq \omega$ and for all $x, y \in \omega$, the A-partial computable function $\Phi_z^A(x, y)$ satisfies $\Phi_z^A(x, y) = \Phi_x^A(y)$.*   □

Theorem 3.3.12 (Relativized s-m-n Theorem).   *For every $m, n \geq 1$ there exists a 1:1 computable (not merely A-computable) function s_n^m of $m + 1$ variables such that for all sets $A \subseteq \omega$ and for all $x, y_1, y_2, \ldots, y_m \in \omega$,*

$$\Phi_{s_n^m(x, y_1, \ldots, y_m)}^A = \lambda z_1, \ldots, z_n \, [\ \Phi_x^A(y_1, \ldots, y_m, z_1, \ldots, z_n)\ ].$$

Proof. Take $m = n = 1$, and let s denote s_n^m. The new program $\widetilde{P}_{s(x,y)}$ on input z consists of applying program $\widetilde{P}_x$ to input (y, z). This defines $s(x, y)$ as a *computable* function since the program $\widetilde{P}_{s(x,y)}$ is independent of the oracle A. Of course, s can be made 1:1 by padding, as in Lemma 1.5.2.   □

Theorem 3.3.13 (Relativized Recursion Theorem, Kleene).   (i) *For all sets $A \subseteq \omega$ and all $x, y \in \omega$, if $f(x, y)$ is an A-computable function, then there is a computable function $n(y)$ such that $\Phi_{n(y)}^A = \Phi_{f(n(y), y)}^A$.*

(ii) *Furthermore, $n(y)$ does not depend upon the oracle A, i.e., if*

$$f(x, y) = \Phi_e^A(x, y)$$

then the computable function $n(y)$ can be found uniformly in e.

Proof. Apply the proof of the Recursion Theorem with Parameters 2.2.6 but notice that $n(y)$ is a computable function (not merely A-computable) because the function $s(x, y)$ in the s-m-n Theorem 1.5.5 is computable, and hence the function $d(x, y)$ is computable, where $d(x, y)$ is obtained as in the proof of Recursion Theorem with Parameters 2.2.6 but with all partial functions relativized to A.   □

Definition 3.3.14. (i) B is *computably enumerable in A* if $B = W_e^A$ for some e, where $W_e^A = \mathrm{dom}(\Phi_e^A)$ as in Definition 3.2.2 (iv).

(ii) B is *in $\Sigma_1^{0,A}$-form* (abbreviated B is Σ_1^A) if $B = \{x : (\exists \overline{y})\, R^A(x, \overline{y})\}$ for some A-computable predicate $R^A(x, \overline{y})$. (By the Quantifier Contraction Theorem 2.1.5, this is equivalent to asserting that $B = \{x : (\exists y)\, R^A(x, y)\}$ for some such R^A.)

All the results of Chapter 2, §2.1 on c.e. sets relativize to A-c.e. sets by virtually the same proofs, where we replace "c.e." and "computable" by "A-c.e." and "A-computable." For example:

Theorem 3.3.15 (Relative Complementation Theorem). $B \leq_{\mathrm{T}} A$ iff B and $\overline{B}$ are c.e. in A.

Theorem 3.3.16. *The following are equivalent:*

(i) B *is c.e. in* A;

(ii) $B = \emptyset$ *or* B *is the range of some* A-*computable total function;*

(iii) B *is* Σ_1^A.

Proof. The proofs of (i) $\Longleftrightarrow$ (ii) and (iii) $\Longleftrightarrow$ (i) are the relativizations to A of the proofs in §2.1. E.g., to prove (i) $\Longrightarrow$ (iii), let $B = W_e^A$. Hence, by the Use Principle Theorem 3.3.9,

$$(3.12) \qquad x \in B \iff (\exists s)\,(\exists \sigma)\,[\,\sigma \prec A \ \& \ x \in W_{e,s}^\sigma\,].$$

Now $x \in W_{e,s}^\sigma$ is a computable relation on (e, σ, x, s) by the Graph Theorem 2.1.9 and $\sigma \prec A$ is an A-computable relation of σ because $\sigma \prec A$ iff $(\forall y < |\sigma|)\,[\sigma(y) = A(y)]$. Hence, (3.12) is of the form $(\exists s)\,(\exists \sigma)\,R(e, \sigma, x, s)$, where R is an A-computable relation.   □

3.3.7 Exercises

Exercise 3.3.17. Give an effective coding of the programs $\{\widetilde{P}_e\}_{e \in \omega}$ using the method of §1.5.1 and prove that under this coding $\theta_e =_{\mathrm{dfn}} \Phi_e^\emptyset$ is an acceptable numbering of the partial computable functions as defined in Definition 1.7.5.

Exercise 3.3.18. (i)   Prove that $\{\langle e, \sigma, x, y, s\rangle \ : \ \Phi_{e,s}^\sigma(x) = y\}$ is computable.

(ii) Prove that $G := \{\ \langle e, \sigma, x, y\rangle \ : \ \Phi_e^\sigma(x) = y\ \}$ is Σ_1, 1-complete, and hence computably isomorphic to K.

Exercise 3.3.19. Given sets B and A prove that $B \leq_{\mathrm{T}} A$ iff there are computable functions f and g such that

$$x \in B \iff (\exists \sigma)\,[\,\sigma \in W_{f(x)} \ \& \ \sigma \prec A\,],$$

$$x \in \overline{B} \iff (\exists \sigma)\,[\,\sigma \in W_{g(x)} \ \& \ \sigma \prec A\,].$$

Exercise 3.3.20. Given c.e. sets A and B prove that $B \leq_{\mathrm{T}} A$ iff there is a computable function h such that for D_y as in Definition 2.3.6 we have:

$$x \in \overline{B} \iff (\exists y)\,[\,y \in W_{h(x)} \ \& \ D_y \subseteq \overline{A}\,].$$

Exercise 3.3.21. In the Relativized *s-m-n* Theorem 3.3.12 we obtained a *computable* (rather than an A-computable) function s. Usually, relativization of the theorems of Chapter 1 and 2 produce only A-*computable* functions. For example, exhibit a set A and an A-computable relation $R^A(x, y)$ which cannot be uniformized by any partial computable function in the sense of Theorem 2.1.8.

Exercise 3.3.22. Prove that if θ is partial computable in $Y \subseteq \omega$, then there is a *computable* function f such that

$$(\forall x \in \mathrm{dom}(\theta)) [\, W^Y_{f(x)} = W^Y_{\theta(x)} \,].$$

Furthermore, an index for f can be found uniformly computably from an e such that $\theta = \Phi^Y_e$.

3.4 * Turing Degrees and the Jump Operator

3.4.1 The Structure of the Turing Degrees

Definition 3.4.1. (i) $A \equiv_T B$ if $A \leq_T B$ and $B \leq_T A$. (Note that $\leq_T$ is reflexive and transitive, so $\equiv_T$ is an equivalence relation.)

(ii) The *Turing degree* (also called *degree of unsolvability*) of A is the equivalence class $\deg(A) = \{B : B \equiv_T A\}$.

(iii) Lower-case boldface letters **a**, **b**, **c** denote degrees, and **D** denotes the class of all degrees.

(iv) The degrees **D** form a partially ordered set $(\mathbf{D}, \leq)$ under the relation $\deg(A) \leq \deg(B)$ iff $A \leq_T B$. We write $\deg(A) < \deg(B)$ if $A <_T B$, i.e., if $A \leq_T B$ and $B \nleq_T A$.

(v) $\deg(A) \vee \deg(B) = \deg(A \oplus B)$. We also write $\deg(A) \cup \deg(B)$ for $\deg(A) \vee \deg(B)$. With $\vee$ the degrees $(\mathbf{D}, <, \vee)$ form an *upper semi-lattice* (as defined in the Notation section) with a supremum but not an infimum.

(vi) A degree **a** is *computably enumerable* if it contains a c.e. set. Let **C** denote the class of c.e. degrees with the same ordering as for **D**.

(vii) A degree **a** is *computably enumerable in* **b** if **a** contains some set A c.e. in some set $B \in \mathbf{b}$.

(viii) A degree **a** is *computably enumerable in and above* **b** (written **a** is c.e.a. in **b**) if $\mathbf{a} \geq \mathbf{b}$ and **a** is c.e. in **b**. The terminology of (vii) and (viii) applies to *sets* as well as degrees.

Two sets of the same degree should be thought of as coding the same information and therefore as equally difficult to compute, while $\mathbf{a} < \mathbf{b}$ asserts that sets of degree **b** are more difficult to compute than those of degree **a**. Furthermore, $\deg(A \oplus B)$ is clearly the least upper bound for $\deg(A)$ and $\deg(B)$ in this partial ordering (see Exercise 3.4.6). Unfortunately, the infimum of two degrees need not always exist for the degrees **D** or even for the c.e. degrees **C**. Hence, these are upper semi-lattices, but neither forms a lattice.

3.4.2   The Jump Theorem

In 1931 Gödel introduced the diagonal set "eine Klasse K" (although he defined K for the Π_1^0 (co-c.e.) sets $\overline{W}_e$, not for the Σ_1 sets, as done today). In 1936 Kleene defined $K = \{e : e \in W_e\}$ as we did in Definition 1.6.3 and showed that: K is not computable, but K is computably enumerable and hence is computably approximable, as we showed in Theorems 1.6.4 and 1.6.5. With the development of computability relative to an oracle A, which we have presented in §3.2, in 1954 Kleene and Post defined this same diagonal operator K^A, but relative to an oracle A, and noted that the previous proofs showed that $K^A >_T A$ and K^A is c.e. in A.

Definition 3.4.2. (i) Let $K^A = \{x : \Phi_x^A(x)\downarrow\} = \{x : x \in W_x^A\}$. K^A is called the *jump* of A and is denoted by A' (pronounced "A prime"). We also use the notation H^A for K^A because this represents the *halting problem* relativized to A of whether $\Phi_x^A(x)$ halts.

(ii) $A^{(n)}$, the nth jump of A, is obtained by iterating the jump n times, namely $A^{(0)} = A$, $A^{(n+1)} = (A^{(n)})'$. Note that $A^{(1)} = A'$.

It follows from the relativization to A of Theorem 2.4.2 that $K^A \equiv K_0^A \equiv K_1^A$ where $K_0^A = \{\langle x, y\rangle : \Phi_x^A(y)\downarrow\}$ and $K_1^A = \{x : W_x^A \neq \emptyset\}$. These alternative characterizations of the jump are useful. The crucial properties of the jump operator are the following. (Recall by Definition 3.4.1 (viii) that B is c.e.a. in A if B is c.e. in A and $B \geq_T A$.)

Theorem 3.4.3 (Jump Theorem). *(i) A' is c.e.a. in A.*

(ii) $A' \not\leq_T A$.

(iii) B is c.e. in A iff $B \leq_1 A'$.

(iv) If A is c.e. in B and $B \leq_T C$ then A is c.e. in C.

(v) $B \leq_T A$ iff $B' \leq_1 A'$.

(vi) If $B \equiv_T A$ then $B' \equiv_1 A'$ (and therefore $B' \equiv_T A'$).

(vii) A is c.e. in B iff A is c.e. in $\overline{B}$.

Proof. Parts (i)–(iii) follow by relativizing the proofs in Chapters 1 and 2. Now $K_0^A := \{\langle x, y\rangle : x \in W_y^A\}$ and $K^A \equiv K_0^A$ by relativizing to A Definition 1.6.6 and Exercise 1.6.22. Therefore, $A' = K^A \equiv K_0^A$. However, A itself is clearly c.e. in A. Hence, $A = W_y^A$ for some y, but $W_y^A \leq_1 K_0^A$, and therefore $A \leq_1 K^A$.

Note that (iii) ($\implies$) uses $K^A \equiv K_0^A$.

(iv) If $A \neq \emptyset$, then A is the range of some B-computable function, and hence of some C-computable function, since $B \leq_T C$.

(v) ($\implies$). If $B \leq_T A$ then B' is c.e. in A by (iv), because B' is c.e. in B by (i). Hence, $B' \leq_1 A'$ by (iii).

(v) ($\impliedby$). If $B' \leq_1 A'$, then both B and $\overline{B}$ are c.e. in A by (iii) (because $B, \overline{B} \leq_1 B'$). Hence, $B \leq_T A$ by the Complementation Theorem 2.1.14.

(vi) follows immediately from (v).

(vii) follows immediately from (iv). $\square$

Let $\mathbf{a}' = \deg(A')$ for $A \in \mathbf{a}$. Note that $\mathbf{a}' > \mathbf{a}$ and $\mathbf{a}'$ is c.e. in $\mathbf{a}$. By the Jump Theorem 3.4.3 (vi) the jump is well defined on degrees. Let $\mathbf{0}^{(n)} = \deg(\emptyset^{(n)})$. Thus, we have an infinite hierarchy of degrees,

$$\mathbf{0} < \mathbf{0}' < \mathbf{0}'' < \cdots < \mathbf{0}^{(n)} < \cdots .$$

The first few degrees in this hierarchy are of special importance and will be shown to be the degrees of certain unsolvable problems considered in Chapter 1.

$$\mathbf{0} = \deg(\emptyset) = \{B : B \text{ is computable}\};$$

$$\mathbf{0}' = \deg(\emptyset'), \text{ where } \emptyset' := K^\emptyset \equiv K \equiv K_0 \equiv K_1;$$

$$\mathbf{0}'' = \deg(\emptyset'') = \deg(\text{Fin}) = \deg(\text{Tot}) = \deg(\text{Inf});$$

$$\mathbf{0}''' = \deg(\emptyset''') = \deg(\text{Cof}) = \deg(\text{Rec}) = \deg(\text{Ext}).$$

By definition, degree $\mathbf{0}$ is the least degree and consists precisely of the computable sets. Degree $\mathbf{0}'$ (read "zero prime") is the degree of the halting problem and also the degree of the problem $K_1 = \{x : W_x \neq \emptyset\}$ because $K \equiv K_1 \equiv K_0$ by Theorem 2.4.2. The above characterizations for $\mathbf{0}''$ and $\mathbf{0}'''$ will be shown in Chapter 4. In §3.6.3 we use the jump to define the *low* and *high* sets according to their information content, and we develop further lowness and highness properties in §4.4 and §4.7.

Definition 3.4.4. Fix $A \subseteq \omega$ and let $\mathbf{a} = \deg(A)$. Relativizing Definition 2.6.1 to A we define a sequence of sets $\{V_n\}_{n \in \omega}$ to be *uniformly computable in A* (*uniformly of degree $\leq \mathbf{a}$*) if there is an A-computable function $g(x, n)$ such that $\lambda x \, [g(x,n)]$ is the characteristic function of V_n, for all n.

3.4.3 Exercises

Exercise 3.4.5. Prove that there are at least $2^{\aleph_0}$ degrees. Conclude that there are *exactly* $2^{\aleph_0}$ degrees.

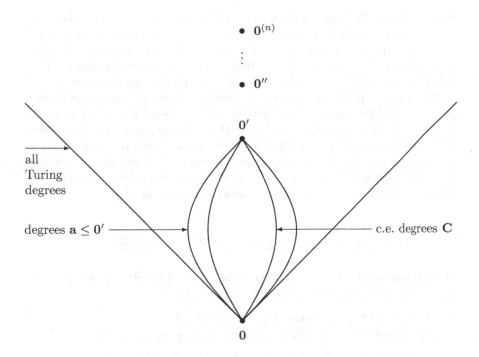

Figure 3.1. Turing Degrees $(\mathbf{D}, \leq)$

Exercise 3.4.6. Show that $\deg(A \oplus B)$ is the least upper bound for $\deg(A)$ and $\deg(B)$ in $(\mathbf{D}, \leq, \cup)$.

Exercise 3.4.7. (a) Let $\{A_y\}_{y \in \omega}$ be any countable sequence of sets. Define the infinite join $\oplus_y A_y$ as in (2.29). Prove that $\deg(\oplus_y A_y)$ is the *uniform* least upper bound for $\{\deg(A_y)\}_{y \in \omega}$ in the sense that if there exist a set C and a computable function f such that $A_y = \Phi^C_{f(y)}$ for all y, then $\oplus_y A_y \leq_T C$.

(b) Prove that this operation is not well defined on degrees. Namely, define $\{A_y\}_{y \in \omega}$ and $\{B_y\}_{y \in \omega}$ such that $A_y \equiv_T B_y$ but $A \not\equiv_T B$ for $A = \oplus_y A_y$ and $B = \oplus_y B_y$.

Exercise 3.4.8. Prove that $\emptyset' \equiv K$. (Recall that $\Phi^\emptyset$ is an acceptable numbering by Exercise 3.3.17.)

3.5 * Limit Computable Sets and Domination

The most interesting and important classes include not only the class of *computable* sets, but also those sets which can be *approximated* by a computable sequence. There are various levels of effectiveness in these approximations. We began with a c.e. set (Σ_1 set) $A = \cup_{s \in \omega} A_s$ where A_s is a monotonic computable sequence. Now we consider the more general case $A = \lim_s A_s$ sometimes called a "trial and error" predicate because membership of $x \in A_s$ can change finitely often as s goes to infinity. This is a *limit computable* approximation also called a Δ_2-*approximation*.

With these approximations we also study the modulus function of Definition 3.5.4 and the computation function of Definition 5.6.5, which measure how quickly the approximation has settled. Closely connected is the idea of *domination* in Definition 3.5.2, which we relate to the modulus function in Propositon 3.5.5. We develop more properties of domination in Definition 4.5.1, along with its negation, called *escape*. We relate these two key ideas of domination and escape to classes of degrees in §4.7.

3.5.1   Domination and Quantifiers $(\forall^\infty x)$ and $(\exists^\infty x)$

Definition 3.5.1. (i) $(\exists^\infty x)\, R(x)$ abbreviates $(\forall y)(\exists z > y)\, R(z)$, i.e., "there exist infinitely many x such that $R(x)$."

(ii) $(\forall^\infty x)\, R(x)$ abbreviates $(\exists y)(\forall z > y)\, R(z)$, i.e., "for almost every x, $R(x)$," often written as $(a.e.\ x)\, R(x)$.

These quantifiers are dual because $(\forall^\infty x)\, R(x)$ iff $\neg(\exists^\infty x)\neg R(x)$.

Definition 3.5.2. (i) A function g *dominates* f, denoted by $f <^* g$, if

$$(3.13) \qquad\qquad (\forall^\infty x)\,[\, f(x) < g(x)\,].$$

(ii) If g does *not* dominate f then we say that f *escapes* g, i.e.,

$$(3.14) \qquad\qquad (\exists^\infty x)\,[\, g(x) \le f(x)\,].$$

3.5.2   Uniformly Computable Sequences

In analysis, a function f on the real numbers is *continuous* at a point x_0 if

$$(\forall \epsilon > 0)(\exists \delta > 0)\,[\ |x - x_0| < \delta \quad\Longrightarrow\quad |f(x) - f(x_0)| < \epsilon\ ].$$

A *sequence* of functions $f_n(x)$ on the reals is *uniformly continuous* if the δ may be chosen to be independent of n, i.e.,

$$(\forall \epsilon > 0)(\exists \delta > 0)(\forall n)[\ |x - x_0| < \delta \quad\Longrightarrow\quad [\ |f_n(x) - f_n(x_0)| < \epsilon\,].$$

We often build a sequence of computable functions $\{f_s(x)\}_{s \in \omega}$ or sets in stages $s \in \omega$. Each function f_s is computable by some individual algorithm, but we usually need to *uniformly* compute the sequence with a *single* computable function $g(x, s) = f_s(x)$.

Definition 3.5.3. A sequence of computable functions $\{f_s(x)\}_{s\in\omega}$ or sets is *uniformly computable* if there is a computable function g such that $g(x, s) = f_s(x)$ for all s and x. We also call this a *computable sequence* because the function g defines the sequence and the elements in its members.

If $\{A_s\}_{s\in\omega}$ is a sequence of computable sets, then the property of being computable is shared by every member of the sequence, i.e., every A_s has a computable characteristic function $h_s(x)$. However, we do not call this a *uniformly* computable sequence unless there is a *single* computable function $g(x, s)$ such that $g(x, s) = h_s(x)$. This is the same concept as that in the Enumeration Theorem 1.5.3. There we had every partial computable function φ_e identified with a Turing machine P_e which computed it. However, in virtually all constructions we also needed to *uniformly* list them with a *single* p.c. function $\psi(e, x) = \varphi_e(x)$ guaranteed by Theorem 1.5.3. Similarly, in Definition 2.6.1 we defined the concept of sequence $\{V_e\}_{e\in\omega}$ of c.e. sets as being *uniformly c.e.* if there is a single computable function $g(e)$ such that $V_e = W_{g(e)}$. Recall that $A \upharpoonright x$ denotes $\{A(y) : y \leq x\}$.

3.5.3   Limit Computable Sets

Definition 3.5.4. (i) A set A is *limit computable* if there is a computable sequence $\{A_s\}_{s\in\omega}$ such that for all x,

(3.15) $$A(x) = \lim_s A_s(x).$$

By the Limit Lemma 3.6.2 (ii), we call $\{A_s\}_{s\in\omega}$ a Δ_2-*approximation* for A.

(ii) Given $\{A_s\}_{s\in\omega}$, any function $m(x)$ is a *modulus (of convergence)* if

(3.16) $$(\forall x)(\forall s \geq m(x))\,[\, A \upharpoonright x = A_s \upharpoonright x\,].$$

We define the *least* function:

(3.17) $$m_A(x) = (\mu s)\,[\, A \upharpoonright x = A_s \upharpoonright x\,].$$

(iii) If A is c.e., then a computable sequence $\{A_s\}_{s\in\omega}$ is a Σ_1-*approximation* to A if $A = \cup_s A_s$ and $A_s \subseteq A_{s+1}$. In this case $m_A(x)$ *is* a modulus and is called the *least modulus*.

We could have alternatively defined the least modulus as the first stage after which the approximation is always correct, i.e.,

(3.18) $$q_A(x) = (\mu s)(\forall t \geq s)\,[\, A \upharpoonright x = A_t \upharpoonright x\,].$$

The advantage is that q_A is a modulus in the sense of (3.16). The disadvantage is that $q_A \not\leq_T A$ in the Δ_2 case, because for a Δ_2-sequence the approximation may attain the correct value, change to an incorrect value, and later return to the correct value.

In the Σ_1 case, $m_A = q_A$ and $m_A \leq_T A$, so we have the best of both worlds. We shall explore modulus properties in the next Proposition 3.5.5, the Limit Lemma 3.6.2 and the Modulus Lemma 3.6.3. We shall later define the closely related *computation function* in Definition 5.6.5.

Proposition 3.5.5. *(i) If A is limit computable via $\{A_s\}_{s\in\omega}$ with any modulus $m(x)$, then $A \leq_T m$.*

(ii) Conversely, if A is c.e. via a Σ_1-approximation $\{A_s\}_{s\in\omega}$ or is Δ_2 via a Δ_2-approximation $\{A_s\}_{s\in\omega}$ with least function $m_A(x)$, then $m_A(x) \leq_T A$.

(iii) Let $\{A_s\}_{s\in\omega}$ be a Σ_1-approximation to a c.e. set A and $m_A(x)$ be the least modulus. If $g(x)$ dominates $m_A(x)$ then $A \leq_T g$. If $g \leq_T \emptyset$ then $A \leq_T \emptyset$.

(iv) Let $\{A_s\}_{s\in\omega}$ and $\{D_s\}_{s\in\omega}$ be Σ_1-approximations of A and D with least modulus functions $m_A(x)$ and $m_D(x)$ respectively. If $m_D(x)$ dominates $m_A(x)$, then $A \leq_T D$.

Proof. (i) $A(x) = A_{m(x)}(x)$.

(ii) From A and $\{A_s\}_{s\in\omega}$ we can compute the least function $m_A(x)$.

(iii) Note that $(\forall^\infty x)[A_{g(x)}(x) = A(x)]$. Therefore, $A \leq_T g$.

(iv) $(\forall^\infty x)[x \in A \iff x \in A_{m_D(x)}]$. $\square$

Proposition 3.5.5 (iii) fails for a Δ_2-approximation even if the computable function $g(x)$ dominates the *least function* $m_A(x)$ in (3.17). Exercise 3.5.6 shows just how weak the least function m_A is for Δ_2-sequences, although for Σ_1-sequences it is very powerful because $m_A = q_A$ for Σ_1-sequences. However, in the Δ_2-case, if we replace the least function $m_A(x)$ by the closely related *computation function* $c_A(x)$ of Definition 5.6.5 (5.15), then Proposition 3.5.5 (iii) holds, as we shall prove in Theorem 5.6.6.

3.5.4   Exercises

Exercise 3.5.6. Construct a noncomputable Δ_2 set A, a computable function $f(x)$, and a Δ_2-approximation $\{A_s\}_{s\in\omega}$ such that $f(x)$ dominates the least function $m_A(x)$. (Compare with the Σ_1 case in Proposition 3.5.5 (iii).)

3.6   ** The Limit Lemma

We have studied the Σ_1^0, Π_1^0, and Δ_1^0 sets. In Chapter 4 we shall study the Σ_n^0, Π_n^0, and Δ_n^0 sets in the arithmetical hierarchy. However, the following property, Δ_2^0, plays such an important role in the Limit Lemma 3.6.2 below that we introduce it now. We often drop the superscript 0 for convenience.

Definition 3.6.1. (i) A set A is Σ_2 if there is a computable relation R with

$$x \in A \iff (\exists y)(\forall z)\, R(x, y, z).$$

(ii) A set A is Π_2 if $\overline{A}$ is Σ_2.

(iii) A set A is Δ_2 if $A \in \Sigma_2$ and $A \in \Pi_2$.

The following three properties are used very often and completely interchangeably without explanation.[2] Shoenfield in [Shoenfield 1959] proved the equivalence of (i) and (iii). The equivalence of (ii) follows from Post's Theorem 4.2.2 (iv), a part of which is introduced here.

Lemma 3.6.2 (Limit Lemma, Shoenfield, 1959). *TFAE:*

(i) A is limit computable;

(ii) $A \in \Delta_2$;

(iii) $A \leq_T \emptyset'$.

Proof. (i) $\implies$ (ii). Let $A(x) = \lim_s A_s(x)$ with $\{A_s\}_{s \in \omega}$ computable. Then

$$x \in A \iff (\exists s)(\forall t)[\, t > s \implies A_t(x) = 1\,]$$

$$x \in \overline{A} \iff (\exists s)(\forall t)[\, t > s \implies A_t(x) = 0\,]$$

and therefore, $A \in \Sigma_2$ and $\overline{A} \in \Sigma_2$.

(ii) $\implies$ (iii). Assume there are computable relations R and S such that

$$x \in A \iff (\exists s)(\forall t)R(x, s, t) \quad \& \quad x \in \overline{A} \iff (\exists s)(\forall t)S(x, s, t).$$

The predicate $(\forall t)R(x, s, t)$ is Π_1 and therefore computable in $\emptyset'$. Hence, the predicate $(\exists s)(\forall t)R(x, s, t)$ is Σ_1 in $\emptyset'$ and therefore c.e. in $\emptyset'$, and likewise for $(\exists s)(\forall t)S(x, s, t)$. Therefore, A and $\overline{A}$ are each c.e. in $\emptyset'$. Thus, $A \leq_T \emptyset'$.

---

[2]Some authors introduce the property of limit computable as above, but use the *name* Δ_2^0 for it. It is not correct to *define* a set A to be Δ_2^0 if it has the limit computable property. However, after proving the equivalence of the three properties in the Limit Lemma 3.6.2, authors often mention one term (such as, A is Δ_2^0 or $A \leq_T \emptyset'$) and then immediately use another property, such as A limit computable, without explanation.

(iii) $\implies$ (i).   Fix a computable sequence $\{K_s\}_{s\in\omega}$ with $\cup_s K_s = K \equiv \emptyset'$. Assume $A = \Phi_e^K$. For every x and s define

$$f(x,s) = \begin{cases} \Phi_{e,s}^{K_s}(x) & \text{if defined;} \\ 0 & \text{otherwise.} \end{cases}$$

For every x the first clause holds for all but finitely many s. Therefore, $A(x) = \lim_s f(x,s)$. $\square$

3.6.1   The Modulus Lemma for C.E. Sets

Lemma 3.6.3 (Modulus Lemma).   *If B is c.e. and $A \leq_T B$, then $A = \lim_s A_s$ for a computable sequence $\{A_s\}_{s\in\omega}$ with a modulus $m(x) \leq_T B$.*

Proof. Let $B = \cup_s B_s$ be c.e., $A = \Phi_e^B$ and define

$$A_s = \{ \, x \; : \; x < s \quad \& \quad \Phi_e^B(x)[s]\downarrow \; = \; 1 \, \},$$

$$m(x) = (\mu s)(\exists z \leq s)(\forall y \leq x)[\, \Phi_e^{B\upharpoonright z}(y)[s]\downarrow \quad \& \quad B\upharpoonright z = B_s\upharpoonright z\,].$$

Now $\{A_s\}_{s\in\omega}$ is a computable sequence because the defining clause for A_s is computable. Furthermore, $m(x)$ is B-computable because the quantifiers are bounded, the first clause in the matrix defining $m(x)$ is computable, and the second clause is B-computable. Finally, $m(x)$ is a modulus because by the Use Principle 3.3.9 the matrix of $m(x)$ holds for all $s \geq m(x)$. $\square$

Definition 3.6.4. Let $\{A_s(x)\}_{s\in\omega}$ be a uniformly computable sequence of computable sets, such that $(\forall x)[\lim_s A_s(x)$ exists$]$. Define the *change set*,

(3.19) $C \; = \; \{ \, \langle s,x \rangle \; : \; (\exists t)_{s<t}[\, A_s(x) \neq A_t(x) \,] \, \}.$

Note that C is Σ_1 and hence c.e., and that the x-section $A^{[x]}$ is finite for every x. (See Definition 3.8.7 for the definition of the x-section.) We call C the *change set* because if $\langle s,x \rangle \in C$ then the value $A_s(x)$ will later change at some stage $t > s$. The change set C will be used in the proof of the following Theorem 3.6.5 and in Theorem 3.8.8.

Theorem 3.6.5 (Modulus Criterion).   *A has c.e. degree iff $A = \lim_s A_s$ for some computable sequence $\{A_s\}_{s\in\omega}$ with a modulus $m \leq_T A$.*

Proof. ($\implies$).   Let $A \equiv_T B$ with B c.e. Apply the Modulus Lemma 3.6.3 to obtain a computable sequence $\{A_s\}_{s\in\omega}$ with a modulus $m(x) \leq_T B$.

($\impliedby$).   Let $A = \lim_s A_s$ with $m \leq_T A$. The definition of C in (3.19) is equivalent if we replace the quantifier bound $s < t$ by $s < t \leq m(x)$ because $m(x)$ is a modulus. Therefore, $C \leq_T m \leq_T A$. Furthermore, $A \leq_T C$

because, given x, we can find some $\langle s, x \rangle \in \overline{C}$. Now $x \in A$ iff $x \in A_s$. Hence, $A \equiv_{\mathrm{T}} C$. □

3.6.2 The Ovals of Σ_1 and Δ_2 Degrees

Imagine the degrees $\mathbf{d} \leq \mathbf{0}'$ as an oval with top $\mathbf{0}'$ and bottom $\mathbf{0}$. The Limit Lemma 3.6.2 exactly characterizes the Δ_2 degrees as the degrees in this outer oval. Now imagine a smaller oval inside the first oval with the same top $\mathbf{0}'$ and bottom $\mathbf{0}$ but containing only the *computably enumerable* degrees. Theorem 3.6.5, building on the Limit Lemma 3.6.2, characterizes the sets and degrees in this inner oval of c.e. degrees. Let A be limit computable via the Δ_2 approximating sequence $\{A_s\}_{s \in \omega}$ with least function $m_A(x)$. By Proposition 3.5.5 (ii) we always have the least function $m_A \leq_{\mathrm{T}} A$. For a Δ_2-approximation the least function $m_A(x)$ may not always be a modulus. Remarkably, however, if A has c.e. *degree*, then $m \leq_{\mathrm{T}} A$ holds for *some* modulus m of some particular Δ_2 approximation for A, as Theorem 3.6.5 demonstrates. Note that all the above results on approximating a set $A = \lim_s A_s$ hold by the same proofs as those for approximating a *function* $f = \lim_s f_s$.

3.6.3 Reaching With the Jump: Low and High Sets

The jump $A' = K^A$ was defined in Definition 3.4.2. If $A \leq_{\mathrm{T}} B$ then $A' \leq_{\mathrm{T}} B'$ by the Jump Theorem 3.4.3 (v). This asserts that the jump preserves the Turing reducibility. Therefore, if $A \leq_{\mathrm{T}} \emptyset'$, then $\emptyset' \leq_{\mathrm{T}} A' \leq_{\mathrm{T}} \emptyset''$. We now classify sets $A \leq_{\mathrm{T}} \emptyset'$ as *low* or *high* according as to whether A' has the lowest or highest possible value.

Definition 3.6.6. Fix a set $A \leq_{\mathrm{T}} \emptyset'$.

(i) A is *low* if $A' \equiv_{\mathrm{T}} \emptyset'$, the lowest possible value.

(ii) A is *high* if $A' \equiv_{\mathrm{T}} \emptyset''$, the highest possible value.

According to the Limit Lemma 3.6.2 relativized to A, we see which A' has the greatest degree under $\leq_{\mathrm{T}}$ that can be approximated by an A-computable sequence. Think of the jump A' as the *reach* of A as measured with respect to standard mileposts such as $\emptyset'$ and $\emptyset''$. The jump (reach) measures the information content of A. A low set A has low computing power because K^A, the halting problem relativized to A, has the same degree as the unrelativized halting problem K. Hence, A does not extend the reach of the jump over the computable case.

As the terminology suggests, a computable set has zero information content, a low set has low information content (but may be noncomputable), a high set has high information content (but is not necessarily complete), and a complete set $A \equiv \emptyset'$ has complete information content with respect to other c.e. sets. The study of the information content and algebraic structure of low and high sets will be a major theme throughout the book. We shall develop a connection between the high/low sets and domination/escape properties in §4.7.

Low Sets

If A is *low* its reach is only $\emptyset'$, as for a computable set C. It resembles the computable sets in algebraic structure and information content, as we shall see later. In the Low Basis Theorem 3.7.2 we shall prove that every infinite computable binary tree $T \subseteq 2^{<\omega}$ may not have a computable path, but it does have a *low* path which is almost as good.

High Sets

If A is *high* its reach is $A' \equiv_T \emptyset''$. In §4.3.1 we shall prove that $\mathrm{Tot} \equiv_T \emptyset''$, where $\mathrm{Tot} := \{\, e : \varphi_e \text{ total} \,\}$. This means that there is an A-computable sequence $\{A_s\}_{s \in \omega}$ such that $\lim_s A_s(x) = \mathrm{Tot}(x)$, which says that Tot is only a "jump away" from A in the sense that Tot is A-limit computable. This is a very strong property with many consequences, as we shall prove in Chapter 4 and in subsequent chapters. For the rest of this chapter we consider only low sets and in Chapter 4 we consider a spectrum of high and low sets.

3.6.4 Exercises

Exercise 3.6.7. In the Limit Lemma 3.6.2 there are six possible direct implications among the three properties (i), (ii) and (iii) listed there. There we proved (i) $\implies$ (ii) $\implies$ (iii) $\implies$ (i). Give a direct proof of the remaining three implications: (i) implies (iii) (which was also proved directly in Theorem 3.6.5 using the change set C defined in (3.19)); (iii) implies (ii); and (ii) implies (i).

Exercise 3.6.8. (Relativized Limit Lemma). Prove the Limit Lemma 3.6.2 relativized to a set A.

Exercise 3.6.9. (Iterated Limit Lemma). Prove that if $n \geq 1$, then $f \leq_T \emptyset^{(n)}$ iff there is a computable function $\widehat{f}$ of $(n+1)$ variables such that

$$f(x) = \lim_{y_1} \lim_{y_2} \cdots \lim_{y_n} \widehat{f}(x, y_1, y_2, \ldots y_n).$$

3.7 * Trees and the Low Basis Theorem

3.7.1 Notation for Trees

Definition 3.7.1. (i) A *tree* $T \subseteq \omega^{<\omega}$ is a set of strings closed under initial segments, i.e., $\sigma \in T$ and $\tau \prec \sigma$ imply $\tau \in T$. Fix any tree T.

(ii) The class of *infinite paths* through T is:

$$(3.20) \qquad [T] = \{\, f : (\forall n)\, [\, f \restriction n \in T \,] \,\}.$$

(iii) A class $\mathcal{C} \subseteq 2^\omega$ is a Π_1^0-*class* if $\mathcal{C} = [T]$ for some computable binary tree $T \subseteq 2^{<\omega}$.

(iv) For $\sigma \in T$, define the subtree T_σ of nodes $\tau \in T$ *comparable with* σ:

$$(3.21) \qquad T_\sigma = \{\, \tau \in T \; : \; \sigma \preceq \tau \quad \vee \quad \tau \prec \sigma \,\}.$$

(v) Define the subtree of *extendible nodes* $\sigma \in T$:

$$(3.22) \qquad T^{\text{ext}} = \{\, \sigma \in T \; : \; (\exists f \succ \sigma)[\, f \in [T]\,]\,\}$$

The tree T is *extendible* if $T = T^{\text{ext}}$. If $T \subseteq 2^{<\omega}$ is computable, then T^{ext} is co-c.e. (See Exercise 3.7.4.)

3.7.2 * The Low Basis Theorem for Π_1^0 Classes

Let $T \subseteq 2^{<\omega}$ be an infinite computable tree. What are the degrees of members $f \in [T]$? We cannot always find a computable path but we can always find a low path. We shall define and explain bases for Π_1^0-classes in Chapter 9.

Theorem 3.7.2 (Low Basis Theorem, Jockusch-Soare, 1972b). *If $\mathcal{C} \subseteq 2^\omega$ is a nonempty Π_1^0 class, then it contains a member f of low degree ($f' \equiv_T 0'$).*

Proof. Now $\emptyset'$ can decide whether a computable tree $G \subseteq 2^{<\omega}$ is finite because

$$(3.23) \qquad |G| < \infty \quad \Longleftrightarrow \quad (\exists n)(\forall \sigma)_{|\sigma|=n}\,[\, \sigma \notin G\,].$$

The bounded quantifier in front of the computable matrix remains computable, and the $(\exists n)$ quantifier makes the condition Σ_1 hence computable in $\emptyset'$. Let T be a computable tree such that $[T] = \mathcal{C}$. Use $\emptyset'$ to define a sequence of infinite computable trees $T = T_0 \supseteq T_1 \supseteq \ldots$ as follows. Define:

$$(3.24) \qquad U_e = \{\, \sigma : \Phi_{e,|\sigma|}^\sigma (e)\!\uparrow \,\},$$

which is also a computable tree. Given T_e: (1) define $T_{e+1} = T_e \cap U_e$ if $T_e \cap U_e$ is infinite; and (2) define $T_{e+1} = T_e$ otherwise. If (1), then $\Phi_e^g(e)\!\uparrow$ for all $g \in [T_{e+1}]$, and if (2), then $\Phi_e^g(e)\!\downarrow$ for all $g \in [T_{e+1}]$. (Namely, we say that T_{e+1} *forces the jump* as described in Chapter 6.) Choose $f \in \cap_{e \in \omega}[T_e]$, which is an intersection of a descending sequence of nonempty closed sets, and is hence nonempty by the compactness of Cantor space 2^ω. (See the Compactness Theorem 8.3.1 (ii).) Now $\emptyset'$ can decide using (3.23) which of (1) and (2) holds in the definition of T_{e+1} and hence whether $\Phi_e^f(e)\!\downarrow$ or not. Therefore, $f' \leq_T \emptyset'$ and f is low. $\square$

We can also use this method to prove a similar basis theorem that any nonempty Π_1^0 class $\mathcal{C} \subseteq 2^\omega$ contains a member f such that every $g \leq_T f$ is dominated by a computable function.

3.7.3 Exercises

Exercise 3.7.3. (a) Prove that T^{ext} is the smallest subtree of T under inclusion such that $[T^{\text{ext}}] = [T]$, i.e., prove that if $[T_1] = [T]$, then $T^{\text{ext}} \subseteq T_1$.

  (b) Prove that T_σ is infinite iff $[T_\sigma] \neq \emptyset$.

Exercise 3.7.4. (a) Fix computable tree $T \subseteq 2^{<\omega}$. Let $\overline{T}^{\text{ext}} = T - T^{\text{ext}}$. Prove that $\overline{T}^{\text{ext}}$ is Σ_1 and therefore T^{ext} is Π_1. Conclude that $T^{\text{ext}} \leq_T \emptyset'$.

  (b) Prove directly the *Kreisel Basis Theorem* that if T is computable and $[T] \neq \emptyset$ then $[T]$ contains a member $f \leq_T \emptyset'$. (Do not use the Low Basis Theorem 3.7.2.)

Exercise 3.7.5. Prove that there is a computable infinite tree $T \subset 2^{<\omega}$ with no computable infinite paths.

Exercise 3.7.6. Prove that if $\mathcal{C} \subseteq 2^\omega$ is a nonempty Π_1^0 class, then the lexicographically least member $f \in \mathcal{C}$ has c.e. degree.

Exercise 3.7.7. (a) Let $T \subseteq 2^{<\omega}$ be a tree. A point $f \in [T]$ is *isolated* if there exists a $\sigma \in T$ such that $[T_\sigma] = \{f\}$, in which case we say that σ *isolates* f in T. Prove that if T is computable and f is isolated, then f is computable.

  (b) Show that if T is computable and $[T]$ is finite, then all its members are isolated and hence computable.

3.8   Bounded Reducibilities and n-C.E. Sets

3.8.1   A Matrix M_x for Bounded Reducibilities

A *bounded reducibility* is a Turing reducibility $\Phi_e^A(x)$ with a computable function $h(x)$ which bounds the use function, i.e., $\varphi_e^A(x) < h(x)$. Given $h(x)$ imagine a matrix M_x with a *column* y for every $y < h(x)$ corresponding to the question "is $y \in A$?" The *rows* of M_x are all strings σ of length $h(x)$ corresponding to the $2^{h(x)}$ many answers over all such $y < h(x)$.

The action of Φ_e^A on x is entirely determined by the action of $\Phi_e^\sigma(x)$ for these rows σ of M_x. For example, if $B \leq_m A$ via a computable function $f(x)$, then $h(x) = f(x) + 1$ and $x \in B$ iff $\sigma(f(x)) = 1$, where $\sigma \prec A$.

We begin in §3.8.2 with the most general bounded reducibility called *bounded Turing reducibility ($B \leq_{bT} A$)*, where we are given only the computable bound $h(x)$, i.e., only the matrix M_x. In §3.8.3 we study the more restrictive case of *truth-table reducibility ($B \leq_{tt} A$)*, where $\Phi_e^\sigma(x)$ converges for every x and *every* row $\sigma \in M_x$. (See Remark 3.8.3.)

3.8.2   Bounded Turing Reducibility

Definition 3.8.1. (i) A set B is *bounded Turing reducible (bT-reducible)* to a set A $(B \leq_{bT} A)$ (also called *weak truth-table (wtt) reducible*) if there is a Turing reduction $\Phi_e^A = B$ and a computable function $h(x)$ such that the use $\varphi_e^A(x) < h(x)$.[3]

(ii) A set B is *identity bounded Turing reducible* to A $(B \leq_{ibT} A)$ if $B \leq_{bT} A$ with $h(x) = x + 1$.[4]

The bT-reductions and ibT-reductions occur naturally in several parts of the subject.[5] For example, often we are given a noncomputable c.e. set A and we construct a simple set $B \leq_T A$ by ordinary permitting, as in

---

[3]In his 1944 paper Section 6 Post introduced the definition of $B \leq_{tt} A$ which became rapidly understood. Turing reducibility was not well understood in the 1940s and 1950s. Therefore, in 1959 Friedberg and Rogers introduced *bT*-reducibility as *wtt-reducibility* $(B \leq_{wtt} A)$ i.e., as a *weakening* of tt-reducibility which was already a *strengthening* of $\leq_T$. Today we understand Turing reducibility as in this chapter and view *bT*-reducibility as a one-step *strengthening* of $\leq_T$ without the need for any intermediate step. Also *bT* is more recognizable from its name than *wtt*.

[4]Note that *ibT* occurs often, for example, with simple permitting in Theorem 5.2.7.

[5]More recently, *ibT* has occurred in applications of computability to differential geometry by Soare in 2004 and Csima and Soare in 2006, and ibT-reducibility has also been used in applications to algorithmic randomness and Kolmogorov complexity. Barmpalias and Lewis in 2006 have shown the nondensity of the *ibT*-degrees of c.e. sets.

§5.2 where an element x is allowed to enter B at some stage only if some $y \leq x$ has just entered A. When $A \upharpoonright x$ has settled, $B \upharpoonright x$ has settled also, so $B \leq_{ibT} A$.

3.8.3 Truth-Table Reductions

Recall that $|\sigma|$ denotes the length of σ, and σ_y is the string with canonical index y as defined in Definition 2.3.6. We identify σ_y and its index y and write $g(\sigma)$ to mean $g(y)$ in (ii).

Definition 3.8.2. A set B is *truth-table reducible (tt-reducible)* to a set A ($B \leq_{tt} A$) if there are computable functions $h(x)$ and $g(x)$ with the following properties.

(i)   To determine whether $x \in B$ the reduction procedure may ask questions of the form "is $y \in A$?" only for elements $y < h(x)$. (This is exactly as in the bounded Turing Definition 3.8.1.) This determines $2^{h(x)}$ many strings $\{ \sigma_i : i < 2^{h(x)} \}$ of 0's and 1's, each of length $h(x)$, which we call the *rows* of the truth-table.

(ii)   The function $g(x)$ selects which rows for which the answer is "yes" that $x \in B$ as follows.

$$x \in B \iff (\exists i < 2^{h(x)})\, [g(i) = 1 \ \& \ \sigma_i \prec A\}.$$

Remark 3.8.3. Both the bT-reduction and the tt-reduction use the matrix M_x described in §3.8.1, The crucial difference between a bT-reduction and a tt-reduction is that in the latter case $\Phi_e^\sigma(x)$ must be defined for *every* row $\sigma \in M_x$ but in the bT case only *some* rows σ need to have $\Phi_e^\sigma(x)$ defined. Therefore, the *tt*-reduction must not only have a bound $h(x)$ on the information scanned, but in advance of A it must also give the *answer* $\Phi_e^\sigma(x)$ for all rows σ of length $h(x)$. As we range through these rows, $g(\sigma) = 1$ in Theorem 3.8.5 selects exactly those rows σ which have $\Phi_e^\sigma(x) = 1$. See Theorem 5.3.6 for a natural *tt*-reduction $D \leq_{tt} A$. For a bT-reduction $\Phi_e^A = B$ we do not have a complete table M_x as for tt. We have only a c.e. graph G_e as in Definition 3.3.7 consisting of strings $\langle \sigma, x, y \rangle$ except that all the strings σ have $|\sigma| < h(x)$. We cannot decide which rows σ will appear in G_e.

Definition 3.8.4. (i) A *partial functional* Ψ on Cantor space 2^ω is a partial map with $\mathrm{dom}(\Psi)$ and $\mathrm{rng}(\Psi) \subseteq 2^\omega$. (We use the term *functional* for maps whose input and output may be an infinite object $A \in 2^\omega$ to distinguish from *functions* from ω to ω.)

(ii) A Turing functional Φ_e on 2^ω is *total* if $(\forall X)(\forall x)[\, \Phi_e^X(x)$ is defined $]$.

The next theorem gives the very pleasing characterization that Φ_e is a tt-reduction iff Φ_e is total. This is sometimes taken as a definition of the former, but this elegant property is a *theorem*, not a definition.

Theorem 3.8.5 (Truth-Table Theorem, Nerode).   *The following definitions of tt-reducible are equivalent.*

(i) $B \leq_{\mathrm{tt}} A$ as defined in Definition 3.8.2.

(ii) $\Phi_e^A = B$ for some total Φ_e.

(iii) There is a computable function $g(x)$ such that for every x

$$ x \in B \quad \Longleftrightarrow \quad (\exists y \in D_{g(x)})[\,\sigma_y \prec A\,]. $$

Proof. (i) $\implies$ (ii). Obvious because (i) ensures Φ_e total.

(ii) $\implies$ (iii). Uniformly in x, enumerate the set $U_x = \{\,\sigma\,:\,\Phi_e^\sigma(x)\!\downarrow\}$. Since Φ_e is total, we apply the Compactness Theorem 8.3.1 (iv) to get a finite subset $F_x \subseteq U_x$ with $\cup_{\sigma \in F_x}[\![\,\sigma\,]\!] \;=\; 2^\omega$. To find F_x uniformly effectively in x, keep enumerating U_x until F_x appears. Define $h(x) = \max\{\,|\sigma| \,:\, \sigma \in F_x\}$ and define $g(x)$ by

$$ D_{g(x)} = \{\,y\,:\,|\sigma_y| = h(x) \;\;\&\;\; \Phi_e^{\sigma_y}(x) = 1\,\}. $$

Now $g(x)$ satisfies (iii).

(iii) $\implies$ (i). If (iii) holds via $g(x)$, let $h(x) = \max\{\,|\sigma| : \sigma \in D_{g(x)}\,\}$. For every σ of length $h(x)$, define:

$$ \Phi_e^\sigma(x) = \begin{cases} 1 & (\exists \tau \preceq \sigma)[\,\tau \in D_{g(x)}\,]; \\ 0 & \text{otherwise.} \end{cases} $$

Now Φ_e with $h(x)$ satisfies (i). $\square$

3.8.4   Difference of C.E., n-c.e., and ω-c.e. Sets

The Limit Lemma 3.6.2 characterized sets $A \leq_{\mathrm{T}} \emptyset'$ as those where $A = \lim_s A_s$ for a computable sequence $\{A_s\}_{s \in \omega}$. Now consider special cases based on how many times the approximation changes on x. These notions are more general than c.e. sets A, but not the most general case of $A \leq_{\mathrm{T}} \emptyset'$.

Definition 3.8.6. (i) A set D is the *difference of c.e. sets* (*d.c.e.*) if $D = A - B$ where A and B are c.e. sets.

(ii) A set A is *omega-c.e.* (written ω-c.e.) if there is a computable sequence $\{A_s\}_{s \in \omega}$ with $A_0 = \emptyset$ and $A_s(x) \in \{0,1\}$, and a computable function $g(x)$

which bounds the number of changes in the approximation $\{A_s\}_{s\in\omega}$ in the following sense,

(3.25) $A = \lim_s A_s$ & $|\{ s : A_s(x) \neq A_{s+1}(x) \}| \leq g(x).$

(iii) For $n \in \omega$ the set A is n-c.e. if $g(x) \leq n$.

For example, the only 0-c.e. set is $\emptyset$, the 1-c.e. sets are the usual c.e. sets, and the 2-c.e. sets are the d.c.e. sets. The next theorem gives an elegant characterization of ω-c.e. sets.

Definition 3.8.7. For any set $A \subseteq \omega$ define the *y-section* of A,

(3.26) $A^{[y]} = \{ \langle x, z\rangle : \langle x, z\rangle \in A \ \& \ z = y \}.$

Using the pairing function we can identify A with a subset of $\omega \times \omega$ and view $A^{[y]}$ as the yth *row* of A. We use the square bracket notation $A^{[y]}$ to distinguish from the yth jump $A^{(y)}$. We also use this notation in Chapter 6 and other chapters.

Theorem 3.8.8 (Bounded T-Reducibility Theorem). *The following are equivalent.*

(i) $A \leq_{\text{bT}} \emptyset'.$

(ii) A *is ω-c.e.*

(iii) $A \leq_{\text{tt}} \emptyset'.$

Proof. Clearly, (iii) implies (i).

(i) $\implies$ (ii). We use K for $\emptyset'$ for notational clarity. Suppose $\Phi_e^K = A$ with computable bound $g(x) \geq \varphi_e^K(x)$ as in (3.25). Let $\{\widehat{A}_s\}_{s\in\omega}$ be any computable enumeration of A. We *speed up* the given A enumeration $\{\widehat{A}_s\}_{s\in\omega}$ to obtain a new enumeration $\{A_s\}_{s\in\omega}$ of A as follows. Given s find the least $t > s$ such that

$$(\forall x \leq t)\,[\ \widehat{A}_t(x) \ = \ \Phi_{e,t}^{K_t}(x)\ ],$$

and define $A_s = \widehat{A}_t$. Now $A_{s+1}(x) \neq A_s(x)$ only if some element $z \leq g(x)$ enters K, which can happen at most $g(x) + 1$ times, once for each $z \in [0, g(x)]$.

(ii) $\implies$ (iii). Assume $A = \lim_s A_s$ satisfies (3.25) via $g(x)$. Define the c.e. *change set* C as in (3.19). Therefore, $|C^{[x]}|$ is the number of changes on x during the approximation, and $|C^{[x]}| \leq g(x)$ by hypothesis (ii). Furthermore, $A(x) = 1$ iff $|C^{[x]}|$ is odd, because the approximation changes between 0 and 1 starting with 0.

First build a truth-table to demonstrate that $A \leq_{\text{tt}} C$. For a given x the truth-table has rows of width $g(x)$ as in Definition 3.8.2 (ii). For each

$k \leq g(x)$ construct a row beginning with k many 1's followed by all 0's. Now to tt-compute from C whether $x \in A$, compute $k = |C^{[x]}|$. Next, find the row beginning with k many 1's. Now $A(x) = 1$ iff k is odd. Hence, $A \leq_{tt} C$. However, C is c.e. Hence, $C \leq_m \emptyset'$ and therefore $A \leq_{tt} \emptyset'$.   □

This proof shows the difference between T-reductions and tt-reductions. First, the assumption on $h(x)$ ensures that $A \leq_{bT} C$. Second, the approximation always begins with $A_s(x) = 0$ for $s = 0$ and changes between 0 and 1 because all values are in $\{0, 1\}$ rather than in ω. Therefore, we can make up a row in the truth-table which gives a value for $A(x)$ based only on the number of changes. This ensures that $A \leq_{tt} C$. This case is unusual since most reductions we consider produce only $A \leq_T B$ for some set B and sometimes produce $A \leq_{bT} B$, as in permitting arguments. Note that instead of directly building a truth-table we could argue that this process builds a *total* functional $\Phi_e^C = A$, which it clearly does.

Remark 3.8.9. To *speed up* an enumeration of a c.e. set A means to take a given computable enumeration $\{\widehat{A}_s\}_{s \in \omega}$ and produce a new computable enumeration $\{A_s\}_{s \in \omega}$ with $A_s \supseteq \widehat{A}_s$ by using some special hypotheses on A such as above or such as A being low as we shall see later.

3.8.5   Exercises

Exercise 3.8.10. Let f be a one-one computable function with range A. Define the *deficiency set* for this enumeration f to be:

$$D = \{\, s : (\exists t > s)\,[\, f(t) < f(s)\,]\,\}.$$

Prove that $A \leq_T D$ and $D \leq_{tt} A$.

Exercise 3.8.11. Show that the ω-c.e. sets are closed under union and complementation and therefore form a Boolean algebra.

Exercise 3.8.12. (i) Show that A is $(2n+1)$-c.e. iff A is the union of n d.c.e. sets and a c.e. set.

(ii) Show that a set A is $(2n+2)$-c.e. iff A is the union of $n+1$ d.c.e. sets.

Exercise 3.8.13. Show that the m defined in the Modulus Lemma 3.6.3 is not necessarily the least modulus. Explicitly define the least modulus as an A-computable function.

Exercise 3.8.14. Define $C_n = \{\, e : |W_e| = n\,\}$.

(i) For $n \geq 0$, show that C_n is d.c.e.

(ii) For $n \geq 0$, show that C_n is not c.e.

Exercise 3.8.15. Show that the d.c.e. sets are closed under intersection.

Exercise 3.8.16. Show that for each n there is an $(n+1)$-c.e. set which is not n-c.e. Hence, the d.c.e. sets are not closed under union.

Hint. Fix n. First specify a method to effectively list all n-c.e. sets $\{Z_e^n\}_{e\in\omega}$. Next, treat this as a game in which Player I can insert or delete an element x_e from his $(n+1)$-c.e. set D in order to arrange that $D(x_e) \neq Z_e^n(x_e)$. The D-player has one more move for x_e than the Z_e^n-player does. Hence, this closely resembles and generalizes Theorem 1.6.5 that K is not computable because as a 1-c.e. set K has one more move than any computable function φ_e.

Exercise 3.8.17. (Jockusch and others). Show by induction on n that if A is n-c.e. then either A or $\overline{A}$ contains an infinite c.e. set.

Exercise 3.8.18. Suppose that A and B are c.e. sets and A is not computable. Prove that $A \times \overline{A} \not\leq_m B \oplus \overline{B}$.

Hint. (Jockusch). Suppose for a contradiction that f is a computable function such that for all x and y

$$x \in A \ \& \ y \in \overline{A} \iff f(x,y) \in B \oplus \overline{B}.$$

Define $C = \{\, y : (\exists x)\,[\, f(x,y) \in B \oplus \emptyset\,]\,\}$. Show that C is c.e., $C \subseteq \overline{A}$, and that there exists $a \in \overline{A} - C$. Show that for all x,

$$x \in A \iff f(x,a) \in \emptyset \oplus \overline{B}.$$

Hence, A is co-c.e.

Exercise 3.8.19. (i) Show that $K \times \overline{K}$ is d.c.e.

(ii) Show that if D is d.c.e. then $D \leq_1 K \times \overline{K}$. (Hence, $K \times \overline{K}$ is 1- complete with respect to d.c.e. sets.)

(iii) Show that $K \oplus \overline{K}$ is d.c.e.

(iv) Using Exercise 3.8.18 show that it is false that $K \times \overline{K} \leq_m K \oplus \overline{K}$. (Hence, $K \oplus \overline{K}$ is neither 1-complete nor m-complete with respect to d.c.e. sets.

Exercise 3.8.20. Define $A := \{x : W_x = \{0\}\}$. Give an alternative proof that $K \times \overline{K} \not\leq_m K \oplus \overline{K}$ by showing that $K \times \overline{K} \equiv_m A$ and that $A \not\equiv_m \overline{A}$ by the Recursion Theorem.

Exercise 3.8.21. Prove the relativized form of the Modulus Lemma 3.6.3. Show that for any set $B \subseteq \omega$ if A is c.e.a. in B and the function $f \leq_T A$, then $f = \lim_s \widehat{f}(x,s)$ for some B-computable function $\widehat{f}$ with modulus $m \leq_T A$.

4
The Arithmetical Hierarchy

4.1 Levels in the Arithmetical Hierarchy

In addition to the notions of computability and relative computability, the Kleene arithmetical hierarchy is one of the fundamental concepts of computability theory. In §2.1 we showed that a set A is c.e. iff it has the syntactical form Σ_1^0 defined with a string of existential quantifiers. Now we define the more general notion of Σ_n^0 with n alternating blocks of quantifiers. We prove that $\emptyset^{(n)} \in \Sigma_n^0 - \Sigma_{n-1}^0$ for $n > 1$. Therefore, the Σ_n^0 classes do not collapse, but rather form a hierarchy called the *arithmetical hierarchy* because these classes are definable in arithmetic. The relativized form of the hierarchy enables us to define several important special classes of sets and degrees called *high$_n$* and *low$_n$*, some of whose properties we develop now, and more later. The arithmetical hierarchy was introduced in Kleene's paper [Kleene 1943] and was developed in Kleene's book [Kleene 1952].

Convention 4.1.1. We now define the Σ_n^0 and Π_n^0 sets, where the superscript 0 indicates that we are counting *number* quantifiers, not *function* quantifiers as in Σ_1^1. We rarely mention function quantifiers until Part II on open and closed classes in Cantor space. Therefore, in Part I Chapters 1–7 we usually drop the superscript 0 from Σ_n^0, Π_n^0, and Δ_n^0, and abbreviate these by Σ_n, Π_n, and Δ_n. Particularly in the relativized case, we write Σ_n^A rather than $\Sigma_n^{0,A}$.

Definition 4.1.2. (i) A set B is in Σ_0 (Π_0, Δ_0) iff B is computable. As in Definition 2.3.1, a Δ_0-*index* for B is an index e such that $\varphi_e = \chi_B$. (Indices for Σ_n and Π_n sets will be given in Definition 4.2.4.)

(ii) For $n \geq 1$, B is *in* Σ_n (written $B \in \Sigma_n$) if there is a computable relation $R(x, y_1, y_2, \ldots, y_n)$ such that

$$x \in B \iff (\exists y_1)\,(\forall y_2)\,(\exists y_3) \cdots (Qy_n)\,R(x, y_1, y_2, \ldots, y_n),$$

where Q is $\exists$ for n odd, and $\forall$ for n even.

(iii) Likewise, B is Π_n ($B \in \Pi_n$) if

$$x \in B \iff (\forall y_1)\,(\exists y_2)\,(\forall y_3) \cdots (Qy_n)\,R(x, y_1, y_2, \ldots, y_n),$$

where Q is $\exists$ or $\forall$ according to whether n is even or odd.

(iv) Similarly, B is Δ_n ($B \in \Delta_n$) if $B \in \Sigma_n$ and $B \in \Pi_n$.

(v) B is *arithmetical* if $B \in \bigcup_n (\Sigma_n^0 \cup \Pi_n^0)$.

Note that B is arithmetical iff B can be obtained from a computable relation by finitely many applications of projection and complementation. (See Exercise 4.1.10.)

Definition 4.1.3. Fix a set A. If we replace everywhere "computable" in Definition 4.1.2 by "A-computable" then we have the definition of B being Σ_n *in* A (written $B \in \Sigma_n^A$), B being Π_n *in* A ($B \in \Pi_n^A$), $B \in \Delta_n^A$, and B being *arithmetical in* A.

4.1.1   Quantifier Manipulation

We say that a *formula* is Σ_n (Π_n) if it is Σ_n (Π_n) as a relation of its free variables. We assume familiarity with the usual rules of quantifier manipulation from elementary logic for converting a formula to an equivalent one in prenex normal form, consisting of a string of quantifiers (*prefix*) followed by a formula with no quantifiers (*matrix*), which will in our case be a computable relation. Using these rules we can show the following facts, which will be frequently used to prove that a particular set is in Σ_n or Π_n. The only nontrivial fact is (vi), concerning bounded quantifiers. A *bounded quantifier* is one of the form $(Qx \leq y)F$, which abbreviates $(\forall x)\,[x \leq y \implies F]$ if Q is $\forall$, and $(\exists x)\,[x \leq y \ \& \ F]$ if Q is $\exists$. Part (vi) asserts that bounded quantifiers may be moved to the right past ordinary quantifiers and thus may be ignored in counting quantifier complexity.

Theorem 4.1.4. (i) $A \in \Sigma_n \iff \overline{A} \in \Pi_n$;

(ii) $A \in \Sigma_n(\text{ or } \Pi_n) \implies (\forall m > n)\,[\,A \in \Sigma_m \cap \Pi_m\,]$;

(iii) $A, B \in \Sigma_n(\Pi_n) \implies A \cup B, A \cap B \in \Sigma_n(\Pi_n);$

(iv) $[R \in \Sigma_n \ \& \ n > 0 \ \& \ A = \{x : (\exists y)\, R(x, y)\}] \implies A \in \Sigma_n;$

(v) $[B \leq_m A \ \& \ A \in \Sigma_n] \implies B \in \Sigma_n;$

(vi) If $R \in \Sigma_n(\Pi_n)$, and A and B are defined by

$$\langle x, y \rangle \in A \iff (\forall z < y)\, R(x, y, z),$$

$$\langle x, y \rangle \in B \iff (\exists z < y)\, R(x, y, z),$$

then $A, B \in \Sigma_n(\Pi_n)$.

Proof. (i)   If $A = \{x : (\exists y_1)\,(\forall y_2) \cdots R(x, \vec{y})\}$, then

$$\overline{A} = \{x : (\forall y_1)\,(\exists y_2) \cdots \neg R(x, \vec{y})\}.$$

(ii)   For example, if $A = \{x : (\exists y_1)\,(\forall y_2)\, R(x, y_1, y_2)\}$, then

$$A = \{x : (\exists y_1)\,(\forall y_2)\,(\exists y_3)\, [R(x, y_1, y_2) \ \& \ y_3 = y_3]\}.$$

(iii)   Let $A = \{x : (\exists y_1)\,(\forall y_2) \cdots R(x, \vec{y})\}$, and

$$B = \{x : (\exists z_1)\,(\forall z_2) \cdots S(x, \vec{z})\}.$$

Then

$$x \in A \cup B \iff (\exists y_1)\,(\forall y_2) \cdots R(x, \vec{y}) \quad \lor \quad (\exists z_1)\,(\forall z_2) \cdots S(x, \vec{z}),$$

$$\iff (\exists y_1)\,(\exists z_1)\,(\forall y_2)\,(\forall z_2) \cdots [\, R(x, \vec{y}) \quad \lor \quad S(x, \vec{z})\,],$$

$$\iff (\exists u_1)\,(\forall u_2) \cdots [\, R(x, (u_1)_0, (u_2)_0, \dots) \quad \lor \quad S(x, (u_1)_1, (u_2)_1, \dots)\,],$$

where $(u)_0$ is the prime power coding as in the Notation section. Likewise, this holds for $A \cap B$.

(iv)   Immediate by quantifier contraction, as in (iii).

(v)   Let $A = \{x : (\exists y_1)\,(\forall y_2) \cdots R(x, \vec{y})\}$ and $B \leq_m A$ via f. Then

$$B = \{x : (\exists y_1)\,(\forall y_2) \cdots R(f(x), \vec{y})\}.$$

(vi)   The proof is by induction on n. If $n = 0$, then A and B are clearly computable. Fix $n > 0$, suppose $R \in \Sigma_n$ and assume (vi) for all $m < n$. Then $B \in \Sigma_n$ by (iv). Now there exists $S \in \Pi_{n-1}$ such that

$$\langle x, y \rangle \in A \iff (\forall z < y)\, R(x, y, z),$$

$$\iff (\forall z < y)\,(\exists u)\, S(x, y, z, u),$$

$$\Longleftrightarrow (\exists \sigma)\,(\forall z < y)\; S(x,y,z,\sigma(z)),$$

where σ ranges over $\omega^{<\omega}$. Now $(\forall z < y)\; S \in \Pi_{n-1}$ by the inductive hypothesis, so $A \in \Sigma_n$. The case $R \in \Pi_n$ follows from the case $R \in \Sigma_n$ by (i). $\qquad\square$

4.1.2   Placing a Set in Σ_n or Π_n

Proposition 4.1.5. Fin $\in \Sigma_2$.

Proof.

$$
\begin{aligned}
x \in \text{Fin} &\iff W_x \text{ is finite}\\
&\iff (\exists s)\,(\forall t)\,[\, t \le s \;\lor\; W_{x,t} - W_{x,s} \,].
\end{aligned}
$$

The bracketed relation of x,s,t is clearly computable. $\qquad\square$

Proposition 4.1.6. Cof $\in \Sigma_3$.

Proof.

$$
\begin{aligned}
x \in \text{Cof} &\iff \overline{W}_x \text{ is finite}\\
&\iff (\exists y)\,(\forall z)\,[\, z \le y \;\lor\; z \in W_x \,]\\
&\iff (\exists y)(\forall z)(\exists s)\,[\, z \le y \;\lor\; z \in W_{x,s} \,].
\end{aligned}
$$

$\qquad\square$

Since the final prefix depends only on the type and relative position of the quantifier symbols and sentential connectives, we frequently abbreviate these calculations by replacing previously identified predicates with strings of quantifiers indicating the classes to which they belong.

Proposition 4.1.7. $\{\, \langle x,y \rangle : W_x \subseteq W_y \,\} \in \Pi_2$.

Proof.

$$
\begin{aligned}
W_x \subseteq W_y &\iff (\forall z)\,[z \in W_x \implies z \in W_y]\\
&\iff (\forall z)\,[\, z \notin W_x \;\lor\; z \in W_y \,]\\
&\iff (\forall z)\,[\, \forall \;\lor\; \exists \,]\\
&\iff \forall\forall\exists\,[\dots]\\
&\iff \forall\exists\,[\dots].
\end{aligned}
$$

$\qquad\square$

Corollary 4.1.8 (Classification of Tot). (i) $\{\, \langle x,y \rangle : W_x = W_y \,\} \in \Pi_2$.

(ii) Tot $= \{\, y : W_y = \omega \,\} \in \Pi_2$.

Proof. (i) follows by Proposition 4.1.7 and Theorem 4.1.4 (iii), and (ii) follows from Proposition 4.1.7 with $W_x = \omega$. $\qquad\square$

Corollary 4.1.9. $\mathrm{Rec} \in \Sigma_3$. ($\mathrm{Rec} := \{e : W_e \equiv_{\mathrm{T}} \emptyset\}$ in Definition 1.6.15.)

Proof.

$$
\begin{aligned}
x \in \mathrm{Rec} \quad &\Longleftrightarrow \quad W_x \text{ is computable} && \text{(i.e., recursive)} \\
&\Longleftrightarrow \quad (\exists y)\,[\, W_x = \overline{W}_y \,] \\
&\Longleftrightarrow \quad (\exists y)\,[\, W_x \cap W_y = \emptyset \ \ \& \ \ W_x \cup W_y = \omega \,] \\
&\Longleftrightarrow \quad \exists \ [\, \forall \ \& \ \forall\exists \,] && \text{by Corollary 4.1.8} \\
&\Longleftrightarrow \quad \exists\forall\exists\,[\ldots].
\end{aligned}
$$

$\square$

4.1.3 Exercises

Exercise 4.1.10. Prove that A is arithmetical, i.e., that $A \in \bigcup_n (\Sigma_n \cup \Pi_n)$, iff A can be obtained from a computable relation by a finite number of applications of projection and complementation.

Exercise 4.1.11. Prove that $\mathrm{Ext} \in \Sigma_3$ for Ext as defined in Definition 1.6.15.

Exercise 4.1.12. Prove that

$$\{\langle x, y \rangle : W_x \text{ and } W_y \text{ are computably separable}\} \in \Sigma_3.$$

(Recall from Remark 2.4.15 and Exercise 1.6.26 that W_x and W_y are *computably separable* if $W_x \subseteq R$ and $W_y \subseteq \overline{R}$ for some computable set R, and W_x and W_y are *computably inseparable* otherwise.)

Exercise 4.1.13. Define $A \subseteq^* B$ if $A - B$ is finite, i.e., if $A \subseteq B$ except for at most finitely many elements. Define $A =^* B$ if $A \subseteq^* B$ and $B \subseteq^* A$. Prove that the following are two sets are Σ_3:

$$\{\langle x, y \rangle : W_x \subseteq^* W_y\};$$

$$\{\langle x, y \rangle : W_x =^* W_y\}.$$

Exercise 4.1.14. Show that $\{\, x : W_x \text{ is creative} \,\} \in \Sigma_3$.

4.2 ** Post's Theorem and the Hierarchy Theorem

Definition 4.2.1. A set A is Σ_n-*complete* (Π_n-*complete*) if $A \in \Sigma_n(\Pi_n)$ and $B \leq_1 A$ for every $B \in \Sigma_n(\Pi_n)$. (By Exercises 4.2.6 and 4.2.7 it makes no difference whether we use "$B \leq_m A$" or "$B \leq_1 A$" in the definition of Σ_n–complete and Π_n-complete.)

Note that A is Σ_1-complete iff A is 1-complete as defined in Definition 2.4.1. Hence, K is Σ_1-complete and $\overline{K}$ is Π_1-complete. The following fundamental theorem relates the jump hierarchy of degrees from §3.4 to the arithmetical hierarchy.

4.2.1 Post's Theorem Relating Σ_n to $\emptyset^{(n)}$

Theorem 4.2.2 (Post's Theorem). *For every $n \geq 0$,*

(i) $B \in \Sigma_{n+1}$ $\Longleftrightarrow$ *B is c.e. in some Π_n set*

$\Longleftrightarrow$ *B is c.e. in some Σ_n set*

by Theorem 3.4.3 (vii).

(ii) $\emptyset^{(n)}$ is Σ_n-complete for $n > 0$;

(iii) $B \in \Sigma_{n+1}$ $\Longleftrightarrow$ B is c.e. in $\emptyset^{(n)}$;

(iv) $B \in \Delta_{n+1}$ $\Longleftrightarrow$ $B \leq_T \emptyset^{(n)}$.

Proof. (i) ($\Longrightarrow$). Let $B \in \Sigma_{n+1}$. Then $x \in B \iff (\exists y)\, R(x, y)$ for some $R \in \Pi_n$. Hence B is Σ_1 in R and therefore c.e. in R by Theorem 3.3.16.

(i) ($\Longleftarrow$). Suppose B is c.e. in some Π_n set C. Then for some e,

$$x \in B \iff x \in W_e^C$$

$$x \in B \iff (\exists s)\,(\exists \sigma)\,[\, \sigma \prec C \ \ \& \ \ x \in W_{e,s}^\sigma \,].$$

Clearly, $x \in W_{e,s}^\sigma$ is computable by Oracle Graph Theorem 3.3.8. Hence, by Theorem 4.1.4 (iv) it suffices to show that $\sigma \prec C$ is Σ_{n+1}. Now

$$\sigma \prec C \iff (\forall y < lh(\sigma))\,[\sigma(y) = C(y)]$$

$$\iff (\forall y < lh(\sigma))\,[[\sigma(y) = 1 \ \& \ y \in C] \vee [\sigma(y) = 0 \text{ and } y \notin C]]$$

$$\iff (\forall y < lh(\sigma))\,[\Pi_n \vee \Sigma_n]$$

because $C \in \Pi_n$. Hence, $\sigma \prec C$ is Σ_{n+1} by Theorem 4.1.4 (ii), (iii), and (vi).

(ii) This is proved by induction on n and is clear for $n = 1$. Fix $n \geq 1$ and assume $\emptyset^{(n)}$ is Σ_n-complete. Hence $\overline{\emptyset^{(n)}}$ is Π_n-complete. Now

$$B \in \Sigma_{n+1} \iff B \text{ is c.e. in some } \Sigma_n \text{ set by (i)}$$

$$\iff B \text{ is c.e. in } \emptyset^{(n)} \text{ by inductive hypothesis}$$

$$\iff B \leq_1 \emptyset^{(n+1)} \text{ by the Jump Theorem 3.4.3 (iii).}$$

Hence, $\emptyset^{(n+1)}$ is Σ_{n+1}-complete.

(iii) Now $\overline{\emptyset^{(n)}}$ is Π_n-complete for $n > 0$ by (ii), and (i) and (ii) imply (iii).

(iv)

$$B \in \Delta_{n+1} \iff B, \overline{B} \in \Sigma_{n+1},$$

$$\iff B, \overline{B} \text{ are c.e. in } \emptyset^{(n)}, \text{ by (iii)},$$

$$\iff B \leq_T \emptyset^{(n)}.$$

$\square$

Corollary 4.2.3 (Hierarchy Theorem). $(\forall n > 0)[\, \Delta_n \subset \Sigma_n \quad \& \quad \Delta_n \subset \Pi_n \,]$. *Clearly,* $\Delta_n \subseteq \Sigma_n$. *The content here is that* $\Sigma_n \not\subseteq \Delta_n$.

Proof. $\emptyset^{(n)} \in \Sigma_n - \Pi_n$, by Post's Theorem 4.2.2 (ii) and (iv), and the Jump Theorem 3.4.3 (ii). Likewise, $\overline{\emptyset^{(n)}} \in \Pi_n - \Sigma_n$. $\square$

Definition 4.2.4. (Σ_n and Π_n Indices).

(i) By Definition 2.1.4, e is a Σ_1-*index* for B if $B = W_e$, and we also say that e is a Π_1-*index* for $\overline{B}$.

(ii) For $n > 0$, by Theorem 4.2.2 (iii), $B \in \Sigma_n$ iff $B \leq_1 \emptyset^{(n)}$, say via φ_e. Then e is a Σ_n-*index* for B and a Π_n-*index* for $\overline{B}$.

(iii) As in Definition 2.3.1 and Definition 4.1.2 (i), a Δ_0-*index* for B is an index e such that $\varphi_e = \chi_B$. For $n \geq 1$, a Δ_n-*index* for B is a pair $\langle e, i \rangle$ where e is a Σ_n index for B and i is a Π_n index for B. (These definitions relativize to an oracle A.)

4.2.2   Exercises

Exercise 4.2.5. In the Limit Lemma 3.6.2, prove that we can pass effectively from an index for any one characterization (i), (ii), or (iii) to any other. An index for (i) is an e such that $\varphi_e(s, x) = A_s(x)$ and $A(x) = \lim_s A_s(x)$; An index for (ii) is a Δ_2-index for A. An index for (iii) is an e such that $A = \Phi_e^K$.

Exercise 4.2.6. Prove that if $B \leq_m A$ and $A = \emptyset^{(n)}$ for $n \geq 1$ then $B \leq_1 A$. *Hint.* Use the Padding Lemma 1.5.2. An alternative proof is to show that $B \in \Sigma_n$ and hence B is c.e. in $\emptyset^{(n-1)}$ by Post's Theorem 4.2.2. Therefore, we can apply the Jump Theorem 3.4.3.

Exercise 4.2.7. Prove that if $B = \emptyset^{(n)}$ for $n \geq 1$, and $B \leq_m A$, then $B \leq_1 A$. *Hint.* Use the method of Theorem 2.3.9. (By Exercise 4.2.7, in order to prove that A is Σ_n-complete it suffices to prove that $\emptyset^{(n)} \leq_m A$ rather than proving $\emptyset^{(n)} \leq_1 A$.)

4.3 * Σ_n-Complete Sets and Π_n-Complete Sets

We have shown that $\emptyset^{(n)}$ is Σ_n-complete for all n. (Following Convention 4.1.1 we normally drop the superscript 0 from now on.) However, there are other Σ_n-complete sets with natural definitions which will be useful in later applications. For example, we know that K, K_0 and K_1 are all Σ_1-complete and we shall now show that Fin is Σ_2-complete and Cof and Rec are Σ_3-complete. Once we have classified a set A as being in Σ_n by the method of §4.1, we attempt to show that the classification is the best possible by proving that $B \leq_1 A$ for some known Σ_n-complete set B, thus showing that A is Σ_n-complete. Recall from Definition 2.4.9 that $(A, B) \leq_m (C, D)$ via f computable if $f(A) \subseteq C$, $f(B) \subseteq D$, and $f(\overline{A \cup B}) \subseteq \overline{C} \cup \overline{D}$. We write "$\leq_1$" if f is 1:1.

Definition 4.3.1. For $n \geq 1$ define $(\Sigma_n, \Pi_n) \leq_m (C, D)$ if $(A, \overline{A}) \leq_m (C, D)$ for some Σ_n-complete set A, and similarly for $\leq_1$ in place of $\leq_m$. In this case we also write $\Sigma_n \leq_m C$ and $\Pi_n \leq_m D$. (By the same remark as that in Definition 4.2.1, it makes no difference whether we write "$\leq_m$" or "$\leq_1$" here.)

(This notation seems strange because Σ_n and Π_n are *classes* not sets. It is justified because if $(\Sigma_n, \Pi_n) \leq_m (C, D)$ then $(A, \overline{A}) \leq_m (C, D)$ and $(\overline{B}, B) \leq_m (C, D)$ for *any* Σ_n set A and Π_n set B.)

4.3.1   Classifying Σ_2 and Π_2 Sets: Fin, Inf, and Tot

Theorem 4.3.2. $(\Sigma_2, \Pi_2) \leq_1$ (Fin, Tot). *Therefore,* Fin *is Σ_2-complete,* Inf *and* Tot *are Π_2-complete, and* Inf $\equiv_1$ Tot. *Hence,* Inf *and* Tot *are computably isomorphic, written* Inf $\equiv$ Tot.

Proof. By Proposition 4.1.5 and Corollary 4.1.8, Fin $\in \Sigma_2$ (so Inf $\in \Pi_2$) and Tot $\in \Pi_2$. Fix $A \in \Sigma_2$. Therefore, $\overline{A} \in \Pi_2$, and there is a computable relation R such that

$$x \in \overline{A} \quad \Longleftrightarrow \quad (\forall y)(\exists z)\, R(x, y, z).$$

Using the *s-m-n* Theorem 1.5.5, define a 1:1 computable function f by

$$\varphi_{f(x)}(u) = \begin{cases} 0 & \text{if } (\forall y \leq u)\,(\exists z)\, R(x, y, z); \\ \uparrow & \text{otherwise.} \end{cases}$$

Now

$$x \in \overline{A} \quad \Longrightarrow \quad W_{f(x)} = \omega \quad \Longrightarrow \quad f(x) \in \text{Tot}, \quad \text{but}$$

$$x \in A \quad \Longrightarrow \quad W_{f(x)} \text{ is finite} \quad \Longrightarrow \quad f(x) \in \text{Fin}.$$

$\square$

4.3.2 Constructions with Movable Markers

Most of the definitions of c.e. sets so far have been *static* in the sense of §2.6.2, but from now on we often give *dynamic* definitions. For example, we may define a c.e. set B by a construction using a computable sequence of stages s where B_s represents the set of elements enumerated in B by the end of stage s and $B = \cup_s B_s$. To construct B we concentrate on the stage s approximation to the *complement* $\overline{B}$ because these are the only elements over which we still have control. Those *already* in B are irretrievable. Given B_s, define the element b_y^s for $y \in \omega$ as follows:

$$(4.1) \qquad \overline{B} = b_0 < b_1 < b_2 < \ldots \qquad \& \qquad \overline{B}_s = b_0^s < b_1^s < b_2^s < \ldots$$

To define B_s it is often useful to imagine a sequence of *markers* $\{\Gamma_y\}_{y \in \omega}$ such that the marker Γ_y is associated with element b_y^s at the end of stage s. Now $b_y^s \leq b_y^{s+1}$. Therefore, we may imagine marker Γ_y as moving upwards and being associated with a nondecreasing (possibly finite) sequence of elements $\{b_y^s\}_{s \in \omega}$ among the integers. Hence, the name *movable markers* is used in the literature.[1]

The advantage of concentrating on the marker Γ_y rather than the element $z = b_y^s$ it is currently resting on is that for applications we may have an additional c.e. *kicking set* V_y which is coordinated with the marker Γ_y. Whenever V_y receives a new element, the current position of Γ_y is enumerated in B. Hence, Γ_y comes to a limit and b_y exists iff V_y is finite. Therefore, $\overline{B}$ is infinite iff every V_y is finite. We have already implicitly used this method for $y = 0$ in Exercise 2.5.3 to prove that Inf $\leq_1$ Cof. We now illustrate the movable marker method in the following Theorem 4.3.3 but with a movable marker for every y not only for $y = 0$.

4.3.3 Classifying Cof as Σ_3-Complete

Theorem 4.3.3. *Cof is Σ_3-complete.*

Proof. Fix $A \in \Sigma_3$. Now for some relation $R \in \Pi_2$, $x \in A$ iff $(\exists y)R(x, y)$. Since $R \in \Pi_2$ there is a computable function g by Theorem 4.3.2 such that $R(x, y)$ iff $W_{g(x,y)}$ is infinite. Therefore,

$$(4.2) \qquad x \in A \quad \Longleftrightarrow \quad (\exists y)[\, W_{g(x,y)} \text{ is infinite }].$$

---

[1] For more sophisticated applications, it is better to think of the markers as fixed *boxes* or *windows* sometimes arranged in some geometrical pattern, such as a matrix, a tree, or simply a line as here, through which the integers move downwards. From this point of view the boxes are fixed and the integers are moving among them, but we still have box Γ_y associated with a nondecreasing sequence of elements $\{b_y^s\}_{s \in \omega}$.

We shall define a c.e. set B^x uniformly in x such that $x \in A$ iff B^x is cofinite. Fix x. For notational convenience we drop the superscript x. We enumerate $B = \cup_{s \in \omega} B_s$ by stages s in the following computable construction. Use the notation of §4.3.2 and (4.1). We think of $W_{g(x,y)}$ as a *kicking set* so that each new element entering $W_{g(x,y)}$ "kicks" the marker Γ_y and forces it to move once more.

Stage $s = 0$. Set $B_0 = \emptyset$.

Stage $s+1$. Let $\overline{B}_s = \{b_0^s < b_1^s < \cdots < b_y^s < \cdots\}$. For each $y \leq s$ such that $W_{g(x,y),s} \neq W_{g(x,y),s+1}$, enumerate b_y^s in B_{s+1}. If no such y exists, define $B_{s+1} = B_s$. This ends the construction.

Case 1. $x \in A$. By (4.2), choose the least y such that $W_{g(x,y)}$ is infinite. Now marker Γ_y is moved infinitely often. Therefore, $\lim_s b_y^s = \infty$, and $|\overline{B}| \leq y$.

Case 2. $x \notin A$. By induction, fix y, and choose s such that $W_{g(x,y),s} = W_{g(x,y)}$ and, for all $z < y$ such that $b_z^s = b_z$. Now Γ_y never moves again after s. Hence, every marker comes to rest on $\overline{B}$, which is therefore infinite. $\square$

4.3.4   Classifying Rec as Σ_3-Complete

Definition 4.3.4. (i) Cpl $= \{x : W_x \equiv_T K\}$, indices of *complete* c.e. sets.

(ii) Rec $= \{x : W_x \equiv_T \emptyset\}$, indices of *computable (recursive)* sets.

Theorem 4.3.5. $(\Sigma_3, \Pi_3) \leq_1 (\text{Cof}, \text{Cpl})$, *and* $(\Sigma_3, \Pi_3) \leq_1 (\text{Rec}, \text{Cpl})$.

Corollary 4.3.6 (Rogers).   Rec *is Σ_3-complete.*

Proof. By Corollary 4.1.9 and Theorem 4.3.5 because Cof $\subseteq$ Rec and because Rec $\cap$ Cpl $= \emptyset$. $\square$

Proof. (Theorem 4.3.5). Let A be Σ_3. We define a c.e. set B^x uniformly in x such that

(4.3) $x \in A \iff (\exists y)[\, W_{g(x,y)}$ is infinite $] \iff B^x$ is cofinite,

(4.4) $x \notin A \implies B^x \equiv_T K$.

Fix x. For notational convenience we can drop the x. Let $\{K_s\}_{s \in \omega}$ be a computable enumeration of K. The construction is now exactly the same as that of Theorem 4.3.3 except that at Stage $s+1$ we replace the second sentence by the following:

"For each $y \leq s$ such that *either* $W_{g(x,y),s} \neq W_{g(x,y),s+1}$
or $y \in K_{s+1} - K_s$, enumerate b_y^s in B_{s+1}."

Now if $x \in A$ then some $W_{g(x,y)}$ is infinite and it causes $\overline{B}$ to be finite as before. If $x \notin A$ then the extra clause generates at most one extra move for marker Γ_y. Therefore, all markers move finitely often and $\overline{B}$ is infinite. The extra coding ensures that $K \leq_T B$. Choose a stage s such that marker Γ_y has settled on b_y^s by the end of stage s. Then $y \in K$ iff $y \in K_s$ because if y enters K at some stage $t > s$ then marker Γ_y must move at stage t, which it cannot. $\qquad\square$

Remark 4.3.7. Theorem 4.3.5 also implies the previous Theorem 4.3.3 that Cof is Σ_3-complete, and it shows that $(\Pi_3, \Sigma_3) \leq_m (\mathrm{Cpl}, \overline{\mathrm{Cpl}})$. This does not imply that Cpl is Π_3-complete. It says exactly that Cpl is Π_3-hard, namely that a Π_3-complete set is m-reducible to it. Indeed Cpl is Σ_4-complete.

Remark 4.3.8. An alternative coding is to move the markers to prove that if $x \in \overline{A}$, then $\overline{B}$ dominates all p.c. functions and therefore $K \leq_T B$ by Theorem 4.5.4 (ii). In Theorem 4.3.5 we have two strategies. The primary strategy S_1 uses $W_{g(x,y)}$ to show that if $x \in A$ then $\overline{B}$ is finite. If $x \in \overline{A}$, this primary strategy guarantees only that $\overline{B}$ is infinite. In this case we can simultaneously play the *secondary strategy* S_2, which ensures $B \equiv_T K$. In the Π_3, where $\overline{B}$ is infinite, we can code various other properties into $\overline{B}$. For example, in Chapter 5 Exercise 5.2.10 we prove that $\{e : W_e \text{ simple}\}$ is Π_3-complete.

One may imagine that the Π_3 alternative on $\overline{B}$ is an expert woodsman who goes through the forest chopping down only *certain* trees to code information. If the Σ_3 alternative holds, then the logging company comes through, cutting *all* the trees and erasing any coding done by the woodsman.

In Exercise 4.3.12 we shall prove that Ext is Σ_3-complete by defining a p.c. function $\varphi_{f(x)}$ and having a strategy S_2 for marker Γ_y which guarantees that $\varphi_{f(x)}$ is not extendible to a total function φ_y and that indeed Γ_y bounds a counterexample z. In Exercise 5.2.10, the markers Γ_y, $y < e$, allow some $b_y^s \in W_e$ to enter B to achieve $B \cap W_e \neq \emptyset$, so B^x will be simple (see §5.2). The only restriction on the secondary strategy S_2 is that it must cause the marker Γ_y to move at most finitely often so as not to *accidentally* cause $\overline{B}$ to be finite even though $x \notin A$ which is the Π_3 case.

4.3.5 Σ_3-Representation Theorems

The following are probably the most useful characterizations for approximating a Σ_3 set A, i.e., for "guessing" whether $x \in A$, and should be viewed as refinements of (4.2).

Theorem 4.3.9 (First Σ_3-Representation Theorem). *If $A \in \Sigma_3$ then there is a computable function g such that*

$$(4.5) \qquad x \in A \iff (\forall^\infty y)\,[\,W_{g(x,y)} = \omega\,]\quad and$$

$$(4.6) \qquad x \in \overline{A} \iff (\forall y)\,[\,W_{g(x,y)} \text{ is finite }].$$

Proof. Since $A \in \Sigma_3$, let $A \leq_1$ Cof via f using Theorem 4.3.3. Define g by

$$z \in W_{g(x,y)} \iff (\forall u)\,[y \leq u \leq z \implies u \in W_{f(x)}]. \qquad \text{Hence,}$$

$$x \in A \implies W_{f(x)} \text{ cofinite} \implies (\exists y)(\forall z \geq y)\,[\,z \in W_{f(x)}\,]$$

$$\implies (\exists y)\,(\forall z \geq y)\,[W_{g(x,z)} = \omega]; \qquad \text{and}$$

$$x \in \overline{A} \implies W_{f(x)} \text{ coinfinite} \implies (\forall y)\,(\exists z \geq y)\,[z \notin W_{f(x)}]$$

$$\implies (\forall y)\,[W_{g(x,y)} \text{ finite}].$$

$\square$

Remark 4.3.10. (*Guessing About a Σ_3 Set A*). To "guess" about membership in a Σ_2 set A, we have a computable function f such that $x \in A$ iff $W_{f(x)}$ is finite. For a Σ_3 set A, Theorem 4.3.9 is the two-dimensional analogue where $W_{g(x,y)}$ is viewed as the yth row of a matrix. If $x \in A$, then almost all rows are ω, and the others are finite. If $x \notin A$ then all rows are finite. The next corollary says that in the first case we may redefine the matrix so that there is a *unique* row which is infinite and that row is ω.

Theorem 4.3.11 (Second Σ_3-Representation Theorem-Uniqueness). *If $A \in \Sigma_3$ then there is a computable function h such that the following lines hold:*

$$(4.7)\quad x \in A \iff (\exists!\,y)\,[\,W_{h(x,y)} = \omega \ \& \ (\forall z \neq y)[\,W_{h(x,z)} =^* \emptyset\,]\,],$$

$$(4.8) \qquad x \in \overline{A} \iff (\forall y)\,[\,W_{h(x,y)} =^* \emptyset\,],$$

where $(\exists!\,y)R(y)$ denotes that there exists a unique *y such that $R(y)$.*

Proof. A is Σ_3. Choose $g(x,y)$ satisfying (4.5) and (4.6). Define

$$f(x,y,s) \;=\; y \,+\, \Sigma_{z<y}\,|\,W_{g(x,z),s}\,|.$$

(Think of $f(x,y,s)$ as the position at the end of stage s of a movable marker Γ_y^x which moves along the h rows trying to represent row $W_{g(x,y)}$ on some h row but which is bumped whenever an element appears in some $W_{g(x,z)}$ for some $z < y$.)

Stage $s+1$. Let $z = f(x,y,s)$. Enumerate in $W_{h(x,z)}$ all $w \in W_{g(x,y),s}$.

Verification.

Case 1. $x \in A$. Choose the least y such that $W_{g(x,y)} = \omega$. Then $z = \lim_s f(x,y,s)$ exists, and $W_{h(x,z)} = W_{g(x,y)} = \omega$. Also, $\lim_s f(x,v,s) = \infty$ for all $v > y$ and hence $W_{h(x,u)}$ is finite for all $u > z$.

Case 2. $x \notin A$. For each z there are at most finitely many y such that $\lim_s f(x,y,s) = z$ because of the clause "$y +$" in the definition of $f(x,y,s)$. But each g row $W_{g(x,y)}$ is finite. Hence, every h row $W_{h(x,z)}$ is finite.   □

4.3.6   Exercises

Exercise 4.3.12.$^\diamond$ Prove that $(\Sigma_3, \Pi_3) \leq_1 (\text{Cof}, \overline{\text{Ext}})$ and hence that Ext is Σ_3-complete. *Hint.* Use the notation and method of Theorem 4.3.3 to construct $\varphi_{f(x)}$ such that if $x \in A$, then $f(x) \in \text{Cof} \subset \text{Ext}$, and if $x \notin A$, then $f(x) \in \overline{\text{Ext}}$.

Exercise 4.3.13. Show $\{\langle x,y \rangle : W_x \text{ and } W_y \text{ are computably separable}\}$ is Σ_3-complete. *Hint.* Make $\varphi_{f(x)}$ of Exercise 4.3.12 take values $\subseteq \{0,1\}$.

Exercise 4.3.14. Prove that $\{\langle x,y \rangle : W_x \subseteq^* W_y\}$ and $\{\langle x,y \rangle : W_x =^* W_y\}$ are each Σ_3-complete.

Exercise 4.3.15. Show that if A is a c.e. set, then $G_m(A) \in \Sigma_3$ where

$$G_m(A) := \{ x : W_x \equiv_m A \}.$$

Exercise 4.3.16.$^\diamond$ (Lerman). Let ζ (zeta) denote the order type of the integers $\mathbb{Z}$ (both positive and negative in their natural order). Hence, ζ has order type $\omega^* + \omega$. A ζ-representation for a set $A \subseteq \omega$ is a linear ordering

$$L_A^\zeta = \zeta + a_o + \zeta + a_1 + \ldots,$$

where $A = \{a_0, a_1, \ldots\}$ is not necessarily in increasing order and possibly with repetitions.

(i) Prove that if L_A^ζ is a computable linear ordering, i.e., the $<$ relation on it is computable, then $A \in \Sigma_3$.

(ii)$^\diamond$ Prove that if $A \in \Sigma_3$ then there is a computable ordering L of order type L_A^ζ.

4.4   Relativized Hierarchy: Low$_n$ and High$_n$ Sets

Definition 4.4.1. The definition of Σ_n^A (Π_n^A) is the same as Definition 4.1.2 for Σ_n (Π_n) except that the matrix R is A-computable instead of computable. If $\mathbf{a} = \deg(A)$, we use the notation $\Sigma_n^{\mathbf{a}}$ in place of Σ_n^A since the class Σ_n^A is independent of the particular representative $A \in \mathbf{a}$.

Everything in this chapter can be relativized to an arbitrary set A with virtually the same proofs, and with Σ_n^A, Π_n^A and $A^{(n)}$ in place of Σ_n, Π_n and $\emptyset^{(n)}$, respectively.

4.4.1 Relativized Post's Theorem

Theorem 4.4.2 (Relativized Post's Theorem). *For every $n \geq 0$,*

(i) $A^{(n)}$ *is* Σ_n^A*-complete if* $n > 0$*;*

(ii) $B \in \Sigma_{n+1}^A$ $\iff$ B *is c.e. in* $A^{(n)}$*;*

(iii) $B \leq_T A^{(n)}$ $\iff$ $B \in \Delta_{n+1}^A := \Sigma_{n+1}^A \cap \Pi_{n+1}^A$*;*

(iv) $B \leq_T A^{(n+1)}$ $\iff$ $(\exists f \leq_T A^{(n)})\,[\,B(x) = \lim_s f(x,s)\,]$*.*

Define Fin^A, Tot^A, and Cof^A as before but with W_e^A in place of W_e. The proofs in §4.3 relativize to A and establish that Fin^A is Σ_2^A-complete, Tot^A is Π_2^A-complete, and Cof^A and Rec^A are Σ_3^A-complete, where Rec^A is the set of e's such that W_e^A is A-computable (A-recursive). Hence, if $\mathbf{a} = \deg(A)$, then $\mathbf{a}' = \deg(A')$, $\mathbf{a}'' = \deg(\mathrm{Fin}^A)$, and $\mathbf{a}''' = \deg(\mathrm{Cof}^A)$.

4.4.2 Low$_n$ and High$_n$ Sets

In Definition 3.6.6 we introduced the low and high sets as those sets $A \leq_T \emptyset'$ whose jump A' has the lowest value $\emptyset'$ and highest value $\emptyset''$. In Definition 3.4.2 (ii) we also defined the nth jump $A^{(n)}$ by iterating the jump n times, where $A^{(0)} = A$, $A^{(1)} = A'$ and $A^{(n+1)} = (A^{(n)})'$. If $A \leq_T \emptyset'$, then by iterating the Jump Theorem 3.4.3 we know $\emptyset^{(n)} \leq_T A^{(n)} \leq_T \emptyset^{(n+1)}$.

Definition 4.4.3. Fix a set $A \leq_T \emptyset'$.

(i) A is *low$_n$* if $A^{(n)} \equiv_T \emptyset^{(n)}$, the lowest possible value.

(ii) A is *high$_n$* if $A^{(n)} \equiv_T \emptyset^{(n+1)}$, the highest possible value.

(iii) Let $\mathbf{D}$ denote the Δ_2 degrees and $\mathbf{C}$ the c.e. degrees. A Turing degree $\mathbf{d} \in \mathbf{D}$ is *low$_n$* or *high$_n$* according to whether it contains a *low$_n$* or *high$_n$* set, since this property is degree invariant. For every $n \geq 0$, define the following subclasses of $\mathbf{D}$:

$$\mathbf{H}_n = \{\, \mathbf{d} : \mathbf{d} \in \mathbf{D} \quad \& \quad \mathbf{d}^{(n)} = \mathbf{0}^{(n+1)} \,\}$$

$$\mathbf{L}_n = \{\, \mathbf{d} : \mathbf{d} \in \mathbf{D} \quad \& \quad \mathbf{d}^{(n)} = \mathbf{0}^{(n)} \,\}.$$

(iv) A set or degree which is not *low$_n$* or *high$_n$* for any n is *intermediate*.

Clearly, $\mathbf{L}_n \subseteq \mathbf{L}_{n+1}$ and $\mathbf{H}_n \subseteq \mathbf{H}_{n+1}$ for every n. Even restricted from $\mathbf{D}$ to $\mathbf{C}$ there is an intermediate c.e. degree and that the classes are strictly increasing,

$$(\forall n)\,[\,\mathbf{L}_n \subset \mathbf{L}_{n+1} \quad \& \quad \mathbf{H}_n \subset \mathbf{H}_{n+1}\,].$$

Often we replace the Δ_2 degrees $\mathbf{D}$ by the c.e. degrees $\mathbf{C}$ and use the same low/high notation, $\mathbf{L}_n/\mathbf{H}_n$, as above. Which one is intended will be clear from the context.

4.4.3   Common Jump Classes of Degrees

The most common jump classes of degrees are the following, with their complements (some of which are not given). In §4.7 we relate several of these classes to domination and escape properties.

$$
\begin{array}{lll}
\mathbf{H}_0 & = & \{\mathbf{0'}\} & \text{the complete degree} \\[4pt]
\mathbf{L}_0 & = & \{\mathbf{0}\} & \text{the degree of } \emptyset \\[4pt]
\mathbf{L}_1 & = & \{\mathbf{d} \in \mathbf{D} : \mathbf{d}' = \mathbf{0'}\} & \text{low}_1 \\[4pt]
\mathbf{L}_2 & = & \{\mathbf{d} \in \mathbf{D} : \mathbf{d}'' = \mathbf{0''}\} & \text{low}_2 \\[4pt]
\overline{\mathbf{L}_2} & = & \{\mathbf{d} \in \mathbf{D} : \mathbf{d}'' > \mathbf{0''}\} & \text{nonlow}_2 \\[4pt]
\mathbf{H}_1 & = & \{\mathbf{d} \in \mathbf{D} : \mathbf{d}' = \mathbf{0''}\} & \text{high}_1 \\[4pt]
\overline{\mathbf{H}_1} & = & \{\mathbf{d} \in \mathbf{D} : \mathbf{d}' < \mathbf{0''}\} & \text{nonhigh}_1.
\end{array}
$$

4.4.4   Syntactic Properties of High$_n$ and Low$_n$ Sets

We now develop a syntactic characterization of high and low in terms of arithmetical quantifiers. This is often useful in applying the hypothesis of high or low.

Theorem 4.4.4 (High Theorem). *For any set $A \subseteq \omega$ TFAE:*

(i)   *A is high*   *(i.e., $\emptyset'' \leq_T A'$, whether $A \leq_T \emptyset'$ or not);*

(ii) $\Sigma_2 \subseteq \Delta_2^A$;

(iii) $\Sigma_2 \subseteq \Pi_2^A$;

(iv) $\overline{\emptyset^{(2)}} \leq_1 A^{(2)}$ *(i.e., Tot $\leq_1$ Fin^A).*

Proof.

$$
\begin{aligned}
A \text{ is high} \quad &\Longleftrightarrow \quad \emptyset'' \leq_T A' \\
&\Longleftrightarrow \quad \emptyset'' \in \Delta_2^A && \text{by Post's Theorem 4.4.2} \\
&\Longleftrightarrow \quad \Sigma_2 \subseteq \Delta_2^A && \text{because } \emptyset'' \text{ is } \Sigma_2\text{-complete} \\
&\Longleftrightarrow \quad \Sigma_2 \subseteq \Pi_2^A && \text{because } \Sigma_2 \subseteq \Sigma_2^A \text{ trivially} \\
&\Longleftrightarrow \quad \emptyset^{(2)} \leq_1 \overline{A^{(2)}} && \text{because } \overline{A^{(2)}} \text{ is } \Pi_2^A\text{-complete} \\
&\Longleftrightarrow \quad \text{Tot} \leq_1 \text{Fin}^A && \text{because } \overline{\emptyset^{(2)}} \equiv_1 \text{Tot.} \qquad \square
\end{aligned}
$$

Theorem 4.4.5 (Low Theorem). *For any set $A \subseteq \omega$ TFAE:*

(i) *A is low*   *(i.e., $A' \leq_T \emptyset'$);*

(ii) $\Sigma_1^A \subseteq \Delta_2$,

(iii) $\Sigma_1^A \subseteq \Pi_2$;

(iv) *$A' \leq_1 \overline{\emptyset^{(2)}}$*   *(i.e., iff $K_1^A \leq_1 \text{Tot}$).*

Proof.

$$
\begin{aligned}
A \text{ is low} \quad &\Longleftrightarrow \quad A' \leq_T \emptyset' \\
&\Longleftrightarrow \quad A' \in \Delta_2 && \text{by Post's Theorem 4.2.2} \\
&\Longleftrightarrow \quad \Sigma_1^A \subseteq \Delta_2 && \text{because } A' \text{ is } \Sigma_1^A\text{-complete} \\
&\Longleftrightarrow \quad \Sigma_1^A \subseteq \Pi_2 && \text{because } \Sigma_1^A \subseteq \Sigma_1^{\emptyset'} = \Sigma_2 \\
&\Longleftrightarrow \quad A' \leq_1 \overline{\emptyset^{(2)}} && \text{because } \overline{\emptyset^{(2)}} \text{ is } \Pi_2\text{-complete.} \qquad \square
\end{aligned}
$$

4.4.5 Exercises

Exercise 4.4.6. State and prove classifications for $high_2$ and low_2 similar to those in Theorems 4.4.4 and 4.4.5 for $high_1$ and low_1.

4.5 * Domination and Escaping Domination

Recall the Definition 3.5.1 of the quantifiers $(\forall^\infty x)$ and $(\exists^\infty x)$, and Definition 3.5.2 of domination and escape, which we now repeat and extend.

Definition 4.5.1. (i) A function g *dominates* f, denoted by $f <^* g$, if

(4.9) $(\forall^\infty x)\,[\, f(x) < g(x) \,]$.

A *partial* function $\theta(x)$ dominates a *partial* function $\psi(x)$ if

$$(\forall^\infty x)\,[\, \psi(x){\downarrow} \quad \Longrightarrow \quad \psi(x) < \theta(x){\downarrow} \,].$$

(ii) A function f *escapes* (domination by) g if $f \not\leq^* g$, i.e., if

(4.10) $(\exists^\infty x)\,[\,g(x) \leq f(x)\,]$.

(iii) A function g *majorizes* f, denoted by $f < g$, if

(4.11) $(\forall x)\,[\,f(x) < g(x)\,]$.

(iv) Functions f and g are *almost equal*, denoted by $f =^* g$, if

$$(\forall^\infty x)\,[\,g(x) = f(x)\,].$$

(v) A class $\mathcal{C}$ of functions is *closed under finite differences* if

$$[\,g \in \mathcal{C} \quad \& \quad g =^* h\,] \quad \Longrightarrow \quad h \in \mathcal{C}.$$

Proposition 4.5.2. *Let $\mathcal{C}$ be a class of functions closed under finite differences, such as the computable functions or the A-computable functions for some A. Then for every f,*

$$(\exists g \in \mathcal{C})\,[\,g >^* f\,] \quad \Longleftrightarrow \quad (\exists h \in \mathcal{C})\,[\,h > f\,].$$

Proof. One direction is obvious. For the other direction, assume $g >^* f$, and find $h =^* g$ such that $h > f$.   □

By Proposition 4.5.2, given such a $\mathcal{C}$ and $g \in \mathcal{C}$ with $g >^* f$, we shall assume that $g > f$. In particular, if we have a computable $g >^* f$ then we shall assume we have computable $g > f$.[2]

4.5.1 Domination Properties

Definition 4.5.3. Let $\{A_s\}_{s \in \omega}$ be a computable enumeration of c.e. set A.

(i) The *stage function* is the partial computable function

$$\theta_A(x) = \begin{cases} (\mu s)\,[\,x \in A_s\,] & \text{if } x \in A \\ \text{undefined} & \text{otherwise.} \end{cases}$$

(ii) The *least modulus* as in (3.17) of Definition 3.5.4 is

$$m_A(x) = (\mu s)\,[\,A_s \upharpoonright x = A \upharpoonright x\,].$$

Note that $\theta_A(x)$ is *partial* but partial *computable*, while $m_A(x)$ is *total* but not computable (unless A is computable).

---

[2]Dominate and majorize are very similar. We normally prefer *dominate* because by Proposition 4.5.2 if a computable function g *dominates* f then a computable function h *majorizes* f. The negation of dominate is *escape* which gives a rich structure of nonlow$_2$ degrees in §4.5 and §4.6, but the negation of "g majorizes f" is simply $(\exists x)\,[f(x) \geq g(x)]$, which is not useful.

Theorem 4.5.4 (Domination Properties). *Let $\{A_s\}_{s\in\omega}$ be an enumeration of a c.e. set A and f a total function.*

(i)   If f dominates $\theta_A(x)$ then $A \leq_T f$.

(ii)   For any $D \leq_T \emptyset'$,

$$D \equiv_T \emptyset' \iff (\exists f \leq_T D)[\, f \text{ dominates every partial computable function}\,].$$

(iii)   If f dominates $m_A(x)$ then $A \leq_T f$.

(iv)   If $\{B_s\}_{s\in\omega}$ is an enumeration of a c.e. set B and $m_A(x)$ dominates the least modulus function $m_B(x)$, then $B \leq_T A$.

Proof. (i) $(\forall^\infty x)[\, x \in A \iff x \in A_{f(x)} \,]$.

(ii) $(\Longleftarrow)$ By (i) because f dominates $\theta_K(x)$.

(ii) $(\Longrightarrow)$ Build $f \leq_T \emptyset'$ by using $\emptyset'$ to determine for a given input x which $\varphi_e(x)$ converge for $e \leq x$. Then define $f(x)$ to exceed all these values.

(iii) $(\forall^\infty x)[\, x \in A \iff x \in A_{f(x)} \,]$.

(iv) $(\forall^\infty x)[\, x \in B \iff x \in B_{m_A(x)} \,]$. $\square$

These are only the simplest facts about domination. In §4.5.2 and throughout the book we develop many more domination properties, and extend (ii) to an elegant characterization by Martin in Theorem 4.5.6 of functions which dominate all *total* computable functions. Escape properties are more subtle, but in Theorem 4.6.2 we characterize functions which escape $\emptyset'$-computable functions and we use these in computable model theory.

4.5.2   Martin's High Domination Theorem

The first few levels of the high/low degree hierarchy, especially the high$_1$, low$_1$, and low$_2$ degrees and their complements, have many important applications. In addition to the syntactic characterization of high degrees in Theorem 4.4.4, we now give the very useful characterization (Theorem 4.5.6) by Martin in terms of dominating functions. The following characterization of high degrees gives useful characterizations in §4.7 and §4.8 for uniform enumerations of the computable functions and properties of those Δ_2 sets which are low$_2$ or nonlow$_2$. Later we consider low$_2$ and nonlow$_2$ sets. We now extend the domination notions from Definition 3.5.2.

Definition 4.5.5. f is *dominant* if f dominates every (total) computable function; an *infinite set* $A = \{a_0 < a_1 < \cdots\}$ is *dominant* if its principal function p_A dominates every (total) computable function, where $p_A(n) = a_n$.

Theorem 4.5.6 (High Domination Theorem, Martin, 1966b). *A set A is high ($\emptyset'' \leq_T A'$) iff there is a dominant function $f \leq_T A$.*

Proof. By Theorem 4.3.2 we know that $\text{Tot} \equiv_T \emptyset''$. Hence, by the Limit Lemma 3.6.8 relativized to A, we have $\emptyset'' \leq_T A'$ iff there is an A-computable $\{0,1\}$-valued function $g(e,s)$ such that $\lim_s g(e,s) = \text{Tot}(e) := \chi_{Tot}(e)$.

($\Longrightarrow$). Assume $\emptyset'' \leq_T A'$. Given $g(e,s)$ as above we define a dominant function $f \leq_T A$ as follows:

Stage s. (To define $f(s)$). For all $e \leq s$ define $t(e)$ and $f(s)$ as follows:

$$t(e) = (\mu t > s)\,[\,g(e,t) = 0 \quad \lor \quad (\forall x \leq s)\,[\,\varphi_{e,t}(x)\!\downarrow\,]\,],$$

$$f(s) = \max\{\,t(e) : e \leq s\,\}.$$

Note that $t(e)$ exists because if φ_e is not total, then $\lim_t g(e,t) = 0$. If φ_e is total, then $\lim_t g(e,t) = 1$, and therefore $f(s) > \varphi_e(s)$ for a.e. s. (Recall by Definition 1.6.17 that if $\varphi_{e,t}(x) = y$ then $e, x, y < t$.)

($\Longleftarrow$). Assume $f \leq_T A$ is dominant. Define an A-computable function $g(e,s)$ such that $\lim_s g(e,s) = \text{Tot}(e)$ as follows:

$$(4.12) \qquad g(e,s) = \begin{cases} 1 & \text{if } (\forall z \leq s)\,[\,\varphi_{e,f(s)}(z)\!\downarrow\,]; \\ 0 & \text{otherwise.} \end{cases}$$

Note that if φ_e is total, then so is $\theta_e(y) = (\mu s)\,(\forall z \leq y)\,[\,\varphi_{e,s}(z)\!\downarrow\,]$. Thus, $f(y)$ dominates $\theta_e(y)$. Therefore, $g(e,s) = 1$ for a.e. s. If φ_e is not total, then $\varphi_e(y)$ and $\theta_e(y)$ diverge for some y, and $g(e,s) = 0$ for all $s \geq y$. $\square$

4.5.3 Exercises

Exercise 4.5.7. Give another proof of Martin's Theorem 4.5.6. *Hint.* Assume $A' \geq_T \emptyset''$. Using Theorem 4.4.4 (iv) fix a computable function g such that φ_e is total iff $W^A_{g(e)}$ is finite. Use an A-computable construction to define $f \leq_T A$. To define $f(s)$ first wait for all $e \leq s$ until either $\varphi_e(s)\!\downarrow$ or $W^A_{g(e)}$ receives a new element.

Exercise 4.5.8. Let A be coinfinite, nonhigh, and c.e. Prove that A has a computable enumeration $\{A_s\}_{s \in \omega}$ that is *diagonally correct*, that is, $(\exists^\infty s)\,[\,a^s_s = a_s\,]$, where $\overline{A}_s = \{a^s_0 < a^s_1 < \cdots\}$ and $\overline{A} = \{a_0 < a_1 < \cdots\}$.

4.6   Characterizing Nonlow$_2$ Sets $A \leq_T \emptyset'$

Fix $A \leq_T \emptyset'$ and relativize the previous proof to the cone $\{B : A \leq_T B\}$ with base A in place of $\emptyset$ and with $\emptyset' \geq_T A$ as a set in this cone. We obtain the following useful escape property characterizing nonlow$_2$ sets $A \leq_T \emptyset'$.

Theorem 4.6.1 (Relativized Domination Theorem, Martin, 1966b). *Fix $A \leq_T \emptyset'$. Then $A'' \leq_T \emptyset''$ (i.e., A is low$_2$) if and only if there is a function $g \leq_T \emptyset'$ which dominates every total function $f \leq_T A$.*

Proof. Fix $A \leq_T \emptyset'$. Relativize Martin's Theorem 4.5.6 to the cone of sets $\{X : X \geq_T A\}$. Now A is low$_2$ $(A'' \equiv_T \emptyset'')$ iff $\emptyset'$, viewed as a member of this cone, is high in the cone, namely iff one jump of $\emptyset'$, that is, $\emptyset''$, reaches A'' in Turing degree, because A'' is the double jump of the base A of the cone. By Martin's Theorem 4.5.6 this occurs iff there is a function $g \leq_T \emptyset'$ which is dominant relative to A-computable functions, so that g dominates every total function $f \leq_T A$.   □

Corollary 4.6.2 (Nonlow$_2$ Escape Theorem). *Fix $A \leq_T \emptyset'$. Then A is nonlow$_2$ $(A'' > \emptyset'')$ iff for every function $g \leq_T \emptyset'$ there is a function $f \leq_T A$ which escapes g in the sense of* (4.10), *i.e.,*

(4.13)
$$(\text{Nonlow}_2\ Escape) \qquad (\forall g \leq_T \emptyset')\,(\exists f \leq_T A)\,(\exists^\infty x)\,[\,g(x) \leq f(x)\,].$$

Proof. This is the contrapositive of Theorem 4.6.1.   □

We have stated (4.13) separately for the sake of the list of properties in Theorem 4.7.1.

4.6.1   Exercises

Exercise 4.6.3.$^{\diamond\diamond}$ (Csima, Hirschfeldt, Knight, Soare, 2004). Identify a string σ_y with its code number y. A set A satisfies the *isolated path property* if for every computable tree $T \subseteq 2^{<\omega}$ with no terminal nodes and with isolated paths dense,

$$(\exists g \leq_T A)\,(\forall \sigma \in T)\,[\,g_\sigma \in [T_\sigma]\quad \&\quad g_\sigma \text{ is isolated}\,],$$

i.e., for every $x \in T$, $g_\sigma = \lambda y\,[g(\sigma, y)]$ is a path extending σ, which is an isolated path of the closed set $[T]$. Prove that every nonlow$_2$ set $A \leq_T 0'$ satisfies the isolated path property.

Exercise 4.6.4.$^{\diamond\diamond}$ (Csima, Hirschfeldt, Knight, Soare, 2004). A set A satisfies the *tree property* if for every computable tree $T \subseteq 2^{<\omega}$ with no terminal nodes, and every uniformly Δ_2 sequence of subsets $\{S_i\}_{i \in \omega}$ all dense in $[T]$,

$$(\exists g \leq_T A)\,(\forall \sigma \in T)\,(\forall i)\,(\exists \tau \in S_i)\,[\,\sigma \prec g_\sigma\quad \&\quad \tau \prec g_\sigma\quad \&\quad g_\sigma \in [T]\,].$$

Prove that every nonlow$_2$ set $A \leq 0'$ satisfies the tree property.

4.7 Domination, Escape, and Classes of Degrees

Martin's Theorem 4.5.6 gave a remarkable connection between high degrees $\mathbf{H}_1$ and dominant functions, and the Nonlow$_2$ Escape Theorem 4.6.2 produced an escape characterization for $\overline{\mathbf{L}}_2$ degrees. Now we summarize the previous properties in the following Theorem 4.7.1.

Recall the Definition 4.5.1 of dominate and escape and the common jump classes in §4.4.3. The contrapositive of Martin's High Domination Theorem 4.5.6 is

$$(4.14) \quad A' \not\geq_{\mathrm{T}} \emptyset'' \quad \Longleftrightarrow \quad (\forall g \leq_{\mathrm{T}} A)(\exists f \leq \emptyset)(\exists^\infty x)\,[\,g(x) \leq f(x)\,].$$

In this case we say "f escapes g." We say that a set A satisfying the right-hand side has the *escape property*. Martin's equation (4.14) says that the degrees satisfying the escape property are exactly the nonhigh$_1$ degrees.

However, this definition does not require that we be able to *uniformly* find an index i with $\varphi_i = f$ given an index e with $g = \Phi_e^A$. Roughly, if we can uniformly find i from e, then the A satisfies the *Uniform Escape Property (UEP)*. We now summarize the domination and escape characterizations so far. (The redundancy of these properties is intensional, *e.g.*, our stating a property on one line and its negation on the next, so that we can later refer to a specific property by its line number here, because we intend to further develop both domination and escape.)

Theorem 4.7.1. *Fix a degree* $\mathbf{d} \leq \mathbf{0}'$.

(i) $\mathbf{d} = \mathbf{0}' \quad\Longleftrightarrow\quad (\exists g \leq \mathbf{d})[\,g$ *dominates all p.c. functions* $]$.

(ii) $\mathbf{d} < \mathbf{0}' \quad\Longleftrightarrow\quad (\forall g \leq \mathbf{d})(\exists \theta$ *p.c.* $)[\,\theta$ *escapes* $g\,]$.

(iii) $\mathbf{d} \in \mathbf{H}_1 \quad\Longleftrightarrow\quad (\exists g \leq \mathbf{d})(\forall f \leq \mathbf{0})[\,g$ *dominates* $f\,]$.

(iv) $\mathbf{d} \in \overline{\mathbf{H}}_1 \quad\Longleftrightarrow\quad (\forall g \leq \mathbf{d})(\exists f \leq \mathbf{0})[\,f$ *escapes* $g\,]$.

(v) $\mathbf{d} \in \mathbf{L}_2 \quad\Longleftrightarrow\quad (\exists g \leq \mathbf{0}')(\forall f \leq \mathbf{d})[\,g$ *dominates* $f\,]$.

(vi) $\mathbf{d} \in \overline{\mathbf{L}}_2 \quad\Longleftrightarrow\quad (\forall g \leq \mathbf{0}')(\exists f \leq \mathbf{d})[\,f$ *escapes* $g\,]$.

Proof. Theorem 4.5.4 (ii) establishes (i) and (ii), Martin's Domination Theorem 4.5.6 establishes (iii) and (iv), the NonLow$_2$ Escape Theorem 4.6.2 proves (v) and (vi). □

4.8 Uniform Enumerations of Functions and Sets

Theorem 4.8.2 will relate nicely to the previous Martin Theorem 4.5.6 on dominant functions and high degrees. Also, the notions we now introduce in Definition 4.8.1 have proved useful in other areas of computability theory, computable model theory, and models of arithmetic.

Definition 4.8.1. (i) If $f(x,y)$ is a binary function then

(4.15) f_y denotes $\lambda x \, [f(x,y)]$.

As in analytic geometry, we imagine a two-dimensional plane with horizontal coordinate x and vertical coordinate y. We view $\lambda x, y \, [f(x,y)]$ as specifying a *matrix* with entry $f(x,y)$ at the location (x,y). For vertical coordinate $y \in \omega$ we view f_y as the yth *row* according to our notation (4.15).

(ii) Let $\mathcal{C}$ be a class of (unary) functions and $\mathbf{a}$ be a degree. Then $\mathcal{C}$ is called $\mathbf{a}$-*uniform* ($\mathbf{a}$-*subuniform*) if there is a binary function $f(x,y)$ of degree $\leq \mathbf{a}$ such that

$$\mathcal{C} = \{f_y\}_{y \in \omega} \qquad (\text{respectively}, \ \mathcal{C} \subseteq \{f_y\}_{y \in \omega}).$$

Therefore, f *uniformly* lists the rows $\{f_y\}_{y \in \omega}$. In the uniform case these are exactly the rows of $\mathcal{C}$. In the subuniform case $\mathcal{C}$ may be a proper subclass of these rows.

4.8.1 Limits of Functions

Given $f(x,y)$ as in (4.15), we may need to take limits in both the x and y directions. For example, if $\{A_y\}_{y \in \omega}$ is a uniformly computable sequence of computable sets then the vertical limit $B(x) = \lim_y A_y(x)$ is a Δ_2 set as in the Limit Lemma 3.6.2. Now suppose that $A_y = W_{f(y)}$ where $f(x)$ is the computable function in the proof of Theorem 4.3.2. Hence, $W_{f(y)}$ is finite if W_y is finite and $W_{f(y)} = \omega$ otherwise. Define $C(y) = \lim_x A_y(x)$. Now

$$C(y) \ = \lim_x A_y(x) \ = \ \mathrm{Tot}(y).$$

Hence, $C' \geq_T 0''$, a useful fact in many infinite injury constructions such as the Thickness Lemma, because any set D thick in A also satisfies $D' \geq_T 0''$.

4.8.2 *A*-uniform Enumeration of the Computable Functions

The next useful characterization follows from Martin's Theorem 4.5.6.

Theorem 4.8.2 (Jockusch, 1972a). *If $\mathbf{d}$ is any degree, then statements* (i)–(iv) *are equivalent:*

(i) $\mathbf{d}' \geq \mathbf{0}''$

(ii) *the computable functions are $\mathbf{d}$-uniform;*

(iii) *the computable functions are $\mathbf{d}$-subuniform;*

(iv) *the computable sets are $\mathbf{d}$-uniform.*

If **d** *is c.e., then* (i)–(iv) *are each equivalent to*

(v) *the computable sets are* **d**-*subuniform.*

Proof. The implications (ii) $\Longrightarrow$ (iii), (ii) $\Longrightarrow$ (iv), and (iv) $\Longrightarrow$ (v) are immediate.

(i) $\Longrightarrow$ (ii). By Martin's Theorem 4.5.6 choose a dominant function g of degree $\leq$ **d**. Define $f(\langle e, i\rangle, x) = \varphi_{e,i+g(x)}(x)$ if $\varphi_{e,i+g(y)}(y) \downarrow$ for all $y \leq x$ and $f(\langle e, i\rangle, x) = 0$ otherwise. Now either $f_{\langle e,i\rangle} = \varphi_e$ is a total function, or $f_{\langle e,i\rangle}$ is finitely nonzero. In either case $f_{\langle e,i\rangle}$ is computable. If φ_e is total then $g(x)$ dominates $\theta(x) = (\mu s)[\,\varphi_{e,s}(x) \downarrow\,]$, so $\varphi_e = f_{\langle e,i\rangle}$ for some i.

(iii) $\Longrightarrow$ (i). Let $f(e,x)$ be a function of degree $\leq$ **d** such that every computable function is an f_e. Define $g(x) = \max\{f_e(x) : e \leq x\}$. Then g is dominant, so $\mathbf{d}' \geq \mathbf{0}''$ by Martin's Theorem 4.5.6.

(iv) $\Longrightarrow$ (i). By Theorem 4.3.2 and Exercise 4.3.12 we have

$$(\mathrm{Tot}, \overline{\mathrm{Tot}}) \leq_m (\mathrm{Tot}, \overline{\mathrm{Ext}})$$

via some computable function g. Assume f has degree $\leq$ **d** and that the f_e's are exactly the computable characteristic functions. Then for all e,

$$
\begin{aligned}
e \in \mathrm{Tot} \quad &\Longleftrightarrow \quad (\exists i)\,[\,f_i \text{ extends } \varphi_{g(e)}\,] \\
&\Longleftrightarrow \quad (\exists i)\,(\forall x)\,(\forall y)\,(\forall s)\,[\,\varphi_{g(e),s}(x) = y \quad \Longrightarrow \quad f_i(x) = y\,].
\end{aligned}
$$

Thus, $\mathrm{Tot} \in \Sigma_2^A$. But $\mathrm{Tot} \in \Pi_2 \subseteq \Pi_2^A$. Therefore, $\mathrm{Tot} \in \Delta_2^A$. Hence, $\mathbf{0}'' \leq \mathbf{d}'$ by the Relativized Post's Theorem 4.4.2.

(v) $\Longrightarrow$ (i). (The following resembles the proof that the computable functions are not uniformly computable.) Assume that **d** is c.e. but (i) is false and $f(e, x)$ is any function of degree $\leq$ **d**. We must construct a $\{0,1\}$-valued computable function $h \neq f_e$ for all e. Since $\deg(f) \leq \mathbf{0}'$ there is a computable function $\hat{f}(e, x, s)$ such that $f(e, x) = \lim_s \hat{f}(e, x, s)$ and a modulus function $m(e, x)$ for $\hat{f}$ which has degree $\leq$ **d** by the Modulus Lemma 3.6.3. Let $p(x) = \max\{m(e, \langle e, x\rangle) : e \leq x\}$. Since $\deg(p) \leq$ **d** and (i) fails, there is a computable function $q(x)$ which $p(x)$ fails to dominate. Define $h(\langle e, x\rangle) = 1 \,\dot{-}\, \hat{f}(e, \langle e, x\rangle, q(x))$. Then h is a computable function and $h(\langle e, x\rangle) \neq f_e(\langle e, x\rangle)$ whenever $x \geq e$ and $q(x) \geq p(x)$. (Exercise 4.9.6 on Π_1^0-classes shows that the hypothesis **d** c.e. is necessary for this part.) $\square$

Corollary 4.8.3 (Jockusch). *If* $\mathbf{d} < \mathbf{0}'$ *is c.e. then the class of c.e. sets of degree* $\leq$ **d** *is not* **d**-*uniform.*

Proof. If **d** is a counterexample, then the computable sets are **d**-subuniform. Therefore, $\mathbf{d}' = \mathbf{0}''$ by (v) $\Longrightarrow$ (i) of Theorem 4.8.2. However, since the c.e. sets of degree $\leq \mathbf{d}$ are $\mathbf{d}$-uniform, they are $\mathbf{0}'$-uniform and so $\mathbf{d}'' = \mathbf{0}''$ by a later result.   □

4.9 $\oslash$ Characterizing Low$_2$ Sets $A \leq_T \emptyset'$

Definition 4.9.1. *The $\emptyset'$-uniform property of A asserts:*

(4.16) $U(A):$ $(\exists f \leq_T \emptyset')\,[\,\{\,Y : Y \leq_T A\,\} = \{\,f_e\,\}_{e \in \omega}\,],$

where $f_e = \lambda x\,[\,f(x,e)\,]$ as in (4.15) and is viewed as the eth row of the matrix with characteristic function $f(x,e)$. (We identify a set Y with its characteristic function χ_Y.)

The uniformity property $U(A)$ asserts that there is a $\emptyset'$-computable matrix $f \leq_T \emptyset'$ whose rows $\{\,f_e\,\}_{e \in \omega}$ are exactly the sets $Y \leq_T A$.

Theorem 4.9.2. *If $A \leq_T \mathbf{0}'$ is low$_2$ then $U(A)$ holds, i.e., the A-computable functions (and hence also A-computable sets) are $\mathbf{0}'$-uniform.*

Proof. Let A be low$_2$ i.e., $A'' \leq_T \emptyset''$. Hence, $\mathrm{Tot}^A \leq_T \emptyset''$. Let $\widehat{g}(e,s)$ be a $\emptyset'$-computable function whose limit $g(e) = \lim_s \widehat{g}(e,s)$ is the characteristic function of Tot^A. Now, using a $\emptyset'$ oracle, find for every e and x

$$(\mu t > x)\,[\,\ \Phi^A_{e,t}(x)\!\downarrow\ \ \vee\ \ \widehat{g}(e,t) = 0\ \ ].$$

If the first case holds, define $h(x,e) = \Phi^A_{e,t}(x)$, and in the second case define $h(x,e) = 0$. This produces $h \leq_T \emptyset'$. Let $\omega^{<\omega}$ be $\{\tau_i\}_{i \in \omega}$. Define $f \leq_T \emptyset'$ by

$$f(x,\langle e,i\rangle) = \begin{cases} \tau_i(x) & \text{if } x < |\tau_i|\,; \\ h(x,e) & \text{if } x \geq |\tau_i|\,. \end{cases}$$

For every e, if Φ^A_e is total, then $\Phi^A_e =^* h_e$ and $\Phi^A_e = f_{\langle e,i\rangle}$ for some i.   □

Corollary 4.9.3. *If $X \leq_T \mathbf{0}'$ is low$_2$, then there is a computable function $\widehat{f}(x,y,s)$ such that the limit $f(x,y) = \lim_s \widehat{f}(x,y,s)$ exists for all x and y, and*

(4.17) $\{Y : Y \leq_T X\} = \{\,f_y : y \in \omega\,\}.$

Proof. Apply Theorem 4.9.2 to see that $f(x,y) \leq_T \emptyset'$ exists and apply the Limit Lemma 3.6.2 to derive $\widehat{f}(x,y,s)$.   □

For a fixed low$_2$ set X, we can think of $f(x,y)$ as a $\emptyset'$-*matrix* with rows $\{\,f_y\,\}_{y \in \omega}$, which is approximated at every stage s in our computable construction by $\lambda\,x\,y\,[\,\widehat{f}(x,y,s)\,]$, and which in the limit correctly gives (4.17). We can often use the *dynamic matrix* approximation

$$\{\,\lambda\,e\,y\,[\,\widehat{f}(e,y,s)\,]\,\}_{s \in \omega}$$

to show that a low₂ set resembles a computable set.

Proposition 4.9.4. *Set A satisfies $U(A)$ iff $A \leq_T \mathbf{0}'$ and A is low₂.*

Proof. ($\Longleftarrow$). Apply Theorem 4.9.2.

($\Longrightarrow$). If f is a computable function satisfying (4.17) then $Y = A$ *itself* is one of the rows f_y for some y, but $f \leq_T \mathbf{0}'$, so $A \leq_T \mathbf{0}'$. Using $f \leq_T \mathbf{0}'$ we can define a $\mathbf{0}'$-function which dominates every A-computable function. Now $A'' \leq_T \mathbf{0}''$ by Theorem 4.6.1. $\square$

4.9.1   Exercises

Exercise 4.9.5. Give another proof of Theorem 4.9.2 using domination. *Hint.* If A is low₂ then $\emptyset'$ is high over A. Relativize Theorem 4.8.2 to A, replacing $\emptyset$ by A and A by $\emptyset'$. By Theorem 4.6.1, choose a $\emptyset'$-function g which dominates every total A-computable function Φ_e^A. Since $A \leq_T \emptyset'$ we can $\emptyset'$-computably define:

$$f(\langle e, i \rangle, x) = \begin{cases} \Phi_{e,\,i+g(x)}^A(x) & \text{if} \quad (\forall y \leq x)[\ \Phi_{e,\,i+g(y)}^A(y)\downarrow\ ] \\ 0 & \text{otherwise.} \end{cases}$$

Either $f_{\langle e,i \rangle} = \Phi_e^A$ is a total function, or $f_{\langle e,i \rangle}$ is finitely nonzero. If Φ_e^A is total then $g(x)$ dominates $c(x) = (\mu s)\,[\Phi_{e,s}^A(x)\downarrow]$.

Exercise 4.9.6. [Jockusch] Show that the hypothesis **d** c.e. in the proof of (v) $\Longrightarrow$ (i) of Theorem 4.8.2 was necessary by proving that there is a (non-c.e.) degree **d** such that $\mathbf{d}' = \mathbf{0}'$ and the computable sets are **d**-subuniform. *Hint.* Apply the Low Basis Theorem 3.7.2 to the Π_1^0 class $\mathcal{C} \subseteq 2^\omega$ defined by

$$f \in \mathcal{C} \iff \text{rng}(f) \subseteq \{0,1\} \quad \& \\ (\forall e)\,(\forall x)\,[\ \varphi_e(x)\downarrow \implies f(\langle e, x \rangle) = \min\{1, \varphi_e(x)\}]$$

to obtain some $f \in \mathcal{C}$ of low degree.

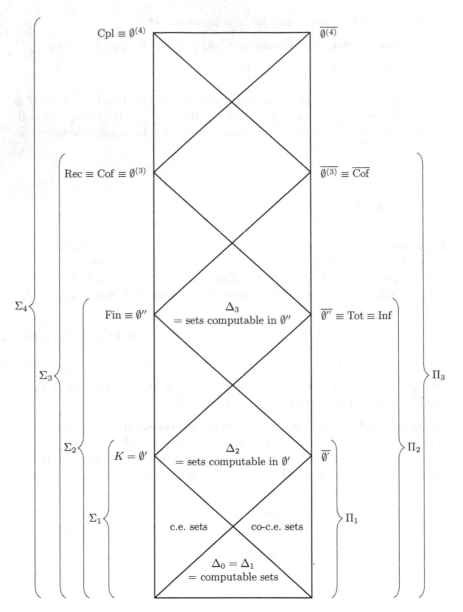

Figure 4.1. Arithmetical hierarchy of sets of integers

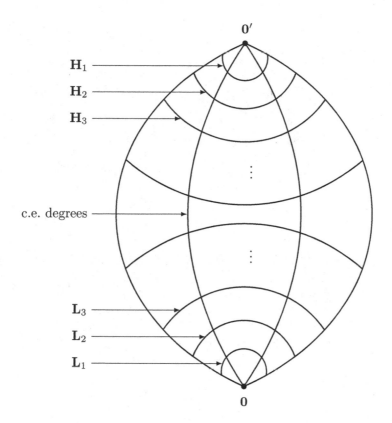

Figure 4.2. High and low degrees

5
Classifying C.E. Sets

5.1 ⋆ Degrees of Computably Enumerable Sets

5.1.1 Post's Problem and Post's Program

In §1.6 we constructed a noncomputable c.e. set K. In §2.4 we studied creative sets and showed that all are computably isomorphic to K. So far the noncomputable c.e. sets we have exhibited are all computably isomorphic to K and therefore of Turing degree $\mathbf{0}'$. Post in [Post 1944] asked whether there is only one noncomputable c.e. set up to Turing equivalence.

Definition 5.1.1. [Post, 1944]. *Post's Problem* was whether there exists a computably enumerable set A with $\emptyset <_{\mathrm{T}} A <_{\mathrm{T}} \emptyset'$.

Post considered this a very significant question because he recognized that important mathematical problems, such as the solvability of Hilbert's tenth problem on Diophantine equations, could be coded as questions of decidability for a c.e. set. Since Post, many more such problems have been discovered and related to c.e. sets, such as problems in number theory, finitely presented groups, differential geometry and other fields. For example, given a computably axiomatizable theory such as Peano arithmetic, the set of theorems can be effectively *enumerated* but not necessarily *decided*. Post recognized that if the answer to his problem is negative, then there is only *one* undecidable problem up to Turing equivalence for these mathematical questions.

Post's *program* for answering his question was to define c.e. sets A whose complements, although infinite, were increasingly thin, such as the simple

107

sets in §5.2 and hypersimple sets in §5.3. He attempted to show that the complete set K cannot be reduced to such an A by one of the strong reducibilities he introduced, such as the many-one reducibility, or the truth-table reduciblility in §3.8.3.

Post's program of thin sets initially succeeded for some bounded reducibilities but not for full Turing reducibility, which was not well understood in 1944 when Post began. From 1944 to 1954 Post gradually developed a deeper understanding of Turing reducibility and information content of c.e. sets and in 1948 he defined the notion of degree of unsolvability (Turing degree). Post's ideas led to our present understanding of Turing functionals as presented in Chapter 3. They were published in the important Kleene-Post paper of 1954, including finite extension oracle constructions of incomparable sets below $\emptyset'$. We shall present these ideas in Chapter 6. Meanwhile, simple sets and bounded reducibilities led to a rich development of many ideas in the subject, as we now explore in this chapter.

5.1.2   Dynamic Turing Reductions on C.E. Sets

The Turing reductions we have considered so far have been *static*, such as $B \leq_m A$ and $B \leq_{tt} A$, meaning they can apply to sets which are not c.e., and even if A and B are c.e. these reductions do not mention any enumeration of them. One of the other main themes in this chapter is the study of *dynamic* Turing reductions, which only apply to c.e. sets and which measure how quickly the enumeration of one settles in comparison with the second.

We begin with the easy notion of *permitting* in §5.2, where we are given a noncomputable c.e. set C and we wish to build a simple set $A \leq_T C$. We say that C *permits* an element x to enter A at some stage only if an element $y \leq x$ simultaneously enters C. This achieves $A \leq_T C$ because $A \upharpoonright x$ settles more quickly than $C \upharpoonright x$, i.e., $m_A(x) \leq m_C(x)$, for these least modulus functions defined in Definition 3.5.4 (iii) for Σ_1-approximations.

We develop this notion of comparing modulus functions much further in §5.4, where we are given a c.e. set A with a certain property such as its being effectively simple or able to compute a fixed point free function. Then we prove that $K \leq_T A$ by giving an enumeration of A which settles more slowly than one for K.

5.2 * Simple Sets and the Permitting Method

Post introduced the notion of a *simple set* in an attempt to solve Post's Problem of Definition 5.1.1. Post succeeded in proving that every simple set A satisfies $K \not\leq_m A$ but not $K \not\leq_T A$ as he had hoped. Indeed, Post himself recognized that there are complete simple sets, and we shall give several proofs of it, including Theorem 5.2.6.

Definition 5.2.1. (i) An infinite set is *immune* if it contains no infinite c.e. set.

(ii) A c.e. set A is *simple* if $\overline{A}$ is immune.

Proposition 5.2.2. *If A is simple then:*
(i) *A is not computable;*
(ii) *A is not creative;*
(iii) *A is not m-complete   (i.e., $K \not\leq_m A$).*

Proof. (i) $\overline{A}$ cannot be c.e. since otherwise it would contain an infinite c.e. set.

(ii) The complement of a creative set does contain an infinite c.e. set by Theorem 2.4.5.

(iii) If $K \leq_m A$, then A is creative by Corollary 2.4.8.   □

Note that in our classsification of the structure of c.e. sets we now have at least two types of noncomputable c.e. sets which cannot be computably isomorphic: creative sets like K, and simple sets.

5.2.1 Post's Simple Set Construction

Theorem 5.2.3 (Post's simple set, 1944). *There exists a simple set A.*

Proof. We construct a coinfinite c.e. set $A = \cup_s A_s$ by stages such that for all e we meet the following *requirement*:

(5.1) $P_e : \quad |W_e| = \infty \quad \implies \quad W_e \cap A \neq \emptyset.$

Construction.
  Stage $s = 0$. Let $A_0 = \emptyset$.
  Stage $s + 1$. Given A_s, choose the least $e < s$ such that

(5.2) $W_{e,s} \cap A_s = \emptyset \quad \& \quad (\exists x > 2e)\,[\,x \in W_{e,s}\,].$

If e exists, choose the least corresponding x and enumerate x in A. If there is no such e, go to stage $s + 2$.
End construction.

Verification. To see that $\overline{A}$ is infinite, note that A contains at most e elements out of the $2e + 1$ elements $\{0, 1, 2, \ldots, 2e\}$, one for each W_i, $0 \leq i < e$, and each such P_i acts at most once because of the first clause of (5.2). Hence, card $(\overline{A} \restriction 2e) \geq e + 1$. Therefore, $\overline{A}$ is infinite.

Next A satisfies requirement P_e. Every P_i receives attention at most once. Choose s_0 such that no P_i, $i < e$, satisfies (5.2) at any stage $s > s_0$. If W_e is infinite, choose $x > 2e$ satisfying (5.2) at some stage $s > s_0$. Hence, P_e acts at stage $s + 1$, and $A \cap W_e \neq \emptyset$ thereafter.   □

Definition 5.2.4. A simple set A is *effectively simple* if there is a computable function f such that

$$(5.3) \qquad (\forall e) [\, W_e \subseteq \overline{A} \quad \Longrightarrow \quad |W_e| \leq f(e) \,].$$

Note that Post's simple set is effectively simple via $f(e) = 2e+1$, because if $W_e \subseteq \overline{A}$ then $W_e \subseteq \{0, 1, \ldots, 2e\}$. Post realized that simple sets are not necessarily Turing incomplete, and indeed he built a complete simple set. Ironically, we shall prove in §5.4 that *all* effectively simple sets are Turing complete. Therefore, effectively simple sets do not help in solving Post's Problem for constructing an *incomplete* c.e. set which is noncomputable.

5.2.2   The Canonical Simple Set Construction

The following construction of a simple set is more flexible than Post's simple set for future constructions and it gives a smaller bound $f(e) = e$ on effective simplicity. For definiteness we refer to it as the *canonical simple set A* because it uses the marker constructions described in §4.3.2 and (4.1), where $\overline{A}_s = \{a_0^s < a_1^s < \cdots \}$.

Theorem 5.2.5 (Canonical simple set). *There is a simple set A which is effectively simple via the function $f(e) = e$.*

Proof.   For every e we meet the requirement P_e of (5.1). We use the notation of §4.3.2, namely $\overline{A}_s = \{\, a_0^s < a_1^s < \ldots \}$.
Construction.   Let $A_0 = \emptyset$.

   Stage $s + 1$. Given A_s. Choose the least $e < s$ such that

$$(5.4) \qquad W_{e,s} \cap A_s = \emptyset \quad \& \quad (\exists i > e) [\, a_i^s \in W_{e,s} \,].$$

Choose the least such i and enumerate a_i^s in A_{s+1}. If there is no such e, go to stage $s + 2$.

End construction.

Verification.   For every e, we act for W_e at most *once*. Therefore,

$$(\forall n)\, (\exists^{<\infty} s)\, (\exists i \leq n) [\, a_i^s \in A_{s+1} - A_s \,].$$

Namely, after $a_j = a_j^s$ for all $j < n$, element a_n^s enters A for the sake of set W_e, $e < n$, at most once for every such $e < n$. Hence, $\lim_s a_n^s = a_n$ exists. Therefore, $\overline{A} = \{a_0 < a_1 < a_2 \ldots\}$ and $|\overline{A}| = \infty$. Now if $W_e \subseteq \overline{A}$ and $|W_e| > e$ then for some $i \geq e$ we have $a_i \in W_e$. Thus, we would have put a_i into A. Therefore, A is effectively simple via $f(e) = e$.   □

5.2.3   Domination and a Complete Simple Set

It will follow that both the Post simple set and the canonical simple set, being effectively simple, are Turing complete. However, by a much shorter and very useful argument, we now construct a complete simple set. It suffices to build a simple set A whose principal function $p_{\overline{A}}$ dominates every p.c. function. This requires only moving the markers slightly more often to achieve domination.

Theorem 5.2.6. *There is a simple set A such that $\overline{A} = \{a_0 < a_1 \ldots\}$ dominates every p.c. function and hence $A \equiv_T K$. (This set is dense simple as defined in Exercise 5.3.12).*

Proof. Perform the construction of the canonical simple set A in Theorem 5.2.5, but, in addition to the former strategy for (5.4), if

$$(5.5) \qquad \varphi_{e,s}(n)\!\downarrow \; \geq \; a_n^s \qquad \& \qquad e \leq n$$

then enumerate a_n^s into A_{s+1} for the sake of φ_e. Now φ_e causes each a_n^s to be enumerated only if $e \leq n$, and even then at most finitely often until $\varphi_e(n) < a_n^s$. Therefore, $a_n = \lim_s a_n^s$ exists. But $\varphi_e(n) < a_n$ for all $n \geq e$ by the construction. Hence, the principal function $p_{\overline{A}}(n) = a_n$ dominates φ_e. It follows from Theorem 4.5.4 (ii) that $p_{\overline{A}} \geq_T \emptyset'$. $\qquad\square$

5.2.4   Simple Permitting and Simple Sets

Theorem 5.2.6 shows that simple sets can have degree $\mathbf{0}'$, and therefore simplicity does not guarantee Turing incompleteness. Now we prove that for every noncomputable c.e. set C we can construct a simple set $A \leq_T C$. Therefore, the structural property of A being simple places no restriction on its degree except that A must be noncomputable. We use the *simple (easy) permitting* method in which an element x is permitted to enter A at a given stage only when an element $y \leq x$ enters C. In the later section §5.5 we consider more general permitting.

Theorem 5.2.7. *For any noncomputable c.e. set C there is a simple set $A \leq_T C$. Indeed, the proof naturally guarantees that $A \leq_{\mathrm{ibT}} C$ for identity bounded Turing reducibility of Definition 3.8.1 (ii). (Furthermore, we can also arrange $A \equiv_T C$ as in Exercise 5.2.12.)*

Proof. Let $\{C_s\}_{s\in\omega}$ be a computable enumeration of C. We modify the Post simple set construction of Theorem 5.2.3 and construct a coinfinite c.e. set A to satisfy for all e the following requirement (5.6) which replaces the Post requirement (5.1).

$$(5.6) \quad P_e: \qquad |W_e| = \infty \qquad \Longrightarrow \qquad [\, W_e \cap A \neq \emptyset \quad \vee \quad C \equiv_T \emptyset \,].$$

Construction. Define $A_0 = \emptyset$.

Stage $s + 1$. Given A_s choose the least e such that

(5.7) $W_{e,s} \cap A_s = \emptyset$ & $(\exists x > 2e)[\, x \in W_{e,s}$ & $C_{s+1} \upharpoonright x \neq C_s \upharpoonright x \,]$.

(Note that (5.7) is the same as the Post simple set (5.2) except that the last clause demands that $(\exists y \leq x)[y \in C_{s+1} - C_s]$, i.e., that C *permits* x at stage $s+1$.) Choose the least such e and enumerate the least corresponding x in A. If there is no such e go to stage $s + 2$.

End construction.

Verification. Clearly, $\overline{A}$ is infinite by the same proof as that of Theorem 5.2.3. Furthermore, $A \leq_T C$ by Proposition 3.5.5 (iii) because $m_A(x) \leq m_C(x)$, with these least modulus functions defined in Definition 3.5.4 (iv).

Lemma 5.2.8. $(\forall e)[\, W_e$ *infinite* $\implies$ $W_e \cap A \neq \emptyset \,]$.

Proof. If not, choose the least e which is a counterexample. Hence, W_e is infinite and $W_e \cap A = \emptyset$. Therefore, there is an increasing c.e. sequence of elements $x_1 < x_2 < \dots$ with $2e < x_1$ and a corresponding c.e. sequence of stages $s_1 < s_2 < \dots$ such that for all i we have $x_i \in W_e[s_i]$ and no element $y \leq x_i$ enters C at any stage $t \geq s_i$. Otherwise, (5.7) would allow $W_e \cap A \neq \emptyset$. Now define a computable function g_e such that $g_e = C$, contradicting the hypothesis that C is noncomputable. When x_i appears in $W_e[s_i]$, define $g_e(y) = C(y)[\, s_i]$ for all $y \leq x_i$. But $C \upharpoonright x_i$ does not change after s_i, and therefore $g_e = C$. Note that we have achieved $A \leq_{ibT} C$. □   □

5.2.5 Permitting as a Game

In this proof we did not introduce the function g_e *during* the construction but only in the verification, using a proof by contradiction. However, it is sometimes more accurate to think of this as a Lachlan game as presented in §2.5 where we define g_e *during* the construction, and we shall present this version of the permitting of a simple set in §16.3.2.

The intuition is that in Theorem 5.2.7 Player I (RED) plays the noncomputable c.e. set C and all the c.e. sets $\{W_e\}_{e \in \omega}$. Player II (BLUE) plays A, attempting to make A simple and $A \leq_T C$ by permitting. While $W_e \cap A$ remains empty, BLUE plays for every e an initial segment of a computable function g_e, threatening to make $g_e = C$. In chess and in Lachlan games the *threat* is often just as effective as the completed action. RED can ignore the threat of g_e only finitely often before he must give C permission for elements to enter A or else A is computable via g_e.

5.2.6 Exercises

Exercise 5.2.9. Let S be the class of simple sets and C the class of cofinite sets. Prove that $S \cup C$ is a filter in $\mathcal{E}$. (In Exercise 5.3.8 we shall see this for the h-simple sets.)

Exercise 5.2.10. $^\diamond$ Show that $\{e : W_e \text{ is simple}\}$ is Π_3-complete. *Hint.* Combine the methods of the canonical simple set construction of Theorem 5.2.5 and the Σ_3 marker construction of Theorem 4.3.3 as follows. Let A be Σ_3-complete and g be as in Theorem 4.3.9. Construct a computable function f such that if $x \in A$ then $W_{f(x)}$ is cofinite, and if $x \notin A$ then $W_{f(x)}$ is simple. Construct $B^x = W_{f(x)}$ as in Theorem 5.2.5 but also move b_n^x whenever a new element is enumerated in $W_{g(x,n)}$.

Exercise 5.2.11. Prove that if $A \leq_m S$ and A is part of a computably inseparable pair (as defined in Exercise 1.6.26) of c.e. sets then S is not simple. Note that $K \equiv \{x : \varphi_x(x) = 0\}$, and hence by Exercise 2.4.18, K is part of an effectively inseparable (and therefore computably inseparable) pair of c.e. sets. (Hence, a simple set is not m-complete.) *Hint.* Let (A, B) be computably inseparable, $A \leq_m S$ via f, and consider $f(B)$.

Exercise 5.2.12. In Theorem 5.2.7 show that we can modify the construction to achieve $C \leq_T A$. *Hint.* In (5.7) replace $2e$ by $3e$. Whenever some x enters C put into A some $y \leq 3x$ currently in $\overline{A}$. Prove that A is simple and that $m_C(x) \leq m_A(3x)$ which guarantees $C \leq_T A$.

5.3 * Hypersimple Sets and Dominating Functions

Although Post proved that simple sets are necessarily m-incomplete, he realized that simple sets could be Turing complete. Therefore, Post continued by defining coinfinite c.e. sets with even thinner complements called hypersimple (h-simple) and hyperhypersimple (hh-simple) sets. Although these sets also failed to solve Post's problem, they were later shown to have interesting characterizations which gave information about the structure of c.e. sets and the relationship between a c.e. set and its degree.

5.3.1 Weak and Strong Arrays of Finite Sets

Definition 5.3.1. (Recall the finite set D_y with index y in Definition 2.3.6.)

(i) A sequence $\{F_n\}_{n \in \omega}$ of *finite* sets is a *strong (weak) array* if there is a computable function f such that $F_n = D_{f(n)}$ $(F_n = W_{f(n)})$.

(ii) An array is *disjoint* if its members are pairwise disjoint.

(iii) An infinite set B is *hyperimmune* abbreviated *h-immune*, (respectively, *hyperhyperimmune*, abbreviated *hh-immune*) if there is no disjoint strong (weak) array $\{\, F_n\,\}_{n\in\omega}$ such that $F_n \cap B \neq \emptyset$ for all n.

(iv) A c.e. set A is *hypersimple*, abbreviated *h-simple* (*hyperhypersimple*, abbreviated *hh-simple*) if $\overline{A}$ is h-immune (hh-immune).

The intention behind h-simplicity and hh-simplicity is that instead of specifying a c.e. set $\{\, a_n\,\}_{n\in\omega} \subseteq \overline{A}$ we specify a disjoint c.e. array of finite sets $\{\, F_n\,\}_{n\in\omega}$ such that each F_n contains some $x \in \overline{A}$ but we cannot tell which $x \in F_n$ has this property. In a strong array we can explicitly compute $\max(F_n)$ and all its members, but in a weak array we can merely enumerate F_n. It can be easily shown that hh-simple implies h-simple and h-simple implies simple. We shall prove later that these implications are not reversible.

5.3.2 Dominating Functions and Hyperimmune Sets

See Definitions 3.5.2, 4.5.1, and §4.5 for the definitions of domination and escape and their properties.

Definition 5.3.2. A function g (or infinite set A) is *computably bounded (computably majorized)* if $(\exists h \leq_T \emptyset)\,(\forall x)[\, g(x) \leq h(x)\,]$. (Extend the definition to infinite sets A by using the principal function $p_A(x)$.)

Theorem 5.3.3 (Kuznecov, Medvedev, Uspenskii). *An infinite set A is hyperimmune iff A is not computably bounded.*

Proof. ($\Longleftarrow$).   Assume A is not hyperimmune. Let $\{D_{g(x)}\}_{x\in\omega}$ be a disjoint strong array with $D_{g(x)} \cap A \neq \emptyset$ for all x. Define $f(x) = \max(\bigcup\{\, D_{g(y)}\,\}_{y\leq x})$. Then $f(x) \geq p_A(x)$ for all x. Therefore, A is computably bounded by f.

($\Longrightarrow$).   Assume p_A is bounded by a computable function f. Define g by $D_{g(0)} = [\,0, f(0)\,]$, where we define $[\,n, m\,] = \{\, n, n+1, \ldots, m\,\}$. Given $g(0), g(1), \ldots, g(n)$, define $k = 1 + \max(\bigcup\{\, D_{g(i)}\,\}_{i\leq n})$ and define $D_{g(n+1)} = [\,k, f(k)\,]$. Now $k \leq p_A(k) \leq f(k)$ so $p_A(k) \in D_{g(n+1)} \cap A$. This disjoint strong array $\{D_{g(n)}\}_{n\in\omega}$ witnesses that A is not hyperimmune. $\square$

Corollary 5.3.4. *A coinfinite c.e. set A is h-simple iff $\overline{A}$ is not computably bounded.* $\square$

Note that Post's simple set S of Theorem 5.2.3 is not hypersimple since $\overline{S}$ is dominated by $f(x) = 2x$. The canonical simple set construction of Theorem 5.2.5 may or may not produce an h-simple set. For example, the simple set A constructed in Theorem 5.2.6 produced $p_{\overline{A}}$ which dominated all p.c. functions and hence makes A h-simple.

5.3.3   Degrees of Hypersimple Sets and Dekker's Theorem

Definition 5.3.5. Let A be an infinite c.e. set and f a 1:1 computable function with range A (an *enumeration* of A). Let a_s denote $f(s)$ and $A_s = \{a_v : v \leq s\}$.

(i) Define the *deficiency set* D to be the Σ_1 set of *deficiency stages*:

(5.8) $D = \{\, s : (\exists t > s)\,[\, a_t < a_s \,] \,\}.$

(ii) The Π_1 set $T = \overline{D}$ is the set of *true stages (nondeficiency stages)*:

(5.9) $T = \{\, t \,:\, A \restriction a_t = A_t \restriction a_t \,\}.$

   Therefore, at a true stage $t \in T$ the element a_t just enumerated is smaller than any element later enumerated. The intuition is that the deficiency set D measures how far f is from being an order-preserving enumeration of A (which, of course, is not possible if A is not computable). The set T of true stages will play an important role in infinite injury, where we shall use the obvious fact that an *apparent* computation $\Phi_e^A(x)[t] = y$ at a *true* stage $t \in T$ is a *true* computation $\Phi_e^A(x) = y$ because no element $z \leq u$ can later enter A, where $u = \varphi_e^A(x)[t]$, the use of the computation. (Also recall truth-table reducibility $B \leq_{tt} A$ from Definition 3.8.2.) See [Dekker 1954].

Theorem 5.3.6 (Dekker, 1954).   *Let A be any noncomputable c.e. set and D the deficiency set of A with respect to some fixed enumeration of A.*

(i) $A \leq_T D$.

(ii) $D \leq_{tt} A$.

(iii) D *is hypersimple.*

(iv) *Reductions in (i) and (ii) are uniform in any Σ_1 indices of A and D.*

Proof. (i) Now D is Σ_1 and hence c.e. Clearly, $\overline{D}$ is infinite. Note that

(5.10) $x \in A$ iff $x \in \{a_0, a_1, \ldots, a_{p_{\overline{D}}(x)}\}$

by the definition of D and the fact that $x \leq a_{p_{\overline{D}}(x)}$. Hence, $A \leq_T D$.

(ii) Next $D \leq_{tt} A$ since to test whether $s \in D$ we A-computably find the least stage v such that $A_v \restriction a_s = A \restriction a_s$. Now $s \in D$ iff $v > s$, i.e., iff $s \in D_v$.

(iii) Now $\overline{D}$ is h-immune, because if some computable function g dominates $p_{\overline{D}}$ then by (5.10) $x \in A$ iff $x \in \{a_0, a_1, \ldots, a_{g(x)}\}$, which implies that A is computable.

(iv) Clearly, the reductions in (i) and (ii) are uniformly computable in Σ_1-indices for A and D. $\square$

Corollary 5.3.7. *For every noncomputable c.e. set A there is an h-simple set $D \equiv_T A$. Hence, every nonzero c.e. degree contains an h-simple set.* $\square$

5.3.4 Exercises

Exercise 5.3.8. Show that the h-simple sets together with the cofinite sets form a filter in $\mathcal{E}$. (In Exercise 5.2.9 we saw this for the simple sets and later we shall see it for the hh-simple sets.)

Exercise 5.3.9. Give a direct "movable marker" type construction of an h-simple set. *Hint.* Modify the construction of the canonical simple set in Theorem 5.2.5 with requirement P_e of (5.1) replaced by the requirement

$$\widehat{P}_e : \; \{\, D_{\varphi_e(x)} \,\}_{x \in \omega} \; \text{ a disjoint strong array } \quad \Longrightarrow \quad (\exists x)[\, D_{\varphi_e(x)} \subseteq A \,].$$

Exercise 5.3.10. Given a noncomputable c.e. set A, use the preceding exercise and the method of Theorem 5.2.7 to find an h-simple set $H \leq_{\mathrm{ibT}} A$.

Exercise 5.3.11. Show that $\{x : W_x \text{ is h-simple}\}$ is Π_3-complete. (See the hint for Exercise 5.2.10.)

Exercise 5.3.12. A coinfinite c.e. set A is *dense simple* (Martin) if the principal function $p_{\overline{A}}$ dominates every total computable function. We constructed a dense simple set in Theorem 5.2.6.

(i) Prove that a dense simple set is h-simple.

(ii) Prove that a dense simple set is high.

(iii) Construct an example of an h-simple set which is not dense simple.

Exercise 5.3.13. Show that any coinfinite c.e. set B has an h-simple superset A. *Hint.* If B is not already h-simple, then by Theorem 2.3 there is an increasing computable function f such that for all n, $[f(n), f(n+1)) \cap \overline{B} \neq \emptyset$. Let $F_n = [f(n), f(n+1))$. Play the strategy of Exercise 2.9 but with the markers Γ_e each associated with some *set* F_n instead of an *integer* n. When x is enumerated in B, it is also enumerated in A. In addition, when F_n has no marker all its members are put into A. Begin by associating Γ_e with F_e. To meet $\widehat{P}_e$ of Exercise 2.9 wait for some x such that $\varphi_e(x)\downarrow$ and $D_{\varphi_e(x)} \cap F_n = \emptyset$ for each F_n associated currently with a marker Γ_i, $i \leq e$. For each $y \in D_{\varphi_e(x)}$ find n such that $y \in F_n$, enumerate *all* members of F_n into A and move the marker (if any) associated with F_n to some F_m, $m \geq n$, F_m not yet a subset of A. Prove that each Γ_e moves finitely often and comes to rest on some F_n such that $F_n \cap \overline{A} \neq \emptyset$. Hence, $\overline{A}$ is infinite. (Note that A is not obtained *uniformly* from B.)

Exercise 5.3.14. A set S is *introreducible* if $S \leq_T T$ for every infinite set $T \subseteq S$. For the deficiency set D of Definition 5.3.5 prove that $\overline{D}$ is introreducible.

Exercise 5.3.15. A set A is *bounded truth-table reducible* to a set B (written $A \leq_{btt} B$) if there is some n (called the *norm* of the reduction) and there are computable functions f and g such that for all x: (1) $|D_{f(x)}| \leq n$ and (2) $x \in A$ iff $B \cap D_{f(x)} = D_y$ for some y in $D_{g(x)}$. (For example, for any set A, $A \leq_{btt} \overline{A}$.) Let $\mathcal{B}$ be the Boolean algebra generated by the c.e. sets. Prove that $A \in \mathcal{B}$ iff $A \leq_{btt} K$. (Note that $A \leq_m B \implies A \leq_{btt} B \implies A \leq_{tt} B$, and that $\leq_{btt}$ and $\leq_{tt}$ are reflexive and transitive and therefore induce equivalence relations $\equiv_{btt}$ and $\equiv_{tt}$.)

Exercise 5.3.16. Prove that if A and B are c.e. and $\neq \omega, \neq \emptyset$, and neither is $\emptyset$ nor ω, and $A \leq_{btt} B$ with norm 1, then $A \leq_m B$.

Exercise 5.3.17. Let $\mathcal{B}$ be the least Boolean algebra containing $\mathcal{E}$. ($\mathcal{B}$ is the Boolean algebra obtained by closing the c.e. sets under complementation, union and intersection).

(i) Prove that $A \in \mathcal{B}$ iff A is a finite union of d.c.e. sets.

(ii) Prove that $A \in \mathcal{B}$ iff $A \leq_{\mathrm{btt}} K$.

(iii) Show that there exists a set $A \leq_{\mathrm{T}} K$ such that $A \notin \mathcal{B}$.

Exercise 5.3.18. $^\diamond$ (Friedberg and Rogers, 1959). Show that $K \not\leq_{bT} H$ for H h-simple. See [Friedberg-Rogers 1959]. *Hint* (Owings). Assume $K = \Phi_e^H$ with f as above. It suffices to show that for every interval $[0, y]$ which intersects $\overline{H}$ we can find uniformly in y some $z > y$ such that $[y+1, z] \cap \overline{H} \neq \emptyset$. Suppose $F = \overline{H} \cap [0, y]$. Let W_n consist of all x such that

$$(\exists \sigma) [\ \Phi_e^\sigma(x) = 0 \quad \& \quad (\forall y < lh(\sigma) [\ \sigma(y) = 0 \implies y \in F\ ]$$

$$\&\ [\sigma(y) = 1 \implies y \in H]\ ].$$

Apply to W_n the productive function for $\overline{K}$ and then apply f to obtain z^F. Let $z = \max\{\ z^F : F \subseteq [0, y]\ \}$.

5.4 * The Arslanov Completeness Criterion

5.4.1 Effectively Simple Sets Are Complete

Although Post constructed a complete simple set, he did not realize that his original simple set and indeed all effectively simple sets (as in Definition 5.2.4) are complete.

Theorem 5.4.1 (Martin, 1966a). *If A is effectively simple then $K \leq_{\mathrm{T}} A$.*

Proof. Let $\{K_s\}_{s \in \omega}$ be a computable enumeration of K and $\theta(x)$ be a p.c. function such that $\theta(x) = (\mu s) [\ x \in K_s\ ]$ if $x \in K$ and $\theta(x)$ diverges otherwise. Let A be effectively simple via f, $\{A_s\}_{s\in\omega}$ be a computable enumeration of A, and $\overline{A}_s = \{ a_0^s < a_1^s < a_2^s < \cdots \}$. By the Recursion Theorem with Parameters 2.2.6, define the computable function h by

$$(5.11) \quad W_{h(x)} = \begin{cases} \{ a_0^{\theta(x)} < a_1^{\theta(x)} < \ldots < a_{fh(x)}^{\theta(x)} \} & \text{if } x \in K; \\ \emptyset & \text{otherwise.} \end{cases}$$

Set $r(x) = (\mu s)\,[a_{fh(x)}^s = a_{fh(x)}]$. Then $r \leq_T A$ because f and h are computable. (Here $fh(x)$ denotes $f(h(x))$.) Now if $x \in K$ and $r(x) \leq \theta(x)$, then $W_{h(x)} \subseteq \overline{A}$ and $|W_{h(x)}| = fh(x) + 1$, contrary to the hypothesis on f. Hence, $r(x) > \theta(x)$ for all $x \in K$. Therefore, $x \in K$ iff $x \in K_{r(x)}$. Hence, $K \leq_T A$. $\qquad\square$

5.4.2   Arslanov's Completeness Criterion for C.E. Sets

Definition 5.4.2. A function f is a *fixed point free (FPF)* function if $(\forall x)\,[\, W_{f(x)} \neq W_x \,]$.

Theorem 5.4.3 (Arslanov Completeness Criterion, 1981).   *A c.e. set A is complete iff there is a fixed point free function $f \leq_T A$.*

Proof. ($\Longrightarrow$).   Assume $A \equiv_T \emptyset' \equiv_T K_1 = \{e : W_e \neq \emptyset\}$ by Exercise 1.6.22. Define

$$W_{f(x)} = \begin{cases} \emptyset & \text{if } W_x \neq \emptyset; \\ \{0\} & \text{otherwise.} \end{cases}$$

($\Longleftarrow$).   Assume $(\forall x)[\, W_{f(x)} \neq W_x \,]$, where $f \leq_T A$. By the Modulus Lemma there is a computable function $\widehat{f}(x,s)$ such that $f(x) = \lim_s \widehat{f}(x,s)$ for every x, and the sequence $\{ \lambda x\,\widehat{f}(x,s) \}_{s\in\omega}$ has a modulus $m \leq_T A$. Let $\{K_s\}_{s\in\omega}$ be a computable enumeration of K. Let $\theta(x) = (\mu s)[x \in K_s]$ if $x \in K$, and let $\theta(x)$ diverge otherwise. By the Recursion Theorem with Parameters 2.2.6 define the computable function h by

$$W_{h(x)} = \begin{cases} W_{\widehat{f}(h(x),\,\theta(x))} & \text{if } x \in K; \\ \emptyset & \text{otherwise.} \end{cases}$$

Now if $x \in K$ and $\theta(x) \geq m(h(x))$, then $\widehat{f}(h(x), \theta(x)) = f(h(x))$ and $W_{f(h(x))} = W_{h(x)}$, contrary to the hypothesis on f. Hence, for all x,

$$x \in K \quad\Longleftrightarrow\quad x \in K_{m(h(x))},$$

and therefore $K \leq_T A$. $\qquad\square$

Corollary 5.4.4 (Arslanov, 1981). *Given a c.e. degree* $\mathbf{a}$, $\mathbf{a} < \mathbf{0'}$ *iff for every function* $f \in \mathbf{a}$ *there exists* n *such that* $W_n = W_{f(n)}$. □

Note that the hypothesis of "A is c.e." in Theorem 5.4.3 is necessary by Exercise 5.4.14. Secondly, if f does not have a fixed point n such that $W_n = W_{f(n)}$, we might at least hope for a *-*fixed point* (*almost fixed point*), namely an n such that $W_n =^* W_{f(n)}$. Exercise 5.4.11 is analogous to the Recursion Theorem since it shows that every function $f \leq_T \emptyset'$ has a *-fixed point, while Exercise 5.4.15 gives a $\emptyset''$-completeness criterion, which is the analogue for *-fixed points of Theorem 5.4.3 for fixed points. Finally, if f has no almost fixed points we might hope that f has a *Turing fixed point*, namely an n such that $W_n \equiv_T W_{f(n)}$. In a later chapter we shall prove that this is true if $f \leq_T \emptyset^{(2)}$.

5.4.3 Exercises

Exercise 5.4.5. (Jockusch). Prove that the following are equivalent for an arbitrary set A:

(i) $(\exists f \leq_T A)\,(\forall e)\,[W_e \neq W_{f(e)}]$,

(ii) $(\exists g \leq_T A)\,(\forall e)\,[\varphi_e \neq \varphi_{g(e)}]$,

(iii) $(\exists h \leq_T A)\,(\forall e)\,[\,h(e) \neq \varphi_e(e)\,]$.

Hint. For (ii) implies (iii), let d be the computable function defined in the proof of the Recursion Theorem, and let $h(e) = g(d(e))$. For (iii) implies (i), fix a p.c. function θ such that if $W_e \neq \emptyset$ then $\theta(e) \in W_e$. Choose a computable function q such that for every e, $\varphi_{q(e)} = \lambda y\,[\,\theta(e)\,]$. Let $W_{f(e)} = \{\,h(q(e))\,\}$ for all e. (Note that the implication (i) implies (iii) is a generalization of the Recursion Theorem.)

Exercise 5.4.6. A set A is *effectively immune* if A is infinite and there is a computable function f such that if $W_x \subseteq A$ then $|W_x| < f(x)$. Prove that if A is effectively immune, B is c.e., and $A \leq_T B$, then $B \equiv_T \emptyset'$. *Hint.* Use the Modulus Lemma 3.6.3 and modify the proof of Theorem 5.4.1.

Exercise 5.4.7. (Smullyan-Martin). (i) Prove that if A is effectively simple and $A \subseteq B \subset_\infty \omega$ and B is c.e. then B is effectively simple. (Recall that $X \subset_\infty Y$ denotes that $X \subset Y$ and $Y - X$ is infinite.)

(ii) Show that there is an effectively simple hypersimple set. *Hint.* Combine the positive requirements of Theorem 5.2.5 and Exercise 5.3.9. The negative requirements assert only that $\lim_s a_{n,s}$ exists, and this will be satisfied if the positive requirements at most finitely often cause $a_{n,s} \neq a_{n,s+1}$, which they do. Note that Post's simple set S of Theorem 1.3 is effectively simple and *not* hypersimple.

Exercise 5.4.8. (Shoenfield, 1957). A is *quasi-creative* if A is c.e. and

$$(\exists f \leq_T \emptyset)\,(\forall x)\,[\,W_x \subseteq \overline{A} \quad \Longrightarrow \quad [D_{f(x)} \subset \overline{A} \ \& \ D_{f(x)} \not\subseteq W_x\,]\,].$$

Prove that A quasi-creative implies A complete. *Hint.* Use the Recursion Theorem in the manner of Theorem 5.4.1 or Theorem 2.4.6.)

Exercise 5.4.9. For each of the following properties define f appropriately and apply Theorem 5.4.3 to show that A is complete.

(i) A creative.

(ii) A quasi-creative (as defined in Exercise 5.4.8).

(iii) A effectively simple.

(iv) B of Exercise 5.4.7.

Exercise 5.4.10. (Morris-Gill). A c.e. set A is *subcreative* (M. Blum) iff there is a computable function g such that for every x if $W_x \cap A$ is finite (possibly empty), then $A \subset W_{g(x)} \subseteq A \cup \overline{W}_x$. Apply Theorem 5.4.3 to prove that every subcreative set is complete. *Hint.* Find the least s such that either $(\exists y)\,[y \in W_{g(x),s} - A]$ or $W_{x,s} \neq \emptyset$. In the former case set $W_{f(x)} = \{y\}$, and otherwise set $W_{f(x)} = \emptyset$. Show that $f \leq_T A$, and f satisfies Theorem 5.4.3.

Exercise 5.4.11. (Arslanov, Nadirov, and Solov'ev, 1977). An integer n is an *almost fixed point* for a function f if $W_{f(n)} =^* W_n$.

(i) Show that for any function θ partial computable in $\emptyset'$ there is a computable function f such that $(\forall x \in \text{dom}(\theta))\,[W_{\theta(x)} =^* W_{f(x)}]$. *Hint.* Choose a total computable function $\widehat{f}(x,s)$ such that $\theta(x) = \lim_s \widehat{f}(x,s)$ for every $x \in \text{dom}(\theta)$. Define

$$W_{f(x)} \;=\; \bigcup \{\, W_{\widehat{f}(x,s),s} \,:\, s \in \omega \,\}.$$

(ii) Show that any function $f \leq_T \emptyset'$ has an almost fixed point.

(iii) Use (ii) to show that for no coinfinite c.e. set A does there exist a function $f \leq_T \emptyset'$ such that for all x with $A \subseteq W_x$: (1) $W_x \cap W_{f(x)} =^* A$; and (2) $W_x \cup W_{f(x)} =^* \omega$. (The point is that we shall show A is hh-simple iff $\mathcal{L}^*(A)$ is a Boolean algebra, i.e., there is a function f satisfying (1) and (2). By this exercise we cannot have $f \leq_T \emptyset'$.)

Exercise 5.4.12. $^\diamond$ (Jockusch and Soare, 1972a). In Theorem 5.2.7 we were given a noncomputable c.e. set C and constructed a simple set $A \leq_T C$ by permitting with $f(x) = x$ and without the explicit coding of $C \leq_T A$ described in Exercise 5.2.12. Prove that this explicit coding is unnecessary by showing that $C \leq_T A$ automatically by using the method of Theorem 5.4.3. (This is an example of a certain *maximum degree principle* which asserts roughly that a c.e. set A being constructed has the highest degree not explicitly ruled out.)

Exercise 5.4.13. (i) (Lachlan, 1968a). A c.e. set A is *weakly creative* if

$$(\exists f \leq_T A)(\forall x)[\, W_x \subseteq \overline{A} \quad \Longrightarrow \quad f(x) \notin W_x \cup A \,].$$

Prove that a c.e. set A is weakly creative iff A is complete.

(ii) (Martin). A c.e. set S is *weakly effectively simple* if

$$|\overline{S}| = \infty \quad \& \quad (\exists f \leq_T S)(\forall x)[\, W_x \subseteq \overline{S} \quad \Longrightarrow \quad f(x) > |W_x| \,].$$

Show that every weakly effectively simple set is complete.

(iii) Show that any complete simple set is weakly effectively simple.

Exercise 5.4.14. (Arslanov, 1979 and 1981). Show that there exists a set $A \leq_T \emptyset'$, $A' \equiv_T \emptyset'$, and $f \leq_T A$, such that $(\forall e)\,[W_e \neq W_{f(e)}]$. Hint. (Jockusch). Show that the class $\mathcal{C}$ of $\{0, 1\}$-valued functions h such that $(\forall e)\,[\, h(e) \neq \varphi_e(e)\,]$ is a Π_1^0 class as in Definition 8.2.1. Apply the Low Basis Theorem 3.7.2 to get $h \in \mathcal{C}$, h of low degree, and apply Exercise 5.4.5.

Exercise 5.4.15. ($\emptyset''$-Completeness Criterion, Arslanov, 1981). Suppose $A \in \Sigma_2$ and $\emptyset' \leq_T A$. Prove that

$$A \equiv_T \emptyset'' \quad \Longleftrightarrow \quad (\exists f \leq_T A)\,(\forall e)\,[\, W_e \neq^* W_{f(e)} \,].$$

(Hence, this theorem is to Exercise 5.4.11 what Theorem 5.4.3 is to the Recursion Theorem.) *Hint* (Jockusch). First show by the method of Exercise 5.4.11(a) that if ψ is a function partial computable in $\emptyset'$ then there exists a computable function $g(e)$ such that

$$(\forall e \in \text{dom}(\psi))\,[\, W_{\psi(e)} =^* W_{g(e)} \,].$$

Replace the construction of Theorem 5.4.3 by a $\emptyset'$-construction. Let $B = \text{Fin}$, which is c.e. in $\emptyset'$. Let $\{A_s\}_{s \in \omega}$ and $\{B_s\}_{s \in \omega}$ be $\emptyset'$-computable enumerations of A and B. Let $\theta(x) = \mu s[x \in B_s]$ if $x \in B$ and let $\theta(x)$ diverge otherwise. From $f = \Phi_e^A$ define $\widehat{f}$ as in Theorem 5.4.3 except that now $\widehat{f} \leq_T \emptyset'$. Hence, $\widehat{f}(y, \theta(x))$ is a function partial computable in $\emptyset'$ so there is a computable function $g(y, x)$ such that

$$(\forall x)\,(\forall y)\,[\, \widehat{f}(y, \theta(x))\!\downarrow \quad \Longrightarrow \quad W_{\widehat{f}(y,\theta(x))} =^* W_{g(y,x)} \,].$$

Use the Recursion Theorem with Parameters 2.2.6 applied with parameter x to the function g to obtain a computable function $h(x)$ such that for all x, $W_{h(x)} = W_{g(h(x),x)}$, and hence

$$(\forall x \in B)\,[\, W_{h(x)} = W_{g(h(x),x)} =^* W_{\widehat{f}(h(x),\theta(x))} \,].$$

Now complete the proof as in Theorem 5.4.3.

5.5 ⊘ More General Permitting

In Theorem 5.2.7 we were given a noncomputable c.e. set C and we built a simple set $A \leq_{ibT} C$ using the identity function to permit. The proof suggests that instead of the identity we could have fixed in advance any computable function $f(x)$ and permitted x to enter A when an element $y \leq f(x)$ entered C.

5.5.1 Standard and General Permitting

Theorem 5.5.1 (Standard Permitting Theorem).   *Let $f(x)$ be a computable function and let A and C be c.e. sets with computable enumerations $\{A_s\}_{s\in\omega}$ and $\{C_s\}_{s\in\omega}$, respectively, such that*

(5.12) $A_{s+1} \upharpoonright x \neq A_s \upharpoonright x$. $\implies$. $(\exists y \leq f(x))[\, y \in C_{s+1} - C_s \,]$.

Then $A \leq_T C$. Indeed, A is bounded Turing reducible to C, $A \leq_{bT} C$, as in Definition 3.8.1 (i).

Proof. The condition (5.12) guarantees that $m_A(x) \leq m_C(f(x))$.   □

When we use a computable function $f(x)$ and (5.12) as in Theorem 5.5.1, we refer to it as *standard permitting*, and we achieve $A \leq_{bT} C$. The proof suggests the more general possibility of fixing a function $f \leq_T C$ and still achieving $m_A(x) \leq m_C(f(x))$, which would ensure $A \leq_T C$ but not $A \leq_{bT} C$. This does not quite succeed because we assume only that $f \leq_T C$, and we need a *computable* function in order to carry out the permitting which is a computable construction. However, if we replace $f(x)$ by a computable approximation $\widehat{f}(x,s)$ and modify the conditions accordingly, then we can succeed. In the *simple permitting* method, we have $f(x) = x$ and conclude that $A \leq_{ibT} C$.

Theorem 5.5.2 (General Permitting Theorem).   *Let A and C be c.e. sets with computable enumerations $\{A_s\}_{s\in\omega}$ and $\{C_s\}_{s\in\omega}$ and let $\widehat{f}(x,s)$ be a computable function satisfying the following conditions:*

(i) $f(x) = \lim_s \widehat{f}(x,s) > 0$ *exists.*

(ii) $0 < \widehat{f}(x,s) \neq \widehat{f}(x,s+1)$. $\implies$.

$$(\exists y \leq \widehat{f}(x,s))[\, y \in C_{s+1} - C_s \,].$$

(iii) $A_{s+1} \upharpoonright x \neq A_s \upharpoonright x$. $\implies$. $(\exists y \leq \widehat{f}(x,s))[\, y \in C_{s+1} - C_s \,]$.

Then $A \leq_T C$.

Proof. Now $f(x) = \lim_s \widehat{f}(x,s) > 0$ by (i). Next, $f \leq_T C$ by (ii) because if $\widehat{f}(x,s) > 0$, then

$$f(x) = \widehat{f}(x,s) \qquad \Longleftrightarrow \qquad \neg(\exists y \leq \widehat{f}(x,s))[\, y \in C - C_s \,].$$

Finally, $A \leq_T C$ because $m_A(x) \leq m_C(f(x))$, which is C-computable.   □

5.5.2   Reverse Permitting

Theorem 5.5.3 is important because it shows that general permitting is the most general form of reducibility on c.e. sets $A \leq_T C$ in the sense that any Turing reduction among them can be reduced to the general permitting in Theorem 5.5.2.

Theorem 5.5.3 (Reverse Permitting). *If A and C are c.e. and $A \leq_T C$, then there are computable enumerations $\{A_s\}_{s \in \omega}$ and $\{C_s\}_{s \in \omega}$ of A and C, and a computable function $\widehat{f}(x, s)$ satisfying the conditions of the General Permitting Theorem 5.5.2.*

Proof. Fix A and C and e such that $A = \Phi_e^C$. Fix computable enumerations $\{\widehat{A}_t\}_{t \in \omega}$ and $\{\widehat{C}_t\}_{t \in \omega}$ of A and C. Now for every s find the least $t > s$ such that

(5.13) $(\forall x \leq s)\, [\, \Phi_e^{\widehat{C}}(x)\,[t] \downarrow \;\; = \;\; \widehat{A}(x)\,[t]\,]$.

Now define $A_s = \widehat{A}_t$ and $C_s = \widehat{C}_t$ and define

$$\widehat{f}(x, s) = \begin{cases} \sup_{z \leq x}\; \varphi_e^{\widehat{C}}(z)\,[t] & \text{if } x \leq s; \\ 0 & \text{if } x > s. \end{cases}$$

Once $\widehat{f}(x, s)$ is defined for some $x \leq s$, the value $\widehat{f}(x, s)$ persists unless $z \in C_{v+1} - C_v$ for some $z \leq \widehat{f}(x, v)$, in which case the process begins again. The function $\widehat{f}(x, s)$ satisfies conditions (i) and (ii) of Theorem 5.5.2, and it satisfies (iii) by the Use Principle in Theorem 3.3.9.   □

5.5.3   Building a Turing Functional $\Theta^C = A$

Now let us relate the permitting ideas of the General Permitting Theorem 5.5.2 and the movable marker arguments of §4.3.2. Often we are given a noncomputable c.e. set C and we wish to build a c.e. set A with certain properties and an explicit Turing reduction $\Theta^C = A$ with use function $\theta^C(x)$. We view $\theta(x)$ in a dual role, as a movable marker taking the place of marker Γ_x in §4.3.2, and as the final value of the use function for $\theta(x) = \lim_s \theta(x)[s]$.

We begin by placing $\theta(x)$ on -1, indicating that it is not yet defined. Later we may place $\theta(x)$ on some use $u > 0$ at some stage s, making $\theta(x)[s] = u$, and simultaneously defining $\Theta(x)[s] = y$ for some y. By doing so we are placing an axiom $\langle \sigma, x, y \rangle$ into the oracle graph G of Θ as explained in §3.3, where $\sigma = C_s \restriction u$. Since $\sigma \prec C_s$, we are committed to these values of $\Theta(x)[s]$ and $\theta(x)[s]$ so long as $\sigma \prec C_t$ for $t > s$. If some $w \leq u$ enters C at stage $t > s$, then we are relieved of any commitment, and we can begin defining $\Theta(x)$ and $\theta(x)$ anew. Now C is c.e. and can never return to the former neighborhood $\sigma \prec C_s$ at any stage $v > t$. Therefore, the former axiom $\langle \sigma, x, y \rangle$ can never apply to C after stage t, and C is *permitted* to ignore this axiom forever.

5.6 $\oslash$ Hyperimmune-Free Degrees

Martin and Miller in [Miller and Martin 1968] extended the definition of hyperimmune from a single *set* to an entire *degree*.

Definition 5.6.1. [Miller and Martin, 1968]   (i)   A degree **d** is *hyperimmune (h-immune)* if it contains a hyperimmune set.

(ii) If degree **d** contains no hyperimmune set, it is called *hyperimmune-free* (since it is *free* of hyperimmune sets). It is also called *computably dominated* because by Theorem 5.3.3 every set $B \in \mathbf{d}$ is dominated by a computable function.

Corollary 5.3.7 shows that every nonzero c.e. degree is hyperimmune. The next few results will demonstrate that every nonzero degree comparable with $\mathbf{0}'$ is hyperimmune, showing that the hyperimmune-free degrees are scarce. However, by Theorem 9.5.1 at least one nonzero hyperimmune-free degree exists.

5.6.1   Two Downward Closure Properties of Domination

The key point of Definition 5.6.1 (ii) is that a function f or set A is *computably bounded* if there is a *single* computable function h which bounds (majorizes) f, while f being *hyperimmune-free* imposes computable bounding on *every* function $g \equiv_T f$, a very strong condition. Remarkably, it turns out not to matter whether we insist on having this condition for all $g \leq_T f$ or merely for all $g \equiv_T f$. (Recall that p_A is the principal function of A, the function listing all elements in increasing order.)

Theorem 5.6.2 (Miller and Martin, 1968). *Suppose A is hyperimmune-free.*

(i) If $B \leq_T A$, then B is hyperimmune-free.

(ii) If $f \leq_T A$, then f is hyperimmune-free.

Proof. (i) Let $B = \Phi_e^A$. Define $g(x)$ as follows. Let $g(0) = 0$. For every $x \in \omega$ define $g(2x+1) = g(2x) + p_B(x) + 1$ and $g(2x+2) = g(2x+1) + p_A(x) + 1$. Now g is strictly increasing and therefore is the principal function of some set $C \equiv_T A$. Therefore, some computable function h bounds $g = p_C$. But then $h(2x+1)$ bounds $p_B(x)$.

(ii) Let $f = \Phi_e^A$. Define $g(0) = 0$. For every $x \in \omega$ define $g(x+1) = g(x) + f(x) + 1$. Now g is strictly increasing and therefore is the principal function of some set $C \equiv_T f$. Therefore, some computable function h bounds $g = p_C$. But then $h(x+1)$ bounds $f(x)$ by part (i).   □

Corollary 5.6.3. *The hyperimmune degrees are closed upwards and the hyperimmune-free degrees are closed downwards.*

Proof. By Theorem 5.6.2.   □

Theorem 5.6.4. *A set A is hyperimmune-free iff for every total f,*

$$(5.14) \qquad\qquad f \leq_T A \quad\Longrightarrow\quad f \leq_{tt} A.$$

Proof. ($\Longrightarrow$). Suppose that A is hyperimmune-free and $f = \Phi_e^A$. Define

$$g(x) = (\mu s)\,[\,\Phi_{e,s}^A(x){\downarrow}\,],$$

which exists because f is total. Also note that $g \leq_T A$. Therefore, by Theorem 5.6.2 there exists some computable function h which bounds g. Define a Turing functional $\Psi^X(x)$ which, on any input x and oracle X, runs $\Phi_{e,s}^X(x)$ for $s = h(x)$ many steps, and outputs y if $\Phi_{e,s}^X(x){\downarrow} = y$ and 0 otherwise. Then Ψ^X is total for every X and $\Psi^A = \Phi_e^A$.

($\Longleftarrow$) (Jockusch) Assume $f \leq_T A$. To prove that A is hyperimmune-free we must show that f is bounded by some computable function h. Now $f \leq_{tt} A$ by (5.14). Fix a total reduction Φ_e such that $f = \Phi_e^A$. Define a computable function h. Given x, search for a level n such that $\Phi_e^\sigma(x){\downarrow}$ for all σ of length n. Such a level must exist, because otherwise $\{\sigma : \Phi_e^\sigma(x){\uparrow}\}$ would be an infinite tree, and $\Phi_e^g(x)$ would not converge for any path g through it. Let $h(x) = \max\{\Phi_e^\sigma(x) : |\sigma| = n\}$. Clearly, h bounds f.   □

5.6.2 Δ_2 Degrees Are Hyperimmune

Every nonzero c.e. degree is hyperimmune because it contains a hypersimple set. We now prove that nonzero degrees $\mathbf{d} \leq \mathbf{0'}$ are also hyperimmune. We also explore the Σ_2 degrees and other degrees with respect to computable domination. Here we use the following computation function, $c_A(x)$, rather than the least function $m_A(x)$ of (3.17) because having a computable function h which dominates $m_A(x)$ for a Δ_2 approximation gives no information while having one which dominates $c_A(x)$ gives much more information.

Definition 5.6.5. Let A be a Δ_2 set and let $\{A_s\}_{s\in\omega}$ be a computable sequence such that $A = \lim_s A_s$. The *computation function* is

$$(5.15) \qquad c_A(x) = (\mu s > x)\,[\,A_s \upharpoonright x = A \upharpoonright x\,],$$

where $A \upharpoonright x$ denotes the restriction of A to elements $y \leq x$.

Theorem 5.6.6. *Let A be Δ_2 and let $\{A_s\}_{s\in\omega}$ be a Δ_2 approximation to A with computation function $c_A(x)$.*

(i) $c_A \equiv_T A$.

(ii) If $g(x)$ dominates $c_A(x)$ then $A \leq_T g$. Therefore, A is computable iff a computable function g dominates $c_A(x)$.

Proof. (i) $A \leq_T c_A$ because $A(x) = A_s(x)$ for $s = c_A(x)$. Also, $c_A(x) \leq_T A$ because we generate $A_s(x)$ until the first $s > x$ with $A \upharpoonright x = A_s \upharpoonright x$.

(ii) If A is computable then c_A is computable by (i). Conversely, assume $c_A(x) < g(x)$ for all x. Define

$$(5.16) \qquad y = (\mu z > x)\,(\forall t)_{z \leq t \leq g(z)}\,[\,A_t \upharpoonright x = A_z \upharpoonright x\,].$$

By the definition of $c_A(x)$ and the fact that $c_A(x) < g(x)$ we know that for all $z \geq x$ the interval $[\,z, g(z)\,]$ (called a *frame*) must contain at least *one* stage t which is z-*true* in the sense that $A_t \upharpoonright z = A \upharpoonright z$. But $A = \lim_s A_s$ implies that $A_s \upharpoonright x = A \upharpoonright x$ for almost all s. Therefore, almost all z-frames must contain only stages t which are x-true. This proves that $A(x) = A_y(x)$ because all values for x in the y-frame agree by (5.16) and one must agree with $A(x)$ because $y > x$ and $c_A(y) \geq c_A(x)$.   □

Corollary 5.6.7 (Miller and Martin, 1968). *If $\emptyset <_T A \leq_T \emptyset'$ then deg(A) is hyperimmune.*

Proof. Let $\emptyset <_T A \leq_T \emptyset'$. Hence, $A \in \Delta_2$. By Theorem 5.6.6, no computable function can dominate $c_A \equiv_T A$. Therefore, by Theorem 5.6.2, A cannot have hyperimmune-free degree.   □

Corollary 5.6.8. *If a nonzero degree $\mathbf{d}$ is comparable with $\mathbf{0}'$ then $\mathbf{d}$ is hyperimmune.*

Proof. By Corollary 5.6.7, all nonzero degrees $\mathbf{d} \leq \mathbf{0}'$ are hyperimmune. By the upward closure of hyperimmune degrees in Theorem 5.6.2, all degrees $\mathbf{d} > \mathbf{0}'$ are also hyperimmune.   □

The next result generalizes Corollary 5.6.7 and shows that most degrees obtained by iterating the jump are hyperimmune.

Corollary 5.6.9 (Miller and Martin, 1968). *If $B <_T A \leq_T B'$ then deg(A) is hyperimmune.*

Proof. The set A is Δ_2^B and there is a B-computable sequence $\{A_s\}_{s\in\omega}$ such that $A = \lim_s A_s$ by Definition 5.6.5 relativized to B. Define the computation function c_A as there. If any computable (or even B-computable) function h dominates c_A then $A \leq_T B$, contrary to hypothesis. $\square$

5.6.3 Σ_2 Approximations and Domination

Definition 5.6.10. (i)   A computable sequence $\{A_s\}_{s\in\omega}$ is a Σ_2 *approximation* to a Σ_2 set A if

$$(5.17) \qquad x \in A \quad \Longleftrightarrow \quad (\forall^\infty s)[\, x \in A_s \,].$$

(ii) For such a Σ_2 sequence define the Σ_2 *computation function*

$$(5.18) \quad c_A(x) \;=\; (\mu s \geq x)(\forall z \leq x)[\, z \in \overline{A} \quad \Longrightarrow \quad (\exists t)_{x \leq t \leq s}[\, z \notin A_t \,]\,].$$

Theorem 5.6.11. *Let A be a Σ_2 set and let $\{A_s\}_{s\in\omega}$ be a Σ_2 approximation to A with $c_A(x)$ the Σ_2 computation function. If a computable function dominates $c_A(x)$ then A is computably enumerable.*

Proof. Let A be Σ_2. Now assume that $(\forall x)[g(x) \geq c_A(x)]$. From (5.17) and (5.18) we know

$$(5.19) \qquad (\forall x)[\, x \in \overline{A} \quad \Longrightarrow \quad (\forall y > x)(\exists t)_{y \leq t \leq g(y)}[\, x \notin A_t \,]\,],$$

$$(5.20) \qquad (\forall x)[\, x \in A \quad \Longleftrightarrow \quad (\exists y > x)(\forall t)_{y \leq t \leq g(y)}[\, x \in A_t \,]\,].$$

Therefore, A is Σ_1 in g. If g is computable, then A is Σ_1 and hence c.e. $\square$

5.7   Historical Remarks and Research References

In 1959 Friedberg and Rogers introduced the notion of bounded Turing reducibility under the name weak truth-table reducibility, wtt-reducibility, because like tt-reducibility it has a computable bound $h(x)$ on the use, but unlike tt it need not be a total functional.

It follows from Corollary 5.3.7 that hypersimplicity does not guarantee incompleteness. Although hypersimple sets can be easily constructed directly (see Exercise 5.3.9), it is not obvious how to construct an hyperhypersimple set. Indeed, the first and most natural example is a maximal set. Maximal sets have the thinnest possible complement which a coinfinite c.e. set can have. However, in 1965 Yates constructed a complete maximal set, thereby refuting Post's idea that the thinness of complement of a coinfinite c.e. set could guarantee incompleteness.

In the original investigations of computability, attention was focused on *total* computable functions as the most familiar and easily understood from traditional mathematics. The Herbrand-Gödel general recursive functions and the Turing machines demonstrated convincingly that the central concept was that of a *partial* computable function and that the total ones should be viewed as a restriction of this central concept. Likewise, the first to systematically explore *relative* reducibilities was Post who built up reducibilities from the most obvious m-reducibility to btt and tt-reducibilities. All these give a *total* reduction $A = \Phi^B$, and indeed tt is equivalent to reducibility by a total Turing reduction Φ (see Theorem 3.8.5).

These concepts of Post so influenced the first two decades of the subject that when Friedberg and Rogers in 1959 introduced the notion $A = \Phi^B$ with use function $\varphi^B(x)$ bounded by a computable function $h(x)$ they thought of it as a "weakening" of the existing concept of tt, rather than the *strengthening* A bounded Turing reducible to B, written $A \leq_{bT} B$, of the general notion of $A \leq_T B$, which was not well understood in 1959.

Today we see that the way to understand a Turing reducibility on c.e. sets $B \leq_T A$ is through the General Permitting Theorem 5.5.2 because it is the most general. The many examples of $B \leq_T A$ through the standard permitting in Theorem 5.5.1 and the more general permitting in Theorem 5.5.2 illustrate this dynamic approach. Early results such as Myhill's 1955 Theorem in Chapter 2 that $K \leq_m C$ for a creative set C tended to reinforce the role of strong and total reducibilities. However, Martin's result on completeness of effectively simple sets and Arslanov's generalization in §5.4 showed that static and total reductions should be replaced now by *partial* and *dynamic* ones.

5.7.1 ⊘ Δ_2-Permitting

In the general permitting of Theorem 5.5.2 we might be given a noncomputable c.e. set A and we might construct a Σ_1 or even Δ_2 set $B \leq_T A$. At some stage s we might have issued an axiom $\langle \sigma, x, 0 \rangle$ for $\Psi^\sigma(x) = 0 = B_s(x)$ where $\sigma = A_s \restriction x$. At a later stage $s + 1$, if $A_{s+1} \restriction x = \tau \neq \sigma$ then we are released forever from the previous axiom because A, being c.e. , can never return to neighborhood σ. This enables us to issue a new axiom, $\langle \tau, x, 1 \rangle$, for $\Psi^\tau(x) = 1$ and define $B_{s+1} \restriction x = \Psi^A \restriction x \, [s + 1]$. Now A being c.e., can never have entered neighborhood τ before, and therefore there is no previous axiom involving τ.

All this fails if A is Δ_2 instead of Σ_1. Using the above example, there may be a later stage $t > s + 1$ when A returns to neighborhood σ. In this case we have already made a commitment of $\Psi^\sigma(x)$ and must return B to the previous state $B_v \restriction x = B_\sigma \restriction x$. Nevertheless, Posner developed a system of Δ_2-permitting by returning after every such change to the previous configuration and commitment. R. Miller refined the method to prove in [R. Miller 2001] that there exists a linear order whose degree spectrum (degrees of isomorphic copies) includes every nonzero Δ_2 degree but not $\mathbf{0}$. Csima in [Csima 2004] also developed the method to find for any nonzero Δ_2 degree $\mathbf{d}$ and any complete decidable theory a $\mathbf{d}$-decidable model omitting the nonprincipal types. The method is particularly useful if one wants to prove a theorem of the following form: for any nonzero Δ_2 degree there exists such and such in that degree. The rough idea is that if the Δ_2 degree $\mathbf{d}$ is not computable, and if one returns as necessary to previous configurations, there will eventually be a permitted change which is not later reversed.

6

Oracle Constructions and Forcing

6.1 ⋆ Kleene-Post Finite Extensions

We have seen in Chapter 5 how Post tried to solve Post's Problem 5.1.1 by defining c.e. sets A with ever thinner complements. Post himself did not live to see the refutation of this approach by [Friedberg 1958], who constructed a maximal set with the thinnest complement of all, and the construction of a *complete* maximal set by Yates, which refuted Post's approach. Post moved on to understand full Turing reducibility in [Post 1944]. He gave an excellent intuitive description of one set being Turing reducible to another.

From 1944 until his death in 1954 Post worked to understand Turing reducibility and decision problems for c.e. sets. Post in [Post 1944] page 289 introduced and later defined in [Post 1948] degrees of unsolvability (Turing degrees,) as we have presented in Definition 3.4.1. Post thought carefully about the properties of Turing reducibility and wrote extensive notes. Just before his death in 1954 he gave his notes to Kleene, who revised and expanded them and published them as [Kleene and Post 1954]. This fundamental paper clarified the properties of a Turing reduction, including the Use Principle 3.3.9, and used it to construct sets of incomparable degree below $\emptyset'$, as in Theorem 6.1.1. The paper did not directly address Post's Problem, because it was for Δ_2 sets, not Σ_1 sets, but it laid the indispensable foundation for the later solution by Friedberg and Muchnik, presented in Chapter 7, who added a computable approximation to the Kleene-Post method to obtain Turing incomparable Σ_1 sets.

The first major contribution in [Kleene and Post 1954] is the *finite extension oracle construction*. Here we fix some oracle X, such as $X = \emptyset'$ or $X = \emptyset''$, and build a set $A \leq_T X$ by an X-computable construction of finite initial segments $\{\sigma_s\}_{s \in \omega}$ of A with $\sigma_s \prec \sigma_{s+1}$. For example, given σ_s, index e, and argument x, we can ask the $\emptyset'$-question,

$$(\exists \rho \succ \sigma_s)(\exists y)(\exists t)\, [\ \Phi_{e,t}^{\rho}(x)\!\downarrow\ =\ y\ ]?$$

If so, we can define $\sigma_{s+1} = \rho$, which guarantees that $\Phi_e^A(x) = y$ for every $A \succ \sigma_{s+1}$ by the Use Principle 3.3.9. If not, then $\Phi_e^A(x)$ diverges for every $A \succ \sigma_s$ and we can define $\sigma_{s+1} \succ \sigma_s$ in any fashion.

By the finite extension of $\sigma_{s+1} \succ \sigma_s$ we have *decided (forced)* Turing computability properties of an infinite set $A \succ \sigma_{s+1}$ not yet fully constructed. In §6.3 on generic sets we study the general case of forcing conditions which are finite initial segments. In §6.5 on least upper bounds we consider infinite matrices as forcing conditions.

Theorem 6.1.1 (Kleene-Post, 1954). *There exist sets A, $B \leq_T \emptyset'$ such that $A \mid_T B$ (i.e., $A \not\leq_T B$ and $B \not\leq_T A$.) Therefore, $\emptyset <_T A$, $B <_T \emptyset'$.*

Proof. We shall construct functions χ_A and χ_B in stages s so $\chi_A = \cup_s \sigma_s$ and $\chi_B = \cup_s \tau_s$, where σ_s and τ_s are the finite strings constructed by the end of stage s. Since the construction of σ_s and τ_s at stage s is computable in $\emptyset'$, the sequences $\{\sigma_s\}_{s \in \omega}$ and $\{\tau_s\}_{s \in \omega}$ are $\emptyset'$-computable sequences. Therefore, $A, B \leq_T \emptyset'$. It suffices to meet, for each e, the *requirements*:

$$(6.1) \qquad R_e:\ A \neq \Phi_e^B \qquad \& \qquad S_e:\ B \neq \Phi_e^A$$

to ensure that $A \not\leq_T B$ and $B \not\leq_T A$. Hence, $A \mid_T B$.

Stage $s = 0$. Define $\sigma_0 = \tau_0 = \emptyset$.

Stage $s + 1 = 2e + 1$. (We satisfy R_e.) Given $\sigma_s, \tau_s \in 2^{<\omega}$ of length $\geq s$. Let $n = |\sigma_s| = (\mu x)\,[\,x \notin \mathrm{dom}(\sigma_s)\,]$. Using a $\emptyset'$-oracle we test whether

$$(6.2) \qquad (\exists t)\,(\exists \rho)\,[\ \rho \succ \tau_s \quad \& \quad \Phi_{e,t}^{\rho}(n)\!\downarrow\ ].$$

Note that $\rho \succ \tau_s$ is computable as a relation of strings ρ and τ_s. The second clause of (6.2) is computable by the Oracle Graph Theorem 3.3.8 (i). Therefore, (6.2) is a Σ_1 statement and can be decided computably in $\emptyset'$.

Case 1. Suppose (6.2) is satisfied. The matrix of (6.2) is computable. Find the least pair $\langle \rho, t \rangle$ satisfying that matrix. Define $\tau_{s+1} = \rho$ and $\sigma_{s+1}(n) = 1 \dot- \Phi_{e,t}^{\rho}(n)$ so that $\sigma_{s+1}(n) \neq \Phi_{e,t}^{\rho}(n)$.

Case 2. Suppose (6.2) fails. Then define $\sigma_{s+1} = \sigma_s\widehat{\ }0$ and $\tau_{s+1} = \tau_s\widehat{\ }0$.

In either case, $|\sigma_{s+1}|, |\tau_{s+1}| \geq s+1$. Therefore, $\chi_A = \cup_s \sigma_s$ and $\chi_B = \cup_s \tau_s$ are defined on all arguments. In either case, if $f \succeq \sigma_{s+1}$ and $g \succeq \tau_{s+1}$ then $f(n) \neq \Phi_e^g(n)$ by the Use Principle Theorem 3.3.9.

Stage $s + 1 = 2e + 2$. (*We satisfy S_e.*) Proceed exactly as above but with the roles of σ_s and τ_s replaced by τ_s and σ_s, mutatis mutandis. $\qquad \square$

Theorem 6.1.2 (Relativized Version). *For any degree* **c**, *there are degrees* **a, b** *such that* $\mathbf{c} \leq \mathbf{a}, \mathbf{b}$ *and* $\mathbf{a}, \mathbf{b} \leq \mathbf{c}'$ *and* $\mathbf{a} \mid \mathbf{b}$.

Proof. Fix a set $C \in \mathbf{c}$. Relativize the above proof to C, using a C'-oracle to build sets A and B such that $(A \oplus C) \mid_{\mathrm{T}} (B \oplus C)$ and $A, B \leq_{\mathrm{T}} C'$. In place of (6.2) we use a C'-oracle to test whether

$$(6.3) \qquad (\exists t)\,(\exists \tau_1)\,(\exists \tau_2)\,[\,\tau_1 \succ \tau_s \quad \& \quad \tau_2 \prec C \quad \& \quad \Phi^{\tau_1 \oplus \tau_2}_{e,t}(n)\downarrow\,],$$

where $\tau_1 \oplus \tau_2$ is defined to be the shortest string $\rho \in 2^{<\omega}$ such that $\rho(2x) = \tau_1(x)$ and $\rho(2x + 1) = \tau_2(x)$. The obvious modification of Cases 1 and 2 ensures that $A \neq \Phi^{B \oplus C}_e$. At stage $2e + 2$ we ensure that $B \neq \Phi^{A \oplus C}_e$. Finally, let $\mathbf{a} = \deg(A \oplus C)$ and $\mathbf{b} = \deg(B \oplus C)$. $\qquad \square$

Definition 6.1.3. A countable sequence of sets $\{A_i\}_{i \in \omega}$ is *computably independent* if for each i, $A_i \not\leq_{\mathrm{T}} \oplus \{A_j : j \neq i\}$, where the infinite join is defined as in Exercise 3.4.7.

6.1.1   Exercises

Exercise 6.1.4. Modify the proof of Theorem 6.1.1 to build an independent sequence $\{A_j\}_{j \in \omega}$ of sets each computable in $\emptyset'$ (indeed, $\oplus_j A_j \leq_{\mathrm{T}} \emptyset'$). *Hint.* Use a finite extension $\emptyset'$-computable construction to build at stage s strings $\{\rho^s_j\}_{s \in \omega}$ such that if $A_j = \cup_s \rho^s_j$ then we meet for each e and i the requirement

$$R_{\langle e, i \rangle} \;:\; A_i \neq \Phi^{\oplus\{A_j\,:\,j \neq i\}}_e.$$

At stage $s = 0$, set $\rho^0_j = \emptyset$ for all j. At stage $s + 1 = \langle e, i \rangle + 1$ we meet requirement $R_{\langle e, i \rangle}$ as follows. Given $\{\,\rho^s_j\,\}_{j \in \omega}$, only finitely many of which are nonempty, let $n = |\rho^s_i|$, and use a $\emptyset'$-oracle to test whether there exist m and (a code number for) a finite sequence of strings $\sigma_0, \sigma_1, \ldots, \sigma_k$ such that

$$(6.4) \quad \Phi^{\oplus\{\,\sigma_j\,:\,j \neq i\,\}}_e(n)\downarrow \;=\; m \quad \& \quad (\forall j \leq k)\,[\,j \neq i \;\Longrightarrow\; \rho^s_j \prec \sigma_j\,].$$

Now according to whether or not (6.4) holds proceed as in Theorem 6.1.1 Case 1 (letting $\rho^{s+1}_i(n) = 1 \,\dot{-}\, m$, and $\rho^{s+1}_j = \sigma_j$ for $j \neq i$), or as in Case 2 otherwise. (Be sure to make each A_i total.)

Exercise 6.1.5. A partially ordered set $\mathcal{P} = (P, \leq_{\mathrm{P}})$ is *countably universal* if every countable partially ordered set is order isomorphic to a subordering of $\mathcal{P}$. Prove that there is a computable partial ordering $\leq_{\mathrm{R}}$ of ω which is countably universal. *Hint.* This can be done either by considering a computably presented atomless Boolean algebra, or by a direct construction where at stage $s + 1$, given a finite set P_s of elements in $\leq_{\mathrm{R}}$, one obtains P_{s+1} by adding a new point for each possible order type over P_s. A Boolean algebra $\mathcal{B} = (\{ b_i \}_{i \in \omega}; \leq, \vee, \wedge, ')$ is *computably presented* if there exist a computable relation $P(i, j)$ and computable functions f, g and h such that $P(i, j)$ holds iff $b_i \leq b_j$, and such that $b_{f(i,j)} = b_i \vee b_j$, $b_{g(i,j)} = b_i \wedge b_j$, and $b_{h(i)} = b_i'$.

Exercise 6.1.6. Show that for a countable partially ordered set $\mathcal{P} = \langle P, \leq_{\mathrm{P}} \rangle$ there is a 1:1 order-preserving map from P into $\mathbf{D}(\leq \mathbf{0}')$, the degrees $\leq \mathbf{0}'$. *Hint.* By Exercise 6.1.5 we may assume $P = \omega$ and $\leq_{\mathrm{P}}$ is a computable relation. Let $\{ A_i \}_{i \in \omega}$ be as in Exercise 6.1.4. Define $f : \omega \to \mathbf{D}(\leq \mathbf{0}')$ by $f(i) = \mathbf{a}_i = \deg(\oplus A_j : j \leq_{\mathrm{P}} i)$. Show that if $i \leq_{\mathrm{P}} j$ then $\mathbf{a}_i \leq \mathbf{a}_j$ (by definition and the fact that $\leq_{\mathrm{P}}$ is computable), and if $i \not\leq_{\mathrm{P}} j$ then $\mathbf{a}_i \not\leq \mathbf{a}_j$ (by the computable independence of $\{ A_i \}_{i \in \omega}$).

Exercise 6.1.7. Prove that there are $2^{\aleph_0}$ mutually incomparable degrees. *Hint.* Recall Definition 8.2.1 of a *tree* $T \subseteq 2^{<\omega}$ and its associated trees in Definition 3.7.1. Construct a tree $T \subseteq 2^{<\omega}$ such that $f \mid_T g$ for every pair $f, g \in [T]$ with $f \neq g$. Let $T = \cup_e T_e$ where tree $T_{e+1} \supset T_e$ and T_{e+1} is defined by induction as follows. Let $T_0 = \{ \emptyset \}$, the tree with the empty node (root) as its only member. Given T_e define L_e to be the *leaves* of tree T_e, namely

$$L_e = \{ \sigma \, : \, \sigma \in T_e \quad \& \quad (\forall \tau \succ \sigma) \, [\, \tau \notin T_e \,] \}.$$

Next define the *successors* to leaves,

$$S_e = \{ \sigma\hat{\ }0 \, : \, \sigma \in L_e \} \quad \cup \quad \{ \sigma\hat{\ }1 \, : \, \sigma \in L_e \}.$$

Suppose $S_e = \{ \rho_i \, : \, i \leq 2^{e+1} \}$. Fix i, $j \leq 2^{e+1}$, $i \neq j$. Use the method of Theorem 6.1.1 to replace ρ_i and ρ_j by strings $\sigma \succ \rho_i$, $\tau \succ \rho_j$ satisfying

$$(6.5) \qquad (\forall f \succ \sigma) \, (\forall g \succ \tau) \, [\, \Phi_e^f \neq g \quad \& \quad \Phi_e^g \neq f \,].$$

Repeat this procedure for all i, $j \leq 2^{e+1}$ with $i \neq j$.

6.2   Minimal Pairs and Avoiding Cones

Definition 6.2.1. Degrees $\mathbf{a}$ and $\mathbf{b}$ form a *minimal pair* if $\mathbf{a}$, $\mathbf{b} > \mathbf{0}$ and

$$(6.6) \qquad (\forall \mathbf{c}) \, [\, [\, \mathbf{c} \leq \mathbf{a} \quad \& \quad \mathbf{c} \leq \mathbf{b} \,] \implies \mathbf{c} = \mathbf{0} \,].$$

Minimal pairs have played an important role in computability theory. Later we shall construct a minimal pair of *computably enumerable* degrees. In §6.5 we shall modify the minimal pair construction to find an exact pair of degrees for an ascending sequence of degrees as defined in Definition 6.5.2. To simplify the notation now and later we introduce a useful remark of Posner which allows us to replace pairs of indices by a single index.

Remark 6.2.2 (Posner).   *For all sets A and B with $A \neq B$ and all i and j, there exists e such that $\Phi_e^A = \Phi_i^A$ and $\Phi_e^B = \Phi_j^B$.*

Proof. Since $A \neq B$ they differ on some element n, say $n \in A - B$. Define the Turing reduction Φ_e^X for any X by

$$\Phi_e^X(y) = \begin{cases} \Phi_i^X(y) & \text{if } n \in X; \\ \\ \Phi_j^X(y) & \text{otherwise.} \end{cases} \qquad \square$$

Theorem 6.2.3. *There is a minimal pair of degrees* $\mathbf{a}, \mathbf{b} < \mathbf{0}'$.

Corollary 6.2.4 (Theorem 6.1.1). *There exist* $\mathbf{a}, \mathbf{b} < \mathbf{0}'$ *with* $\mathbf{a} \mid \mathbf{b}$.

Proof. If $\mathbf{a}, \mathbf{b}$ is a minimal pair then $\mathbf{a} \mid \mathbf{b}$. If $\mathbf{a} \leq \mathbf{b}$ then $\mathbf{a} \wedge \mathbf{b} = \mathbf{a} > \mathbf{0}$. $\quad\square$

Proof of Theorem 6.2.3. It suffices to construct sets A and B unequal and computable in $\emptyset'$ satisfying for all e the following requirements.

(6.7) $N_e:$ $\quad \Phi_e^A = \Phi_e^B$ total $\quad\Longrightarrow\quad$ $(\exists g \leq_T \emptyset)[\, g = \Phi_e^A \,]$.

(6.8) $\qquad\qquad P_e:$ $\quad A \neq \varphi_e$ $\quad$ and $\quad B \neq \varphi_e$.

We shall use a $\emptyset'$-oracle construction to build increasing sequences of strings, $\{\sigma_s\}_{s\in\omega}$ and $\{\tau_s\}_{s\in\omega}$, and then define $A = \cup_s \sigma_s$ and $B = \cup_s \tau_s$. Define $\sigma_0 = \emptyset$ and $\tau_0 = \emptyset$.

Stage $s + 1 = 2e + 1$. (Satisfy P_e for A and B.) Given σ_s and τ_s let $x = |\sigma_s| = (\mu y)[\sigma_s(y) \uparrow]$. Ask $\emptyset'$ whether $\varphi_e(x) \downarrow$. If so, define $\sigma_{s+1}(x) = 1 \dot{-} \varphi_e(x)$ and otherwise define $\sigma_{s+1}(x) = 0$. Do likewise to ensure that τ_{s+1} and φ_e are not compatible.

Stage $s + 1 = 2e + 2$. (Satisfy N_e.) Ask $\emptyset'$ the Σ_1 question,

(6.9) $\qquad (\exists \rho \succ \sigma_s)\,(\exists \nu \succ \tau_s)\,(\exists x)[\, \Phi_e^\rho(x) \downarrow \;\neq\; \Phi_e^\nu(x)\downarrow \,]$?

If so, define $\sigma_{s+1} = \rho$ and $\tau_{s+1} = \nu$. If not, define $\sigma_{s+1} = \sigma_s{}^\frown 0$ and $\tau_{s+1} = \tau_s{}^\frown 0$.

Lemma 6.2.5. $(\forall e)[\, \Phi_e^A = \Phi_e^B = f \text{ total} \;\Longrightarrow\; f \text{ is computable} \,]$.

Proof. Assume $\Phi_e^A = \Phi_e^B = f$ is total. At stage $s+1 = 2e+2$, equation (6.9) could not have held, else $\Phi_e^A \neq \Phi_e^B$. Hence, for any x we can choose $\rho \succ \sigma_s$ such that $\Phi_e^\rho(x) \downarrow = y$ by the Use Principle 3.3.9 because Φ_e^A is total. Now any other $\xi \succ \sigma_s$ for which $\Phi_e^\xi(x) \downarrow = z$ must have $y = z$, else one of y and z must form a disagreement with $\Phi_e^\nu(x)$ for some $\nu \prec B$, contrary to our hypothesis that (6.9) fails. Therefore, $f(x) = y$ even though we may not have $\rho \prec f$. Since searching for the first such string ρ, which must exist, is a computable procedure, we know that f is computable. $\qquad\square \qquad\quad \square$

So far the degrees we have constructed, such as $\mathbf{0}^{(n)}$ or degrees below $\mathbf{0}'$, are comparable to $\mathbf{0}'$. We now show how to construct a degree $\mathbf{a}$ incomparable with a given degree $\mathbf{b} > \mathbf{0}$. To achieve this, $\mathbf{a}$ must avoid the *lower cone* of degrees $\{\,\mathbf{d} : \mathbf{d} \leq \mathbf{b}\,\}$ and the *upper cone* $\{\,\mathbf{d} : \mathbf{d} \geq \mathbf{b}\,\}$. The strategy for accomplishing the latter (which we play on the even stages) will be used in this chapter, and will be refined and often used in constructions of c.e. degrees, such as in the Sacks Splitting Theorem in Chapter 7.

Theorem 6.2.6 (Avoiding Cones). *For every degree* $\mathbf{b} > \mathbf{0}$ *there exists a degree* $\mathbf{a} < \mathbf{b}'$ *such that* $\mathbf{a} \mid \mathbf{b}$.

Proof. Fix $B \in \mathbf{b}$. Construct f, the characteristic function of A, by a B'-computable finite extension construction, $f = \cup_s \sigma_s$, to meet the requirements R_e and S_e of Theorem 6.1.1.

Stage $s = 0$. Set $\sigma_0 = \emptyset$.

Stage $s + 1 = 2e + 1$. (Satisfy $R_e : A \neq \Phi_e^B$.) Let $n = |\sigma_s|$. With a B'-oracle determine whether $\Phi_e^B(n)$ converges, i.e., whether $\langle n, e \rangle \in K_0^B \equiv B'$. If so, define $\sigma_{s+1}(n) = 1 \dot{-} \Phi_e^B(n)$. If not, define $\sigma_{s+1}(n) = 0$.

Stage $s+1 = 2e+2$. (Satisfy $S_e : B \neq \Phi_e^A$.) Given σ_s, first $\emptyset'$-computably test whether the following equation holds:

(6.10) $(\exists \sigma)(\exists \tau)(\exists x)(\exists y)(\exists z)(\exists t)$

$$[\, \sigma_s \prec \sigma, \tau \ \ \& \ \ \Phi_{e,t}^\sigma(x)\!\downarrow \ = y \ \ \neq \ \ z \ = \ \Phi_{e,t}^\tau(x)\!\downarrow \,].$$

If so, one of the values y or z must differ from $B(x)$. Let σ_{s+1} be the first $\sigma \succ \sigma_s$ such that $\Phi_e^\sigma(x)\!\downarrow \ \neq B(x)$ for some x. (This is B'-computable because $B \oplus \emptyset' \leq_T B'$.) If (6.10) fails, we let $\sigma_{s+1} = \sigma_s \hat{\ } 0$. In this case, we claim that for any $f \succ \sigma_s$ if $\Phi_e^f = g$ is total, then g is computable (and hence $\Phi_e^f \neq B$ because $\emptyset <_T B$). To compute $g(x)$, enumerate G_e of the Oracle Graph Theorem 3.3.8 (ii) until the first $\sigma \succ \sigma_s$ is found such that $\Phi_e^\sigma(x)$ converges. Now $g(x) = \Phi_e^f(x) = \Phi_e^\sigma(x)$, because otherwise for some τ, $\sigma_s \preceq \tau \prec f$, $\Phi_e^\tau(x)\!\downarrow \ \neq \Phi_e^\sigma(x)$, thereby satisfying (6.10).   □

6.2.1   Exercises

Exercise 6.2.7. Construct an infinite sequence of degrees $\mathbf{a}_n \leq \mathbf{0}'$, $n \in \omega$, which pairwise form minimal pairs. *Hint.* Build noncomputable sets $\{A_n\}_{n\in\omega}$ meeting for all i, j, and all $m \neq n$,

$$N_{\langle m,n,i,j \rangle} : \quad \Phi_i^{A_m} = \Phi_j^{A_n} = f \ \text{total} \quad . \implies . \quad f \text{ is computable.}$$

Exercise 6.2.8. (i) Fix a degree $\mathbf{c} > \mathbf{0}$. Build a degree $\mathbf{b}$ which forms a minimal pair with $\mathbf{c}$.

(ii)   Given nonzero degrees $\{\mathbf{c}_n\}_{n\in\omega}$, find a sequence of degrees $\{\mathbf{a}_i\}_{i\in\omega}$ each of which is incomparable with $\mathbf{c}_n$ for every n and which pairwise form minimal pairs.

6.3 * Generic Sets

In the two preceding sections we constructed a sequence of finite functions $\sigma_s \preceq \sigma_{s+1}$ so that σ_{s+1} (and indeed all $\tau \succeq \sigma_{s+1}$) met a particular requirement. The generic construction in this section encompasses all the previous examples. Recall that in the Notation section we gave an effective index y to each string $\sigma_y \in 2^{<\omega}$ and we identify the string σ_y with its index y. Likewise, we identify a c.e. set of strings $V_e \subseteq 2^{<\omega}$ with the corresponding c.e. set of integers and use the same notation, V_e. Definition 2.6.1 defined u.c.e. and s.c.e arrays of c.e. sets. Generic sets were studied by [Jockusch 1980].

6.3.1   1-Generic Sets

Definition 6.3.1. Let $\mathbb{V} = \{V_e\}_{e\in\omega}$ be a u.c.e. sequence of c.e. sets $V_e \subseteq 2^{<\omega}$ as in Definition 2.6.1, with strings identified with their indices.

(i) We say $f \in 2^\omega$ *forces* V_e if it satisfies the *forcing* requirement,

$$(6.11) \qquad F_e: \quad (\exists \sigma \prec f)[\,\sigma \in V_e \quad \vee \quad (\forall \rho \succ \sigma)[\,\rho \notin V_e\,]\,].$$

If σ satisfies the matrix of F_e we say that σ *forces* F_e (written $\sigma \Vdash F_e$) and any $f \succ \sigma$ also *forces* F_e (written $f \Vdash F_e$).

(ii) We say f is *generic with respect to* $\mathbb{V} = \{V_e\}_{e\in\omega}$ (written $\mathbb{V}$-*generic*) if f forces V_e for every $e \in \omega$.

(iii) We say f is *1-generic* if it is generic with respect to $\{W_e\}_{e\in\omega}$. (The term "1-generic" refers to the fact that f is deciding Σ_1 statements.)

If the f satisfies the first clause $\sigma \in V_e$ of the matrix in (6.11), then we say f is *e-white* and otherwise f is *e-black*. For every e the 1-generic function f must be either e-black or e-white.

The point about a 1-generic set is that it is amorphous and difficult to describe. For example, it cannot be computable or even c.e. However, we can construct a 1-generic Δ_2 set.

6.3.2   Forcing the Jump

Occasionally, we build f as the characteristic function of a set A and we wish to control the jump A'. (At a finite stage we decide whether $e \in A'$.) We can accomplish this by meeting for all e the following requirement called *forcing the jump* $\Phi_e^A(e)$:

$$(6.12) \qquad J_e: \quad (\exists \sigma \prec A)[\,\Phi_e^\sigma(e)\downarrow \quad \vee \quad (\forall \tau \succeq \sigma)[\,\Phi_e^\tau(e)\uparrow\,]\,].$$

Theorem 6.3.2 (Jockusch-Posner). *A set A is 1-generic iff A forces the jump, i.e., satisfies every jump requirement $\{J_e\}_{e \in \omega}$ of (6.12).*

Proof. ($\Longrightarrow$). Define $W_{h(e)} = \{\sigma : \Phi_e^\sigma(e) \downarrow\}$. Now A forces $W_{h(e)}$. Therefore, A forces the jump $\Phi_e^A(e)$, i.e., satisfies the requirement J_e.

($\Longleftarrow$). Define a computable function $f(e)$ by

$$\Phi_{f(e)}^\sigma(z) = \begin{cases} 1 & \text{if } (\exists s \leq |\sigma|)\,(\exists \tau \preceq \sigma)\,[\,\tau \in W_{e,s}\,], \\ \text{undefined} & \text{otherwise.} \end{cases}$$

If A meets requirement $J_{f(e)}$ of (6.12), then we can clearly see that A forces $V_e = \{\sigma : \Phi_{f(e)}^\sigma(f(e))\downarrow\}$ and A forces W_e. $\qquad\square$

6.3.3 Doing Many Constructions at Once

In the preceding sections we constructed sets with several different properties: incomparable with another, half of a minimal pair, and avoiding a cone. If we now construct a 1-generic set A, then A automatically has all these properties because each property corresponds to a dense set and a 1-generic set meets every dense set of strings. Dense sets, comeager sets, and Banach-Mazur games are explained in Chapter 14. The Banach-Mazur games described there are very similar to the finite extension strategies presented in this chapter.

In Theorem 14.2.1 we shall consider finite extension arguments in the general setting of the Finite Extension Paradigm which subsumes them. This does not cover the coding argument for the Friedberg Completeness Criterion in Theorem 6.4.1 below. However, we extend our paradigm analysis to the Finite Extension Coding Paradigm in Theorem 14.2.2 which covers these examples.

6.3.4 Exercises

Exercise 6.3.3. Construct a 1-generic set $A \leq_T \emptyset'$. *Hint.* Use a $\emptyset'$-oracle and finite extension construction as in the Kleene-Post Theorem 6.1.1 to meet all the jump requirements J_e in (6.12).

Exercise 6.3.4. (Jockusch-Posner) Prove that if a set A is 1-generic, then $A \oplus \emptyset' \equiv_T A'$. Prove that there is a nonzero low degree.

Exercise 6.3.5. (Jockusch-Posner) Assume A is 1-generic.

(i) Prove that A is immune. *Hint.* Let Z be a c.e. subset of A. Define $V_e = \{\sigma : (\exists x \in Z)\,[\sigma(x) = 0]\}$ and use F_e of (6.11) to prove that Z is finite.

(ii) Prove that A is hyperimmune.

(iii) Prove that there is no noncomputable c.e. set $Z \leq_T A$. *Hint.* Assume $Z = \Phi_i^A$ and define

$$V_e = \{\, \sigma : \, (\exists x)\,[\, \Phi_i^\sigma(x) = 0 \,\&\, x \in Z\,]\,\},$$

and apply requirement F_e of (6.11) to show that $\overline{Z}$ is c.e.

(iv) Prove that A_0 and A_1 are Turing incomparable where $A_0(x) = A(2x)$ and $A_1(x) = A(2x+1)$. *Hint.* To see that $A_0 \neq \Phi_e^{A_1}$ consider the c.e. set of strings

$$V_e = \{\, \sigma : (\exists x)\,[\, \Phi_e^{\sigma_1}(x)\!\downarrow \, \neq \, \sigma_0(x)\,]\,\}$$

where $\sigma_0(x) = \sigma(2x)$ and $\sigma_1(x) = \sigma(2x+1)$.

(v) Prove that there are sets $B_i \leq_T A$, $i \in \omega$, such that for every i, we have $B_i \not\leq_T \oplus \{\, B_j : j \neq i\,\}$.

Exercise 6.3.6. (Shoenfield) Show there is a set $A \leq_T \emptyset'$ which does not have c.e. degree.

Exercise 6.3.7. Given B such that $\emptyset <_T B \leq_T \emptyset'$ find a low set A such that $A \not\geq_T B$. *Hint.* Use a $\emptyset'$-construction to build $A = \cup_s \sigma_s$. For each e designate some stage s at which you: (1) force $A \neq \varphi_e$; (2) make A satisfy the lowness requirement for Φ_e; and (3) look for e-splittings, ρ, τ extending σ_s and some x such that $\Phi_e^\rho(x)\!\downarrow \, \neq \Phi_e^\tau(x)\!\downarrow$. If you do not find them, then either Φ_e^A is not total or is computable.

Exercise 6.3.8. (i) Let $\{A_n\}_{n \in \omega}$ be a sequence of sets uniformly computable in $\emptyset'$, i.e., there is a $\emptyset'$-computable function g such that for all x and n, $g(n,x) = A_n(x)$. Prove that there is a set $B \leq_T \emptyset'$ such that $(\forall n)\,[\, B \neq_T A_n\,]$. *Hint.* Ensure that B is noncomputable and for each e and n, if $\Phi_e^B = A_n$, then A_n is computable.

(ii) Give another proof of Exercise 6.3.6.

(iii) Show there is a degree $\mathbf{d} \leq \mathbf{0}'$ which is not n-c.e. and not even ω-c.e.

6.4 * Inverting the Jump

Note that for any degree $\mathbf{a}$, $\mathbf{0} \leq \mathbf{a}$ and hence $\mathbf{0}' \leq \mathbf{a}'$, i.e., any jump is above $\mathbf{0}'$. Hence, the jump, viewed as a map on degrees, has range contained in $\{\mathbf{b} : \mathbf{b} \geq \mathbf{0}'\}$. The next theorem asserts that this map is *onto* the set $\{\mathbf{b} : \mathbf{b} \geq \mathbf{0}'\}$. A degree $\mathbf{a}$ is called *complete* if $\mathbf{a} \geq \mathbf{0}'$. Hence, the result also gives a criterion for $\mathbf{a}$ being complete.

Theorem 6.4.1 (Friedberg Completeness Criterion). *For every degree* $\mathbf{b} \geq \mathbf{0}'$ *there is a degree* $\mathbf{a}$ *such that* $\mathbf{a}' = \mathbf{a} \cup \mathbf{0}' = \mathbf{b}$.

Proof. Fix $B \in \mathbf{b}$. We shall construct f, the characteristic function of A, by finite initial segments $\{\sigma_s\}_{s\in\omega}$ using a B-computable finite extension construction.

Stage $s = 0$. Set $\sigma_0 = \emptyset$.

Stage $s + 1 = 2e + 1$. (We decide whether $e \in A'$.) We meet the *forcing the jump* requirement J_e of (6.12). If A meets J_e we say that A *forces the jump* on argument e. Given σ_s, use a $\emptyset'$-oracle to test whether

$$(6.13) \qquad (\exists \sigma)\,(\exists t)\,[\ \sigma_s \prec \sigma \quad \& \quad \Phi^\sigma_{e,t}(e)\!\downarrow\ ].$$

(Note that the matrix is computable. Therefore, (6.13) is a Σ_1 condition and is computable in $\emptyset'$.) If (6.13) is satisfied, let σ_{s+1} be the first such σ in the standard enumeration of G_e of the Oracle Graph Theorem 3.3.8. If not, set $\sigma_{s+1} = \sigma_s$.

Stage $s+1 = 2e+2$. (We code $B(e)$ into A.) Let $n = |\sigma_s|$. Define $\sigma_{s+1}(n) = B(e)$. (This completes the construction.)

Now $f = \cup_s \sigma_s$ is total since $|\sigma_{2e}| \geq e$. Let $A = \{x : f(x) = 1\}$, and $\mathbf{a} = \deg(A)$. The construction is B-computable because at odd stages we use a $\emptyset'$-oracle, at even stages we use a B-oracle, and $\emptyset' \leq_T B$. Since $A \oplus \emptyset' \leq_T A'$ for any A, to prove $A' \equiv_T B \equiv_T A \oplus \emptyset'$ it suffices to prove the following two lemmas.

Lemma 6.4.2. $A' \leq_T B$.

Lemma 6.4.3. $B \leq_T A \oplus \emptyset'$.

Proof of Lemma 6.4.2. Since the construction is B-computable, the sequence $\{\sigma_s\}_{s\in\omega}$ is B-computable. To decide whether $e \in A'$, B-computably determine using $\emptyset' \leq_T B$ whether (6.13) holds for σ_{2e}. If so, $e \in A'$, and otherwise $e \notin A$ because no $\sigma \succeq \sigma_s$ has $\Phi^\sigma_e(e)$ defined. $\qquad \square$

Proof of Lemma 6.4.3. We show $\{\sigma_s\}_{s\in\omega}$ is an $(A \oplus \emptyset')$-computable sequence. This suffices because $B(e)$ is the last value of σ_{2e+2}. The proof is by induction on s. Given $\{\sigma_s : s \leq 2e\}$, use a $\emptyset'$-oracle to compute σ_{2e+1}. If $n = |\sigma_{2e+1}|$ then $\sigma_{2e+2} = \sigma_{2e+1}\!\widehat{\ }A(n)$, so σ_{2e+2} is computed from σ_{2e+1} using an A-oracle. $\qquad \square$

This completes the proof of Theorem 6.4.1. $\qquad \square$

Theorem 6.4.4 (Relativized Friedberg Completeness Criterion).
   For every degree $\mathbf{c}$,

$$F_1(\mathbf{c}): \quad (\forall \mathbf{b})\,[\,\mathbf{b} \geq \mathbf{c}' \implies (\exists \mathbf{a})\,[\,\mathbf{a} \geq \mathbf{c} \ \& \ \mathbf{a}' = \mathbf{a} \cup \mathbf{c}' = \mathbf{b}\,]\,].$$

Proof. Do the proof of Theorem 6.4.1 with $\mathbf{c}$ and $\mathbf{c}'$ in place of $\mathbf{0}$ and $\mathbf{0}'$. $\qquad \square$

Corollary 6.4.5. *For every $n \geq 1$, and every degree $\mathbf{c}$,*

$$F_n(\mathbf{c}): \quad (\forall \mathbf{b})\,[\,\mathbf{b} \geq \mathbf{c}^{(n)} \implies (\exists \mathbf{a})\,[\,\mathbf{a} \geq \mathbf{c} \ \& \ \mathbf{a}^{(n)} = \mathbf{a} \cup \mathbf{c}^{(n)} = \mathbf{b}\,]\,].$$

Proof. To prove $(\forall \mathbf{c})F_n(\mathbf{c})$ holds for all $n \geq 1$, use induction on n and the fact that $F_{n+1}(\mathbf{c})$ follows from $F_n(\mathbf{c})$ and $F_1(\mathbf{c}^{(n)})$. $\qquad\square$

Although Theorem 6.4.1 demonstrates a pleasant property of the jump operator, it also demonstrates an unpleasant property, namely that the jump as a map on degrees is not 1:1. To see this, apply Theorem 6.4.1 with $\mathbf{b} = \mathbf{0}''$ to obtain $\mathbf{a}$ such that $\mathbf{a}' = \mathbf{a} \cup \mathbf{0}' = \mathbf{0}''$. Clearly, $\mathbf{a} \mid \mathbf{0}'$, yet they have the same jump. It is also possible to have $\mathbf{a} < \mathbf{b}$ and $\mathbf{a}' = \mathbf{b}'$. (It is easy to see that the jump is 1:1 on *sets*.)

6.4.1 Exercises

Exercise 6.4.6. [Jockusch-Shore, 1983] Prove that for any $i \in \omega$ and any B such that $\emptyset' \leq_T B$ there exists A such that

$$A \oplus W_i^A \equiv_T A \oplus \emptyset' \equiv_T B.$$

Note that Theorem 6.4.1 is a special case of this setting where i is defined by $W_i^X = X'$. *Hint.* Do the proof of Theorem 6.4.1 but in (6.12) replace $\Phi_e^\rho(e)\!\downarrow$ by $e \in W_i^\rho$ for $\rho = \sigma$ or τ. (Note that this construction is uniform in B and in any j such that $\Phi_j^B = \emptyset'$.)

Exercise 6.4.7. Prove that

$$(\forall \mathbf{b} \geq \mathbf{0}')\,(\exists \mathbf{a}_0)\,(\exists \mathbf{a}_1)\,[\,\mathbf{a}_0 \mid \mathbf{a}_1 \ \& \ \mathbf{a}_0' = \mathbf{a}_0 \cup \mathbf{0}' = \mathbf{b} = \mathbf{a}_1 \cup \mathbf{0}' = \mathbf{a}_1'\,].$$

Hint. Combine the constructions of Theorems 6.1.1 and 6.4.1 to handle four types of requirements, the two types from Theorem 6.1.1 and the two from Theorem 6.4.1. As in that theorem at stage $2e + 2$, code $B(e)$ into *both* of A_0 and A_1.

6.5 Upper and Lower Bounds for Degrees

Every nonempty finite set of degrees has a least upper bound (lub). In this section we show that this is false for greatest lower bounds (glb's). Hence, the degrees do not form a lattice, but merely an upper semi-lattice.

Definition 6.5.1. (i) For any set A define the *ω-jump* of A,

$$A^{(\omega)} = \{\,\langle x, n \rangle : x \in A^{(n)}\,\}.$$

In Exercise 6.5.9 we show that this is well-defined on degrees. Therefore, we can define the induced ω-jump on *degrees* $\mathbf{a}^{(\omega)} = \deg(A^{(\omega)})$ for $A \in \mathbf{a}$.

(ii)   An infinite sequence of degrees $\{\mathbf{a}_n\}_{n\in\omega}$ is *ascending* if $\mathbf{a}_n \leq \mathbf{a}_{n+1}$ for all n and *strictly ascending* if $\mathbf{a}_n < \mathbf{a}_{n+1}$ for all n. For example, $\mathbf{0}, \mathbf{0}^{(1)}, \mathbf{0}^{(2)}, \ldots$ is strictly ascending, and $\mathbf{0}^{(\omega)}$ is a natural upper bound for the sequence, although by the next theorem the sequence has no lub.

Definition 6.5.2. If $\{\mathbf{a}_n\}_{n\in\omega}$ is an ascending sequence of degrees then upper bounds $\mathbf{b}$ and $\mathbf{c}$ are an *exact pair* for the sequence if for every degree $\mathbf{d}$,

$$[\,\mathbf{d} \leq \mathbf{b} \quad \& \quad \mathbf{d} \leq \mathbf{c}\,] \quad \Longrightarrow \quad (\exists n)\,[\,\mathbf{d} \leq \mathbf{a}_n\,].$$

Theorem 6.5.3 (Kleene-Post-Spector).   *For every ascending sequence of degrees, $\{\mathbf{a}_n\}_{n\in\omega}$, namely $\mathbf{a}_n \leq \mathbf{a}_{n+1}$, there exist upper bounds $\mathbf{b}$ and $\mathbf{c}$ which form an exact pair for the sequence.*

Corollary 6.5.4.   *No infinite strictly ascending sequence of degrees, i.e., $\mathbf{a}_n < \mathbf{a}_{n+1}$, has a least upper bound.* □

Corollary 6.5.5.   *There are degrees $\mathbf{b}$ and $\mathbf{c}$ with no greatest lower bound.* □

Before proving Theorem 6.5.3 we make some definitions and introduce some new notation.

Definition 6.5.6.   For any set $A \subseteq \omega$ define the *y-section* of A,

(6.14) $A^{[y]} = \{\langle x, z \rangle : \langle x, z \rangle \in A \ \& \ z = y\}$ and

(6.15) $A^{[<y]} = \bigcup\{\, A^{[z]} : z < y \,\}.$

(Using the pairing function we can identify A with a subset of $\omega \times \omega$ and view $A^{[y]}$ as the yth *row* of A. We use the square bracket notation $A^{[y]}$ to distinguish from the yth jump $A^{(y)}$.)

Definition 6.5.7.   (i) Given sets A and B, for every y the *thickness requirement* for y states

(6.16) $T_y: \quad A^{[y]} =^* B^{[y]}$

where $X =^* Y$ denotes that $(X - Y) \cup (Y - X)$ is finite.

(ii) A subset $A \subseteq B$ is a *thick subset* of B, written $A \subseteq_{\text{thick}} B$, if T_y is satisfied for all y.

Thick subsets will be very useful here and in later constructions of c.e. sets and degrees, such as the thickness lemma and infinite injury constructions.

Definition 6.5.8.   Partial functions θ, ψ are *compatible*, which we write as $compat(\theta, \psi)$, if they have a common extension, i.e., if there is no x for which $\theta(x)$ and $\psi(x)$ are defined and unequal. Otherwise, they are *incompatible*.

Proof of Theorem 6.5.3. Choose $A_y \in \mathbf{a}_y$ for each y and then define $A = \{\langle x, y \rangle : x \in A_y\}$, so that $\langle x, y \rangle \in A^{[y]}$ iff $x \in A_y$. We shall construct characteristic functions f and g of sets B and C which are thick in A (so that $A_y \equiv_{\mathrm{T}} B^{[y]} \leq_{\mathrm{T}} B$, and likewise for C). This ensures that $\mathbf{b} = \deg(B)$ and $\mathbf{c} = \deg(C)$ are upper bounds for $\{\mathbf{a}_n\}_{n \in \omega}$. For all y we must meet the thickness requirements,

$$T_y^B : \quad B^{[y]} =^* A^{[y]},$$

$$T_y^C : \quad C^{[y]} =^* A^{[y]}.$$

We must also meet, for all e and i, the exact pair requirements,

$$R_{\langle e,i \rangle} : \quad \Phi_e^B = \Phi_i^C \text{ total} \quad \Longrightarrow \quad (\exists y)\,[\,\Phi_e^B \leq_{\mathrm{T}} A_y\,]$$

by looking for "*e*-splittings" as we did in proving Theorem 6.2.3.

Let σ_s, τ_s, B_s, and C_s be the portions of f, g, B, and C defined by the end of stage s of the following construction.

Stage $s = 0$. Set $\sigma_0 = \tau_0 = \emptyset$.

Stage $s + 1$. Assume that σ_s and τ_s are defined on $\omega^{[<s]}$ and assume that:

(6.17) $(\forall y < s)\,[\,B_s^{[y]} =^* C_s^{[y]} =^* A^{[y]}\,]$;   and

(6.18) $(\mathrm{dom}(\sigma_s) - \omega^{[<s]}) =^* \emptyset =^* (\mathrm{dom}(\tau_s) - \omega^{[<s]})$.

Step 1. (Satisfy $R_{\langle e,i \rangle}$ for $s = \langle e, i \rangle$.) If

(6.19) $(\exists \sigma)\,(\exists \tau)\,(\exists x)\,(\exists t)\,[\,\mathrm{compat}(\sigma, \sigma_s) \quad \& \quad \mathrm{compat}(\tau, \tau_s)$
$$\& \quad \Phi_{e,t}^\sigma(x){\downarrow} \neq \Phi_{i,t}^\tau(x){\downarrow}\,],$$

then let σ and τ be the first such strings and extend σ_s to $\widehat{f} = \sigma_s \cup \sigma$ and τ_s to $\widehat{g} = \tau_s \cup \tau$. Otherwise, let $\widehat{f} = \sigma_s$, and $\widehat{g} = \tau_s$. Note that $\sigma_s \equiv_{\mathrm{T}} A^{[<s]} \equiv_{\mathrm{T}} \tau_s$ by (6.17) and (6.18). Hence, $\mathrm{compat}(\sigma, \sigma_s)$ is an $A^{[<s]}$-computable relation on σ. (Note that for $s > 0$, Step 1 requires an $A_{s-1}' \equiv_{\mathrm{T}} (A^{[<s]})'$ oracle.)

Step 2. (Satisfy T_s^B and T_s^C.)

Let $\sigma_{s+1} = \widehat{f}$ on $\mathrm{dom}(\widehat{f})$. On all $x \in \omega^{[s]} - \mathrm{dom}(\widehat{f})$ define $\sigma_{s+1}(x) = A(x)$. Let $\tau_{s+1} = \widehat{g}$ on $\mathrm{dom}(\widehat{g})$ and $\tau_{s+1}(x) = A(x)$ for all $x \in \omega^{[s]} - \mathrm{dom}(\widehat{g})$. By (6.18), σ_s (and hence $\widehat{f}$) is already defined on at most finitely many elements of $\omega^{[s]}$, and similarly for τ_s, so $B_{s+1}^{[s]} =^* A^{[s]} =^* C_{s+1}^{[s]}$, and f and g are now defined on $\omega^{[\leq s]}$. This ends the construction.

If (6.19) holds, then $\Phi_e^B \neq \Phi_i^C$. If (6.19) fails and $\Phi_e^B = \Phi_i^C = h$ is total, then for $s = \langle e, i \rangle$ we shall show that $h \leq_{\mathrm{T}} A^{[<s]}$. Notice that $A^{[<s]} \leq_{\mathrm{T}} A_s$

because

$$A^{[<s]} \equiv_T A^{[0]} \oplus \cdots \oplus A^{[s-1]} \equiv_T A_0 \oplus \cdots \oplus A_{s-1} \leq_T A_s.$$

To $A^{[<s]}$-computably determine $h(x)$, find the first string σ in some enumeration of $\{\sigma : \Phi_e^\sigma(x)\downarrow\}$ such that compat(σ, σ_s) and set $h(x) = \Phi_e^\sigma(x)$. Now $h(x) = \Phi_e^f(x)$, or else for some $\sigma' \prec f$, compat(σ', σ_s) holds and $\Phi_e^{\sigma'}(x)\downarrow = y \neq \Phi_e^\sigma(x)$, so (6.19) holds for either σ or σ' and for any $\tau \prec C$ such that $\Phi_i^\tau(x)$ converges.   □

6.5.1   Exercises

Exercise 6.5.9. Let the ω-jump $A^{(\omega)}$ be defined as in Definition 6.5.1. Prove that if $A \equiv_T B$, then $A^{(\omega)} \equiv_T B^{(\omega)}$. *Hint.* To show that $B^{(\omega)} \leq_T A^{(\omega)}$ we must prove that $B^n \leq_T A^{(\omega)}$ uniformly in n. Apply the Jump Theorem 3.4.3 (vi) to show that $B^{(n)} \equiv_T A^{(n)}$ uniformly in n.

Exercise 6.5.10. Show that the proof of Theorem 6.5.3 automatically produces sets B and C computable in $\oplus\{A_y'\}_{y \in \omega}$.

Exercise 6.5.11. Show that in the proof of Theorem 6.5.3 if B is any upper bound for the A_y sets then we can modify Steps 1 and 2 to construct C such that B and C satisfy the same requirements as before.

Exercise 6.5.12. Let $\mathbf{I}$ be a countable ideal contained in the Turing degrees $\mathbf{D}$. Prove that there exist degrees $\mathbf{b}, \mathbf{c}$ such that for all $\mathbf{a} \in \mathbf{D}$,

$$\mathbf{a} \in \mathbf{I} \quad \Longleftrightarrow \quad [\, \mathbf{a} \leq \mathbf{b} \ \& \ \mathbf{a} \leq \mathbf{c}\,].$$

We call $\mathbf{b}$ and $\mathbf{c}$ an *exact pair* for the ideal $\mathbf{I}$ as in Definition 6.5.2, and "ideal" is defined in the Notation section.

Exercise 6.5.13. ◦ (K. Lange). Fix an infinite computable tree $T \subseteq 2^{<\omega}$. Fix a set $\mathcal{A} = \{A_n\}_{n \in \omega} \subseteq [T]$ of computable paths through T (not necessarily closed) but *dense* in T in the sense that

$$(\forall \sigma \in T)(\exists A_n \succ \sigma)[\, A_n \in \mathcal{A}\,].$$

For some degree $\mathbf{d}$, a $\mathbf{d}$-*basis* for $\mathcal{A}$ is a sequence of paths $X = \{B_n\}_{n \in \omega} \subseteq [T]$ and a function $f \leq_T \mathbf{d}$ such that $\varphi_{f(n)} = B_n$, i.e., $\mathbf{d}$ can uniformly compute a Δ_0-index for every path in $\mathcal{A}$, viewed as a row in the $\mathbf{d}$-computable matrix $B = \oplus_n B_n$.

(i) If the set $\mathcal{A}$ of isolated paths of T is dense in T, prove that $\mathcal{A}$ has a $\mathbf{0}'$-basis.

(ii) Prove that if $\mathcal{A}$ has a $\mathbf{0}'$-basis $X = \{A_n\}_{n \in \omega}$, then there is a sequence $\{B_n\}_{n \in \omega}$ such that $B = \oplus_n B_n$ is low and the collection of paths $\{B_n\}_{n \in \omega}$

equals the collection of paths $\{A_n\}_{n\in\omega}$, although the sequences may not be the same.

Hint for (ii). Given a $\mathbf{0}'$-basis $X = \{A_n\}_{n\in\omega}$, use a $\mathbf{0}'$-construction to build another basis $Y = \{B_n\}_{n\in\omega}$ having the same rows $[A_n]$ as X but perhaps in a different order. Simultaneously, force the jump of the matrix $B = \oplus_n B_n$ so that B is low. Search only through strings σ such that $(\forall j \le |\sigma|)[\sigma^{[j]} \in T]$, where $\sigma^{[j]}(x) = \sigma(\langle x, j\rangle)$. Now extend these $\sigma^{[j]}$ on the B side by $\mathbf{0}$-effectively choosing some row on the A side extending $\sigma^{[j]}$ and filling this row in on the B side.

Remark 6.5.14. Note that if $A \le_T \emptyset'$ and $S = A'$ then $S \ge_T \emptyset'$ and S is c.e. in $\emptyset'$. Therefore, the jump map takes the degrees $\mathbf{a} \le \mathbf{0}'$ *into* the degrees c.e. in and above $\mathbf{0}'$. The next theorem proves that this map is *onto*.

Exercise 6.5.15. $\diamond$ (Shoenfield Jump Inversion Theorem). Fix S such that $\emptyset' \le_T S$ and S is c.e. in $\emptyset'$, namely such that S is c.e. in and above (c.e.a.) in $\emptyset'$. Construct $A \le_T \emptyset'$ such that $A' \equiv_T S$. *Hint.* Define a $\emptyset'$-sequence $\{\sigma_s\}_{s\in\omega}$ of $\{0,1\}$-valued partial functions such that $\sigma_s \preceq \sigma_{s+1}$ and $\lim_s \sigma_s = \chi_A$. We ensure that $S \le_T A'$ by arranging that for all y, $\lim_x A(\langle x, y\rangle) = \chi_S(y)$. We ensure $A' \le_T S$ by forcing the jump $\Phi_e^A(e)$. Fix a $\emptyset'$-computable enumeration $\{S_s\}_{s\in\omega}$ of S such that $|S_{s+1} - S_s| = 1$. Let $\sigma_0 = \emptyset$. The following is a $\emptyset'$-construction.

Stage $s + 1$. Assume that if $y \in S_s$ then $\sigma_s(\langle x, y\rangle) = 1$ for almost every x, and otherwise $\sigma_s(\langle x, y\rangle)\downarrow$ for at most finitely many x and $\sigma_s(\langle x, y\rangle)\uparrow$ for all other x.

Step 1. Now σ_{s+1} has a computable domain and is computable on its domain. Hence, we can $\emptyset'$-computably test for each $e \le s$ which has not yet been forced in A' whether

$$(6.20) \qquad (\exists t)\,(\exists \sigma)\,[\text{compat}(\sigma, \sigma_s) \quad \& \quad \Phi_{e,t}^\sigma(e)\downarrow$$
$$\& \quad (\forall y < e)\,(\forall x)\,[\,\langle x, y\rangle \notin \text{dom}(\sigma_s) \quad \Longrightarrow \quad \sigma(\langle x, y\rangle) = 0\,]\,].$$

If so, choose the least e and the least corresponding string σ. Define $\tau_{s+1} = \sigma_s \cup \sigma$ and say that e is *forced into* A'. Otherwise, define $\tau_{s+1} = \sigma_s$.

Step 2. Enumerate the next element $z \in S_{s+1} - S_s$. Define

$$\sigma_{s+1}(\langle x, y\rangle) =$$
$$\begin{cases} \tau_{s+1}(\langle x, y\rangle) & \text{if } \langle x, y\rangle \in \text{dom}(\tau_{s+1}); \\ 1 & \text{if } y = z \text{ and } \langle x, y\rangle \notin \text{dom}(\tau_{s+1}); \\ 0 & \text{if } y \notin S_{s+1}, \langle x, y\rangle \le s, \langle x, y\rangle \notin \text{dom}(\tau_{s+1}). \end{cases}$$

The last clause is to ensure that if $y \notin S$ then $\lim_x A(\langle x, y\rangle) = 0$. To see that $A' \le_T S$, fix e, assume that membership of $i \in A'$ has been decided for all $i < e$, and find s such that $S_s \restriction e = S \restriction e$. Show that if e has not been forced into A' by stage s, then $e \notin A'$, i.e., it has been forced out of A'.

7
The Finite Injury Method

7.1 A Solution to Post's Problem

A positive solution to Post's problem was finally achieved by Friedberg in [Friedberg 1957] and independently by Muchnik in [Muchnik 1956], who built c.e. sets A and B of incomparable (Turing) degree. Friedberg then produced other results on c.e. sets in [Friedberg 1957b], [Friedberg 1957c] and [Friedberg 1958].

7.1.1 The Intuition Behind Finite Injury

To achieve this we must meet requirements similar to those in Chapter 6, such as $R_e : \Phi_e^B \neq A$, in Kleene-Post Theorem line (6.1). However, we use a *computable* construction rather than the *oracle* constructions which were used in Chapter 6 to build $A = \cup_s \sigma_s$ for $\sigma_s \prec \sigma_{s+1}$. If we build A by finite extensions in a computable construction, then A must be computable and therefore not interesting. Hence, we now build $A = \cup_s A_s$ as a computably *enumerable* set: a computable union of finite sets $\{A_s\}_{s \in \omega}$.

This presents a new difficulty. In Chapter 6 we normally met a requirement like R_e by finding a string $\rho \in V_e$ (which was a certain c.e. set of strings associated with R_e), and then by extending σ_s to some $\sigma_{s+1} \succ \rho$, thereby satisfying R_e forever. Here we also meet R_e (at least temporarily) by enumerating elements into A so that $A_{s+1} \succ \rho$. Unlike the oracle construction in Kleene-Post Theorem 6.1.1, this does *not* guarantee that $\rho \prec A$ because new elements $x < |\rho|$ may later enter A.

147

Therefore, to ensure $A_t \succ \rho$ for all $t > s$, we must prevent 0's in the characteristic function of A_{s+1} from later changing to 1's. Hence, we issue a *restraint function* $r(e, s+1)$ which with priority R_e prevents any *lower priority* requirement R_i, $i > e$, from enumerating any $x \leq r(e, s+1)$ into A. If this restraint is successful, then $\rho \prec A$ and R_e is satisfied at the end.

However, the requirements $\{R_i\}_{i<e}$ have *higher priority* than R_e, and can override the restraint $r(e, s+1)$ in an effort to satisfy their own conditions. If this happens at stage $s+1$ we say R_e is *injured* at stage $s+1$ and R_e begins anew to satisfy its condition at later stages. This injury may arise because some higher priority requirement R_0 is slow to act. Meanwhile, R_e has acted and has built a wall of restraint to defend its action. Only then does R_0 act, injuring any restraint by R_e. We must allow R_0 to act and injure R_e or else R_0 may never be satisfied.

The solution is the *Golden Rule:* every requirement R_e treats the weaker requirements $\{R_j\}_{j>e}$ as it *itself* wants to be treated by the higher priority requirements $\{R_i\}_{i<e}$. For this chapter on finite injury this means that R_e acts at most finitely often to injure any lower priority requirement R_j, $j > e$. By the Golden Rule the higher priority requirements $\{R_i\}_{i<e}$ injure R_e at most finitely often and R_e is satisfied.

7.1.2   The Injury Set for Requirement N_e

Definition 7.1.1. Let $\{A_s\}_{s\in\omega}$ be a computable enumeration of a c.e. set A and $r(e, s)$ be a computable function representing the restraint for a future requirement such as $R_e : \Phi_e^A \neq B$, as in (7.6) or $N_e : \Phi_e^A \neq C$, as in (7.8). We shall see that the requirement wishes to restrain elements $x \leq r(e, s)$ from entering A at stage $s+1$. We say that element x *injures* the requirement associated with $r(e, s)$ at stage $s+1$ if $x \in A_{s+1} - A_s$ and $x \leq r(e, s)$. Define the *injury set,*

$$(7.1) \qquad I_e = \{\, x \ : \ (\exists s)\,[\, x \in A_{s+1} - A_s \quad \& \quad x \leq r(e, s)\,]\,\}.$$

Remark 7.1.2. In the three sections §7.2, §7.3, and §7.5 we give three different results, each of which yields a solution to Post's problem, and each of which uses a different kind of finite injury priority argument. Each of the three methods is fundamental and will be essential for results in later chapters. The three methods are in ascending order with respect to technical complexity because if $|I_e|$ is the number of times that requirement R_e is injured then $|I_e| \leq e$ in §7.2, $|I_e| < 2^e$ in §7.3, and $|I_e|$ is finite, but not even computably bounded, in §7.5. In Chapter 8 we shall have *infinite* injury with the injury set I_e infinite.

7.2 * Low Simple Sets

Probably the easiest solution to Post's problem is the construction of a low simple set A. Simplicity guarantees that A is c.e. and noncomputable while the lowness of A, i.e., $A' \equiv_T \emptyset'$, guarantees incompleteness, i.e., $A <_T \emptyset'$. Furthermore, in this construction the requirements may be separated into the *positive* requirements P_e, which attempt to put elements *into* A, and the *negative* requirements N_e, which attempt to keep elements *out of* A. Low sets and degrees were studied in §4.4.2. They have several pleasant structural properties which will be explored later.

7.2.1   The Requirements for a Low Simple Set A

Theorem 7.2.1. *There is a simple set A which is low ($A' \equiv_T \emptyset'$).*

Corollary 7.2.2. *There is a noncomputable c.e. set A which is incomplete, i.e., $\emptyset <_T A <_T \emptyset'$.*

Proof of Theorem 7.2.1. Construct A c.e. to meet for all e the requirements:

(simplicity) $\qquad P_e: \quad |W_e| = \infty \qquad \Longrightarrow \qquad W_e \cap A \neq \emptyset;$

(lowness) $\qquad N_e: \quad (\exists^\infty s)\,[\,\Phi^{A_s}_{e,s}(e)\!\downarrow\,] \qquad \Longrightarrow \qquad \Phi^A_e(e)\!\downarrow;$

($\overline{A}$ infinite) $\qquad C_e: \quad a_e = \lim_s a^s_e$ exists,

where A_s consists of the elements enumerated in A by the end of stage s, $A = \cup_s A_s$, $\overline{A}_s = \{a^s_0 < a^s_1 < \cdots\}$, and $\overline{A} = \{a_0 < a_1 < \cdots\}$, as in the canonical simple set construction in Theorem 5.2.5, which this Theorem 7.2.1 generalizes. (The expression $(\exists^\infty x)R(x)$ was defined in Definition 3.5.1.)

The priority ranking of the requirements is $\{N_e \prec C_e \prec P_e\}_{e \in \omega}$, meaning that N_e has higher priority than C_e, which has higher priority than P_e, which has higher priority than N_{e+1}, and so on. Requirements N_e and C_e are both *negative*, tending to restrain elements *out of* A, and thus there is no conflict between them, but P_e is a *positive* requirement, tending to enumerate elements *into* A, and must respect any restraint imposed by N_i or C_i for all $i \leq e$. Note that the requirements $\{N_e\}_{e \in \omega}$ guarantee $A' \leq_T \emptyset'$ as follows.

7.2.2   A Computable $\widehat{g}(e,s)$ with $g(e) = \lim_s \widehat{g}(e,s) = A'(e)$

Define the computable function $\widehat{g}(e,s)$ by

$$\widehat{g}(e,s) := \begin{cases} 1 & \text{if} \quad \Phi_{e,s}^{A_s}(e)\downarrow; \\ 0 & \text{otherwise.} \end{cases}$$

If requirement N_e is satisfied for all e, then $g(e) = \lim_s \widehat{g}(e,s)$ exists for all e. But $g \leq_T \emptyset'$ by the Limit Lemma 3.6.2, and g is the characteristic function of A'. Therefore, $A' \leq_T g \leq_T \emptyset'$.

Recall from Definition 3.2.2 that the *use function* $\varphi_{e,s}^{A_s}(x)$ is the maximum element used in the computation $\Phi_{e,s}^{A_s}(x)$ if the latter converges, and is 0 otherwise. To aid in meeting N_e, given A_s, define for all e the *restraint function*,

(7.2) $r(e,s) := \varphi_{e,s}^{A_s}(e).$

Now $r(e,s)$ is a computable function because $\{A_s\}_{s\in\omega}$ is a computable sequence. To meet N_e we attempt to restrain with priority N_e any elements $x \leq r(e,s)$ from entering A_{s+1}. The point is that if $\Phi_{e,s}^{A_s}(e)\downarrow$, with $r = \varphi_{e,s}^{A_s}(e)$, and N_e succeeds in preventing any $x \leq r$ from later entering A, then $A \upharpoonright r = A_s \upharpoonright r$ and hence $\Phi_e^A(e)\downarrow$. Thus, such elements can only enter A for the sake of some P_i of higher priority (namely $i < e$). The requirement C_e restrains all elements $\{a_j^s\}_{j\leq e}$ from entering A_{s+1}, unless they are enumerated by a higher priority requirement P_i, $i < e$, exactly as in the Canonical Simple Set Theorem 5.2.5.

7.2.3   The Construction of a Low Simple Set A

Stage $s = 0$. Let $A_0 = \emptyset$.

Stage $s + 1$. Given A_s, let $\overline{A}_s = \{a_0^s < a_1^s < \ldots\}$ and compute $r(e,s)$ for all e. Choose the least $i \leq s$ such that

(7.3) $W_{i,s} \cap A_s = \emptyset;$ and

(7.4) $(\exists x)[\, x \in W_{i,s} \quad \& \quad x > a_i^s \quad \& \quad (\forall e \leq i)[\, r(e,s) < x\,]\,].$

(The first clause of (7.4) says x is useful for P_i because $x \in W_{i,s}$. The second clause says x is not restrained by any requirement C_e, $e \leq i$, because $x > a_e^s$. The third clause says that x is not restrained by any requirement N_e, $e \leq i$.) If i exists, choose the least x satisfying (7.4). Enumerate x in A_{s+1}, and say that requirement P_i *receives attention*. Hence, $W_{i,s} \cap A_{s+1} \neq \emptyset$, P_i is satisfied, and (7.3) fails for stages $> s+1$. Therefore, P_i never again receives attention. If i does not exist, do nothing, and define $A_{s+1} = A_s$. Let $A = \cup_s A_s$. This ends the construction.

7.2.4   The Verification of a Low Simple Set A

Lemma 7.2.3. *Define the injury set I_e as in (7.1). For all e,*

(i)   The injury set I_e is finite. (That is, N_e is injured finitely often.)

(ii) $a_e = \lim_s a_e^s$ exists. (C_e is injured finitely often.) Hence, $|\overline{A}| = \infty$.

Proof. Each positive requirement P_i contributes at most one element to A by (7.3). But by the second and third clauses of (7.4), C_e and N_e can be injured by P_i only if $i < e$, and at most once for each such i. Hence, $|I_e| \le e$, and likewise for requirement C_e.   □

Lemma 7.2.4. $(\forall e)\,[\,requirement\ N_e\ is\ met\ \&\ \ r(e) := \lim_s r(e,s)\ exists\,]$.

Proof. Fix e. By Lemma 7.2.3, choose stage s_e such that N_e is not injured at any stage $s > s_e$. If $\Phi_{e,s}^{A_s}(e)$ converges for $s > s_e$, then by induction on $t \ge s$, $\ r(e,t) = r(e,s)$ and $\Phi_{e,t}^{A_t}(e) = \Phi_{e,s}^{A_s}(e)$ for all $t \ge s$. Hence, $A_s \upharpoonright r = A \upharpoonright r$ for $r = r(e,s)$. Therefore, $\Phi_e^A(e)$ is defined by the Use Principle Theorem 3.3.9.   □

Lemma 7.2.5. $(\forall i)\,[\,requirement\ P_i\ is\ met\,]$.

Proof. Fix i such that W_i is infinite. By Lemma 7.2.4, choose s such that

$$(\forall t \ge s)(\forall e \le i)\,[\,r(e,t) = r(e)\ \ \&\ \ a_e = a_e^t\,].$$

Choose $s' \ge s$ such that no P_j, $j < i$, receives attention after stage s'. Choose $t > s'$ such that

$$(\exists x)\,[\,x \in W_{i,t}\ \ \&\ \ x > a_i\ \ \&\ \ (\forall e \le i)[\,r(e) < x\,]\,].$$

Now either $W_{i,t} \cap A_t \ne \emptyset$ or else P_i receives attention at stage $t+1$. In either case, $W_{i,t} \cap A_{t+1} \ne \emptyset$. Therefore, P_i is met by the end of stage $t+1$.   □

7.2.5   The Restraint Functions $r(e,s)$ as Walls

In this and later constructions one should think of $r(e,s)$ as a *wall* imposed by N_e, extending from 0 to $r(e,s)$. For fixed e the wall $r(e,s)$ is not monotonically increasing in s because after a wall $r(e,s)$ is erected, penetration of that wall by some x injuring N_e may cause $\Phi_{e,s+1}^{A_{s+1}}(e)\uparrow$ so that the wall drops to 0. If P_i wishes to contribute some element x to A, then x must lie beyond all walls $r(e,s)$, $e \le i$. The crucial feature of all finite injury constructions is that each wall is penetrated (injured) finitely often and therefore comes to a limit. But then each positive requirement is satisfied because it merely chooses a witness beyond all walls of higher priority.

7.2.6   Exercises

Exercise 7.2.6. In the proof of Theorem 7.2.1, replace the requirements P_e by the requirements of Exercise 5.3.9 to give a direct construction of a low hypersimple set.

Exercise 7.2.7. Combine the method of Theorem 7.2.1 with the permitting method of Theorem 5.2.7 to prove that for any noncomputable c.e. set A there is a low simple (indeed, low hypersimple) set $B \leq_{ibT} A$.

7.3   * The Friedberg-Muchnik Theorem

The Friedberg-Muchnik Theorem states that there exist c.e. sets A and B of incomparable degree. Its proof is the *canonical* finite injury theorem, not only because it was the first example of a finite injury proof, but also because the injury pattern is the most typical. The injury pattern is a bit more complex than that of the low simple set of Theorem 7.2.1, where the injury set I_e satisfies $|I_e| \leq e$, but not as complex as the unbounded injury pattern in the Sacks Splitting Theorem 7.5.1, where we have merely $|I_e| < \infty$. Here we have $|I_e| < 2^e$, which is typical of these arguments.

Theorem 7.3.1 (Friedberg, 1957; Muchnik, 1956). *There exist c.e. sets A and B of incomparable Turing degree.*

7.3.1   Renumbering the Requirements

Convention 7.3.2 (Indexing Convention). *From now on we assume that for every Turing reduction Φ_e there are infinitely many even indices i and odd indices j such that $\Phi_i = \Phi_e = \Phi_j$. We also assume that φ_e has been defined by $\varphi_e := \Phi_e^\emptyset$, so the same holds for the p.c. functions φ_e.*

Proof. Hence, to satisfy all the incomparability requirements it suffices to satisfy merely all even requirements for Φ_e^B and all odd requirements for Φ_e^A. (This enables us to have the subscript of R_e match that of Φ_e, which makes the proof more perspicuous than having $R_{2e} : A \neq \Phi_e^B$.) We construct c.e. sets A and B to meet the same requirements as in (6.1) of the Kleene-Post theorem,

$$(7.5) \qquad\qquad R_e : \quad A \neq \Phi_e^B \qquad\quad \text{for } e \text{ even,}$$

$$(7.6) \qquad\qquad R_e : \quad B \neq \Phi_e^A \qquad\quad \text{for } e \text{ odd.}$$

We say that requirement R_i has *higher priority* than R_j if $i < j$. We first present the *basic module* or *atomic strategy* for meeting a single requirement R_e and then explain how to combine these strategies.

7.3.2   The Basic Module to Meet R_e for e Even

Select an integer x not yet in A. Wait for s such that $\Phi_{e,s}^{B_s \restriction u}(x) \downarrow = 0$, which must happen if $\Phi_e^B = A$. (Now $u < s$ by Definition 3.2.2 (iii).)

Action. At stage $s + 1$ enumerate x in A, and define a restraint function $r(e, s + 1) = s + 1$ which prevents lower priority requirements R_j, $j >$ e, from enumerating any $z \leq r(e, s + 1)$ into B. If the higher priority requirements R_i, $i < e$, have finished acting by stage s then the restraint function $r(e, s + 1)$ ensures that $B \restriction s = B_s \restriction s$ so that

$$\Phi_e^B(x) \downarrow = 0 \quad \neq \quad 1 = A(x)$$

and requirement R_e is met. If some R_i, $i < e$, acts at some stage $t > s + 1$, then we reset the R_e basic module with a fresh witness x' and begin all over. On the other hand $r(e, s) = 0$ indicates that requirement R_e is not currently satisfied and may need to be satisfied again if the opportunity arises.

7.3.3   The Full Construction

We now combine all the strategies as follows. Let $\omega^{[y]} = \{\langle x, y \rangle : x \in \omega\}$. To avoid conflict between requirements, we choose witnesses $x \in \omega^{[e]}$ to meet R_e. For every e we set the restraint function $r(e, s) = 0$ for $s = 0$. We later reset $r(e, s + 1) = 0$ if R_e is injured at stage $s + 1$. Therefore, $r(e, s) > 0$ indicates that R_e has been satisfied at some stage $t \leq s$ and not injured since then. On the other hand, $r(e, s) = 0$ indicates that R_e has either never been satisfied, or satisfied, injured, and never satisfied since then until now.

Stage $s = 0$. Define $r(e, 0) = 0$ for all e.

Stage $s + 1$ even. Choose the least even e such that

$$(7.7) \qquad\qquad\qquad r(e, s) = 0 \quad \& $$

$$(\exists x)[\, x \in \omega^{[e]} - A_s \quad \& \quad \Phi_{e,s}^{B_s}(x) \downarrow = 0 \quad \& \quad (\forall i < e)[\, r(i, s) < x \,] \,].$$

Action. If there is no such e, then do nothing and go to stage $s + 2$. Otherwise, choose the least such e and the least corresponding x. We say R_e *acts* at stage $s + 1$. Perform the following steps.

Step 1. Enumerate x in A. (This makes $A_{s+1}(x) = 1$.)

Step 2. Define $r(e, s + 1) = s + 1$. (This attempts to restrain $B_s \restriction s$ in order to preserve the computation $\Phi_{e,s}^{B_s}(x) \downarrow = 0 \neq 1 = A_{s+1}(x)$.)

Step 3. For all $j > e$, define $r(j, s + 1) = 0$. We say that these lower priority requirements $\{R_j\}_{j>e}$ are *injured* at $s + 1$ and are *reset*.

Step 4. For all $i < e$ define $r(i, s+1) = r(i, s)$. (This leaves the previous action performed by these *higher* priority requirements $\{R_i\}_{i<e}$ untouched.)

Stage $s+1$ odd. Do the same for *odd* e with the roles of A and B reversed.

7.3.4   The Verification

Lemma 7.3.3. *If requirement R_e acts at some stage $s + 1$ and is never later injured, then requirement R_e is met and $r(e, t) = s+1$ for all $t \geq s+1$.*

Proof. Suppose R_e acts at stage $s+1$ and e is even. Then $\Phi_e^{B_s}(x)\downarrow = 0$ for some $x \in A_{s+1}$. Since no R_i, $i < e$, ever acts after stage $s + 1$, it follows by induction on $t > s$ that R_e never acts again and $r(e, t) = s + 1$ for all $t > s$. Hence, no R_j, $j > e$, enumerates any $x \leq s$ into B after stage $s+1$. Therefore, $B \upharpoonright s = B_s \upharpoonright s$ and $\Phi_e^B(x)\downarrow = 0 \neq A(x)$. $\square$

Lemma 7.3.4. *For every e, requirement R_e is met, acts at most finitely often, and $r(e) = \lim_s r(e, s)$ exists.*

Proof. Fix e and assume true for all R_i, $i < e$. Let v be the greatest stage at which some such R_i acts, if ever, and $v = 0$ if none exists. Then $r(e, v) = 0$ and this will persist until some stage $s + 1 > v$, if ever, when R_e acts. If R_e acts at some stage $s + 1$ then by Lemma 7.3.3 that action satisfies R_e, which never acts again, and $r(e, t) = s + 1$ for all $t \geq s + 1$.

Either way, $r(e)$ exists and R_e acts at most finitely often. Now suppose that R_e is not met. Then $\Phi_e^B = A$. Now by stage v at most finitely many elements $x \in \omega^{[e]}$ have been enumerated in A. Choose the least $x \in \omega^{[e]} - A_v$ such that $x > v$. Eventually, there will be a stage s such that $\Phi_{e,s}^{B_s}(x)\downarrow = 0$ since $x \notin A_s$. Hence, R_e acts at stage $s+1$, and by Lemma 7.3.3 this action satisfies R_e forever. $\square$

This completes the proof of the Friedberg-Muchnik Theorem 7.3.1 $\square$

Proposition 7.3.5. *Define the injury set I_e as in (7.1) for A and e even and similarly for B in place of A for e odd. Then in Theorem 7.3.1 we have $|I_e| < 2^e$.*

Proof. Define $f_s(i) = 1$ if $r(i, s) > 0$ and $f_s(i) = 0$ otherwise. If R_e is injured at stage $s + 1$ it is only because R_i acts for some $i < e$. Therefore, $f_s(i) = 0$ and $f_{s+1}(i) = 1$, causing f to increase lexicograpically at stage $s + 1$ on the initial segment $[0, e-1]$. However, $f_s \upharpoonright e$ is a binary string of exactly e bits and can increase lexicographically at most $2^e - 1$ times. $\square$

7.3.5   Exercises

Exercise 7.3.6. (i) Show that there is a u.c.e. sequence of c.e. sets $\{A_i\}_{i \in \omega}$ such that for every i, $A_i \not\leq_T \oplus\{A_j\}_{j \neq i}$. *Hint.* Modify the construction of the Friedberg-Muchnik Theorem 7.3.1 to meet the requirements $R_{\langle e,i \rangle}$ of Exercise 6.1.4.

(ii) Show that any countable partially ordered set can be embedded in the c.e. degrees $(\mathbf{C}, \leq)$ by an order-preserving map. (See the hint for Exercise 6.1.6.)

Exercise 7.3.7. (Cooper-Epstein-Lachlan). Construct a pair of c.e. sets A and B such that the d.c.e. set $D = A - B$ does not have c.e. degree. *Hint.* For each e, i, j meet the requirement

$$R_{\langle e,i,j \rangle} : \quad D \neq \Phi_i^{W_e} \quad \vee \quad W_e \neq \Phi_j^D.$$

To meet a single $R_{\langle e,i,j \rangle}$ choose x not yet in A or B and wait for a stage s such that

$$0 = \Phi_{i,s}^{W_{e,s} \upharpoonright u}(x) \quad \& \quad W_{e,s} \upharpoonright u = \Phi_{j,s}^{D_s \upharpoonright v} \upharpoonright u$$

for some u and v, where $D_s = A_s - B_s$. Now enumerate x in A_{s+1} and restrain D from changing on any other elements $\leq v$. We win $R_{\langle e,i,j \rangle}$ by the first clause unless there is some stage $t \geq s + 1$ and $y < u$ such that $y \in W_{e,t} - W_{e,s}$. In this case we enumerate x in B_t so that $D_t \upharpoonright v = D_s \upharpoonright v$ and $\Phi_{j,t}^{D_t \upharpoonright v}(y) = W_{e,s}(y) \neq W_{e,t}(y)$.

Exercise 7.3.8. A set A is *autoreducible* if there is an e such that for all x, $A(x) = \Phi_e^{A-\{x\}}(x)$. (The idea is that Φ_e determines whether $x \in A$ by using oracle questions "$y \in A$?" only for $x \neq y$.)

(i) Construct a c.e. set A which is not autoreducible. *Hint.* For each e choose a witness x, and attempt to meet $A(x) \neq \Phi_e^{A-\{x\}}(x)$ by waiting for $\Phi_{e,s}^{A_s-\{x\}}(x){\downarrow} = y$ for some s, then (1) enumerating x in A iff $y = 0$; and (2) attempting to restrain elements $z \leq \varphi_e^{A_s}(x,s)$ from entering A.

(ii) For any noncomputable c.e. set B construct a c.e. set $A \leq_T B$ such that A is not autoreducible. (Ladner has shown that one cannot achieve $A \equiv_T B$ because he proved that A is autoreducible iff A is *mitotic*, namely A is the disjoint union of c.e. sets A_0 and A_1 such that $A \equiv_T A_0 \equiv_T A_1$, and that there is a nonzero c.e. degree containing only mitotic c.e. sets.)

Exercise 7.3.9. Recall the definitions of $\leq_{tt}$, $\leq_{bT}$ and $\leq_{ibT}$ from Chapter 5. Note that if A_0 and A_1 are disjoint c.e. sets, then $A_i \leq_{ibT} A_0 \cup A_1$ for $i = 0, 1$. Use a priority argument to construct disjoint c.e. sets A_0 and A_1 such that $A_i \not\leq_{tt} A_0 \cup A_1$, $i = 0, 1$. *Hint.* Pick a witness x to meet the requirement that $A_i \not\leq_{tt} A_0 \cup A_1$ via the eth tt-reduction; wait until the latter converges on $A_{0,s} \cup A_{1,s} \cup \{x\}$, and put x into A_0 or A_1 to achieve a disagreement.

Exercise 7.3.10. Combine the permitting method with the Friedberg-Muchnik construction of Theorem 7.3.1 to show that for any noncomputable c.e. set C there exist Turing incomparable c.e. sets $A, B \leq_T C$.

7.4 * Preservation Strategy to Avoid Upper Cones

In the Friedberg-Muchnik Theorem 7.3.1 we satisfied an incomparability requirement $A \neq \Phi_e^B$ by enumerating an element x into A and simultaneously preserving $B \restriction \varphi_e^B(x)$. Here we are *given* a noncomputable c.e. set C and we must construct a noncomputable c.e. set $A \not\geq_T C$. In the noneffective case of building A such that $B \neq \Phi_e^A$ using an oracle in Theorem 6.2.6, we searched for e-splittings to meet requirement $R_e : A \neq \Phi_e^B$. If we found none, we claimed that Φ_e^A, if total, was computable. Here we adopt an apparently more passive approach of preserving *agreements* between C and Φ_e^A with the intention that if these agreements are sufficiently well preserved and if $C = \Phi_e^A$, then C is computable, contrary to hypothesis.

Theorem 7.4.1 (Cone Avoidance Theorem, Sacks).
   For every noncomputable c.e. set C there is a simple set A such that $C \not\leq_T A$ (and hence $\emptyset <_T A <_T \emptyset'$).

7.4.1   The Notation

It clearly suffices to construct A to be coinfinite and to satisfy, for all e, the requirements:

(7.8) $N_e : \quad \Phi_e^A \neq C$;

(7.9) $P_e : \quad W_e$ infinite $\implies$ $W_e \cap A \neq \emptyset$.

Let $\{C_s\}_{s \in \omega}$ be a computable enumeration of C. We shall give a computable enumeration $\{A_s\}_{s \in \omega}$ of A. Given $\{C_t\}_{t \leq s}$ and $\{A_t\}_{t \leq s}$, define for all e the following length function $\ell(e, s)$, restraint function $r(e, s)$, and the combined restraint function $R(e.s)$.

(7.10) $\ell(e, s) := \max\{ \, x \, : \, (\forall y < x)\,[\, \Phi_e^A(y)[s]\!\downarrow \, = \, C_s(y) \,] \, \}$;

(7.11) $r(e, s) := \max\{ \, \varphi_e^A(x)[s] \, : \, x \leq \ell(e, s) \, \}$;

(7.12) $R(e, s) := \max\{ \, r(i, s) \, : \, i \leq e \, \}$.

   Define the injury set I_e as in (7.1). If $x \leq r(e, s)$ and $x \in A_{s+1} - A_s$ then x represents an *injury* to requirement N_e at stage $s + 1$. As in Lemma 7.2.4, note that each I_e is finite (indeed, $|I_e| \leq e$) because N_e is injured at most once by each P_i, $i < e$, whereupon P_i is satisfied forever as in Theorem 7.2.1.

7.4.2   The Basic Module for Requirement N_e

Requirement N_e restrains with priority N_e any $x \leq r(e, s)$ from entering $A_{s+1} - A_s$.

Lemma 7.4.2. *If N_e is injured at most finitely often then $\Phi_e^A \neq C$.*

Proof. Assume for a contradiction that $\Phi_e^A = C$. Then $\lim_s \ell(e, s) = \infty$. Choose s' such that N_e is never injured after stage s'. We shall define a computable function g_e such that $g_e(x) = C(x)$, contrary to hypothesis. To compute $g_e(p)$ for some $p \in \omega$ find the least $s > s'$ such that $\ell(e, s) > p$. It follows by induction on $t \geq s$ that

(7.13) $(\forall t \geq s) [l(e, t) > p \quad \& \quad r(e, t) \geq \max\{ \varphi(A_s; e, x, s) : x \leq p \}]$,

and hence that $\Phi_{e,s}^{A_s}(p) = \Phi_e^{A_s}(p) = \Phi_e^A(p) = C(p)$, whence C is computable. To prove (7.13), assume it holds for t. Then $r(e, t)$ and $s > s'$ ensure that $A_{t+1} \restriction z = A_t \restriction z$ for all numbers z used in a computation $\Phi_{e,t}^{A_t}(x) = y$ for any $x \leq p$. Thus, $\Phi_{e,t+1}^{A_{t+1}}(x) = y$, so $l(e, t+1) > p$, unless $C_{t+1}(x) \neq C_t(x)$ for some $x < \ell(e, t)$. But if $C_t(x) \neq C_s(x)$ for some $t \geq s$ and $x \leq p$, where x is minimal, then our use of "$\leq \ell(e, t)$" rather than "$< \ell(e, t)$" in the definition of $r(e, t)$ ensures that the *disagreement* $C_t(x) \neq \Phi_{e,t}^{A_t}(x)$ is preserved forever, contrary to our hypothesis that $C = \Phi_e^A$. □

(Note that even though the Sacks strategy is always described as one which preserves agreements, it is crucial that we preserve at least one *disagreement* as well whenever possible. The reason that the disagreement is preserved is that C is c.e. and therefore the approximation $C_s(x)$ can only change from 0 to 1 and never back again. See Theorem 7.6.1 to extend this analysis to the case where C is Δ_2 in place of C c.e.)

Lemma 7.4.3. *If N_e is injured at most finitely often, and $C \neq \Phi_e^A$, then $\lim_s r(e, s) < \infty$.*

Proof. By Lemma 7.4.2, choose $p = (\mu x) [C(x) \neq \Phi_e^A(x)]$. Choose s' sufficiently large such that for all $s \geq s'$,

(1) $(\forall x < p) [\Phi_{e,s}^{A_s}(x) \downarrow = \Phi_e^A(x)]$;

(2) $(\forall x \leq p) [C_s(x) = C(x)]$; and

(3) N_e is not injured at stage s.

Case 1. $(\forall s \geq s') [\Phi_{e,s}^{A_s}(p) \uparrow]$. Then $r(e, s) = r(e, s')$ for all $s \geq s'$.

Case 2. $\Phi_{e,t}^{A_t}(p) \downarrow$ for some $t \geq s'$. Then $\Phi_{e,s}^{A_s}(p) = \Phi_{e,t}^{A_t}(p)$ for all $s \geq t$ because $\ell(e, s) \geq p$, and so, by the definition of $r(e, s)$, and the fact that N_e is not injured after stage t, the computation $\Phi_{e,t}^{A_t}(p)$ is preserved forever. Thus, $\Phi_e^A(p) = \Phi_{e,s}^{A_s}(p)$. But $C(p) \neq \Phi_e^A(p)$. Hence,

$(\forall s \geq t) [C_s(p) \neq \Phi_{e,s}^{A_s}(p) \quad \& \quad \ell(e, s) = p \quad \& \quad r(e, s) = r(e, t)]$.

Therefore, $r(e, t) = \lim_s r(e, s) = r(e)$. □

7.4.3 The Construction of A

Stage $s = 0$. Define $A_0 = \emptyset$.

Stage $s + 1$. Given A_s, define $\ell(e, s)$ and $r(e, s)$ as above for all e. We say P_i *requires attention* at stage $s + 1$ if $i \leq s$, $W_{i,s} \cap A_s = \emptyset$, and

$$(7.14) \qquad (\exists x)\, [\, x \in W_{i,s} \ \ \& \ \ x > 2i \ \ \& \ \ (\forall e \leq i)\, [\, r(e, s) < x \,]\,].$$

(If x exists, then $x < s$ by Definition 1.6.17. Therefore, (7.14) is a computable condition.) If i exists, choose the least such i and the least corresponding x and enumerate x in A_{s+1}. We say P_i *receives attention (acts)* at stage $s + 1$.

7.4.4 The Verification

Lemma 7.4.4. *For every* e, N_e *is met and* $r(e) := \lim_s r(e, s) < \infty$ *exists.*

Proof. Every positive requirement P_i acts at most once. Hence N_e can be injured at most e times. Apply Lemma 7.4.2 and Lemma 7.4.3.   □

Lemma 7.4.5. $(\forall e)\, [\, W_e \text{ infinite} \implies W_e \cap A \neq \emptyset \,]$.

Proof. By Lemma 7.4.4, define $r(i) := \lim_s r(i, s)$ for $i \leq e$, and define $R(e) := \max\{\, r(i) : i \leq e \,\}$. Now if

$$(\exists x)\, [\, x \in W_e \ \ \& \ \ x > 2e \ \ \& \ \ x > R(e) \,],$$

then $W_e \cap A \neq \emptyset$. Note that $\overline{A}$ is infinite by the clause "$x > 2e$" in (7.14) and hence A is simple.   □

7.5 Sacks Splitting Theorem

Shortly after proving Theorem 7.3.1, Friedberg [1958a] proved the Friedberg Splitting Theorem 2.7.1, stating that any c.e. set $B >_T \emptyset$ could be split as the disjoint union of noncomputable c.e. sets A_0 and A_1. Sacks [1963b] then generalized these two Friedberg theorems simultaneously by showing that A_0 and A_1 could be chosen to be not merely noncomputable but even Turing incomparable. Furthermore, they can both be chosen not to be in the cone above a given noncomputable C. To accomplish this Sacks used the method in the preceding section to avoid an upper cone.

Theorem 7.5.1 (Sacks Splitting Theorem, Sacks, 1963b). *Let B and C be c.e. sets such that C is noncomputable. Then there exist low c.e. sets A_0 and A_1 such that:*

(i) $A_0 \sqcup A_1 = B$ *and*

(ii) $C \nleq_T A_i$, *for* $i = 0, 1$.

Proposition 7.5.2. *If the c.e. set B is the disjoint union of c.e. sets A_0 and A_1 then $B \equiv_T A_0 \uplus A_1$.*

Proof. First, note that $A_i \leq_T B$ for $i = 0, 1$. Hence, $A_0 \oplus A_1 \leq_T B$ because to decide whether $x \in A_i$, first ask whether $x \in B$. If so, enumerate A_0 and A_1 until x appears in one of them. In the other direction, obviously $B \leq_T A_0 \oplus A_1$. $\qquad\square$

Corollary 7.5.3. *If $\mathbf{b}$ is any nonzero c.e. degree, there are incomparable low c.e. degrees $\mathbf{a}_0$ and $\mathbf{a}_1$ such that $\mathbf{b} = \mathbf{a}_0 \cup \mathbf{a}_1$.*

Proof. Choose a c.e. set $B \in \mathbf{b}$. Apply the Sacks Splitting Theorem 7.5.1 to B with $C = B$ to obtain A_0 and A_1. Hence, A_0, $A_1 <_T B$. By Proposition 7.5.2, $B \equiv_T A_0 \oplus A_1$. Now A_0 and A_1 cannot have comparable degree because if, say, $A_0 \leq_T A_1$ then $A_1 \equiv_T B$. Let $\mathbf{a}_0 = \deg(A_0)$ and $\mathbf{a}_1 = \deg(A_1)$. $\qquad\square$

Corollary 7.5.4. *The low c.e. degrees generate the c.e. degrees when closed under join.*

Corollary 7.5.5. *No c.e. degree is minimal.*

Corollary 7.5.6. *For any c.e. degree $\mathbf{c}$, $\mathbf{0} < \mathbf{c} < \mathbf{0}'$, there is a c.e. degree incomparable with $\mathbf{c}$.*

Proof. Let $B = K$, let $C \in \mathbf{c}$ be c.e., and apply the Sacks Splitting Theorem 7.5.1. One of A_0, and A_1 must have degree incomparable with $\mathbf{c}$, or else $A_0 \leq_T C$ and $A_1 \leq_T C$. Hence, $K \leq_T C$, contrary to $C <_T \emptyset'$. $\qquad\square$

Note that in Corollary 7.5.6 we do not produce A incomparable with C *uniformly* in C but merely give a pair A_0, A_1 of which *one* succeeds. Later we shall be able to find A uniformly from C.

Proof of Theorem 7.5.1. (We may assume that B is infinite, or else the result is trivial. This, however, does not affect the uniformity of A_0 and A_1 from B and C.) Let $\{B_s\}_{s \in \omega}$ and $\{C_s\}_{s \in \omega}$ be computable enumerations of B and C such that $B_0 = \emptyset$ and $|B_{s+1} - B_s| = 1$ for all s. We shall give computable enumerations $\{A_{i,s}\}_{s \in \omega}$, $i = 0, 1$, satisfying the single positive requirement

$$P: \quad x \in B_{s+1} - B_s \quad \Longrightarrow \quad [\, x \in A_{0,s+1} \ \vee \ x \in A_{1,s+1} \,],$$

and the negative requirements for $i = 0, 1$, and all e,

$$N_{\langle e,i \rangle}: \quad C \neq \Phi_e^{A_i}.$$

(The lowness of A_i will follow from the strategy to achieve the $N_{\langle e,i \rangle}$ requirements, because the restraint function $r^i(e, s)$ exists and is finite.)

Stage $s = 0$. Define $A_{i,0} = \emptyset$, $i = 0, 1$.

Stage $s+1$. Given $A_{i,s}$, define the computable functions $l^i(e, s)$ and $r^i(e, s)$ as in the proof of the Avoiding Cone Theorem 7.4.1 but with $A_{i,s}$ in place of A_s. Let $x \in B_{s+1} - B_s$. Choose $\langle e', i' \rangle$ to be the least $\langle e, i \rangle$ such that $x \leq r^i(e, s)$, and enumerate x in $A_{1-i',s+1}$. (Namely, choose the highest priority requirement $N_{\langle e', i' \rangle}$ which would be injured by enumerating x in $A_{i'}$, and enumerate x on the "*other side*" $A_{1-i'}$.) If $\langle e', i' \rangle$ fails to exist, enumerate x into A_0.

To see that the construction succeeds, define the injury set I_e^i as in the proof of the Avoiding Cone Theorem 7.4.1 but with A_i in place of A. It follows by induction on $\langle e, i \rangle$ that for $i = 0, 1$, and all e,

(1) I_e^i is finite;

(2) $C \neq \Phi_e^{A_i}$; and

(3) $r^i(e) := \lim_s r^i(e, s)$ exists and is finite.

Namely, fix $\langle e, i \rangle$ and assume (1), (2), and (3) hold for all $\langle k, j \rangle < \langle e, i \rangle$. By (3), choose t such that $r^j(k, s) = r^j(k)$ for all $\langle k, j \rangle < \langle e, i \rangle$ and all $s \geq t$.

Choose r greater than all such $r^j(k)$. Choose $v > t$ with $B \restriction r = B_v \restriction r$. Now $N_{\langle e, i \rangle}$ is never injured after stage v, so (1) holds for $\langle e, i \rangle$. Now (2) and (3) hold for $\langle e, i \rangle$ exactly as in the Avoiding Cone Theorem 7.4.1 and Lemmas 7.4.2, and 7.4.3.

Lemma 7.5.7. A_0 *and* A_1 *are both low.*

Proof. To see that each A_i is low, define the computable function g as follows:

$$\Phi_{g(e)}^X(y) = \begin{cases} C(0) & \text{if } y = 0 \ \& \ \Phi_e^X(e)\downarrow; \\ \text{undefined} & \text{otherwise.} \end{cases}$$

Note that for $i = 0, 1$,

$$e \in A_i' \iff \Phi_e^{A_i}(e)\downarrow \iff \Phi_{g(e)}^{A_i}(0) = C(0) \iff \lim_s l^i(g(e), s) > 0,$$

so $A_i' \in \Delta_2^0$ because l^i and g are computable functions, and $\lim_s l^i(g(e), s)$ exists, and hence $A_i' \leq_T \emptyset'$. □

This completes the proof of the Sacks Cone Avoidance Theorem 7.5.1. □

Note that the injury set I_e^i, although finite, has no obvious bounding function as in §7.2 and §7.3. Indeed, in general there is no computable function g such that $|I_e^i| \leq g(i, e)$.

7.6 Avoiding the Cone Above a Δ_2 Set $C >_T 0$

Now suppose that C is not c.e. as in §7.4 but rather is Δ_2. Let $C = \lim_s C_s$ for a computable sequence $\{C_s\}_{s\in\omega}$. We might begin by defining $\ell(e,s)$ and $r(e,s)$ as in §7.4.1, with the same basic module for N_e. The difficulty is that Lemma 7.4.2 no longer holds. Suppose that N_e is never injured and restrains every $x \le r(e,s)$ from entering A at stage $s+1$. When $\ell(e,s) > p$, we define $g_e(p) = \Phi_e^A(p)[s] = q$. Later, C may change, causing $\ell(e,t) < p$ for some $t > s$ which causes $r(e,t) < r(e,s)$, and may allow elements to enter A, thus destroying the A-computation $\Phi_e^A(p)[s] = q$. Now both C and Φ_e^A may later take a different value $q' = \Phi_e^A(p)[t] = C(p) \ne q$. The result is that $\Phi_e^A = C$ but $g_e \ne C$, and we cannot claim that C is computable. The solution is to define a stronger length of agreement function.
(7.15)
(maximum length function) $\max(e,s) := \max\{\,\ell(e,t) : t \le s\,\}$.

Theorem 7.6.1 (Δ_2-Avoiding Cone Theorem). *For every noncomputable Δ_2^0 set C there is a simple set A such that $C \not\le_T A$.*

Proof. Let $C = \lim_s C_s$ for a computable sequence $\{C_s\}_{s\in\omega}$. Let N_e and P_e be as in Theorem 7.4.1. Let $\max(e,s)$ be as in (7.15), and use it in place of $\ell(e,s)$ to define the new *restraint function*,

(7.16) $r(e,s) := \max\{\,\varphi_e^A(y)\,[s]\, : \, y \le \max(e,s)\,\}$.

Now perform the same construction and proof as in Theorem 7.4.1.

Lemma 7.6.2. *If N_e is injured at most finitely often, then $C \ne \Phi_e^A$.*

Proof. Assume that $C = \Phi_e^A$. Then $\lim_s \ell(e,s) = \lim_s \max(e,s) = \infty$. Choose s' such that N_e is never injured after stage s'. We shall define a computable function g_e such that $g_e(x) = C(x)$, contrary to hypothesis. To compute $g_e(p)$ for some $p \in \omega$ find the least $s > s'$ such that $\max(e,s) > p$ and $\Phi_e^A(p)[s]$ is defined. It follows by induction on $t \ge s$ that

(7.17) $(\forall t \ge s)[\,\max(e,t) > p \;\;\&\;\; r(e,t) \ge \max\{\,\varphi_e^A(x)[s] : x \le p\,\}\,]$.

Hence, $g_e(p) = \Phi_e^A(p)[s] = \Phi_e^A(p) = C(p)$ and C is computable. $\square$ $\square$

7.6.1 Exercises

Exercise 7.6.3. Prove that if $\{C_j\}_{j\in\omega}$ is a computable sequence of noncomputable Δ_2^0 sets and B is a c.e. set then there exist disjoint c.e. sets A_0 and A_1 such that $B = A_0 \cup A_1$ and $C_j \not\le_T A_i$ for $j \in \omega$, $i \in \{0,1\}$. Conclude that there is no computable enumeration $\{A_i\}_{i\in\omega}$ of c.e. (Δ_2^0) sets such that $\{\deg(A_i)\}_{i\in\omega}$ consists precisely of the nonzero c.e. (Δ_2^0) degrees. *Hint.* Replace the negative requirements in the Sacks Splitting Theorem 7.5.1 by $N_{\langle e,i,j\rangle} : \; C_j \ne \Phi_e^{A_i}$ and use the same method.

Exercise 7.6.4. Given the sequence $\{C_j\}_{j\in\omega}$ as in Exercise 7.6.3 find an infinite sequence $\{A_i\}_{i\in\omega}$ of pairwise Turing incomparable c.e. sets meeting all the requirements $N_{\langle e,i,j\rangle}$ of Exercise 7.6.3.

Exercise 7.6.5. Prove that there exist low c.e. degrees $\mathbf{a}_0$ and $\mathbf{a}_1$ such that for any c.e. degree $\mathbf{c}$ there exist c.e. degrees $\mathbf{b}_0 \le \mathbf{a}_0$ and $\mathbf{b}_1 \le \mathbf{a}_1$ such that $\mathbf{c} = \mathbf{b}_0 \cup \mathbf{b}_1$. *Hint.* Apply the Sacks Splitting Theorem 7.5.1 to $K_0 = \{\langle x,y\rangle : x \in W_y\}$ to obtain A_0 and A_1. Fix $W_e \in \mathbf{c}$, consider $y = e$, and exhibit B_0 and B_1.

Part II

Trees and Π_1^0 Classes

8
Open and Closed Classes

8.1 Open Classes in Cantor Space

Using ordinal notation identify the ordinal 2 with the set of smaller ordinals $\{0,1\}$. Identify set $A \subseteq \omega$ with its characteristic function $f : \omega \to \{0,1\}$ and represent the set of these functions as 2^ω. Use the conventions of the Notation section, especially the numbering σ_y of strings $\sigma \in 2^{<\omega}$. We use the notation on trees of §3.7 and sometimes use Convention 4.1.1 of dropping the superscript 0 in defining arithmetic classes. We now deal with *classes* $\mathcal{A} \subseteq 2^\omega$, i.e., *second-order* objects, rather than just *first-order* objects like sets $A \subseteq \omega$.[1]

Definition 8.1.1. (i) *Cantor space* is 2^ω with the following topology (collection of open classes). For every $\sigma \in 2^{<\omega}$ define the *basic clopen class* (closed and open class)

$$(8.1) \qquad [\![\sigma]\!] = \{\, f \ : \ f \in 2^\omega \ \ \& \ \ \sigma \prec f \,\}.$$

The *open* classes of Cantor space are unions of basic clopen classes.[2]

---

[1]Some material from the chapters in Part II was modified from that in the paper by Diamondstone, Dzhafarov, and Soare [2011].

[2]The classes $[\![\sigma]\!]$ are called *clopen* because they are both closed and open. Cantor space and Baire space are both *separable*. They have a countable base of open classes as above. Therefore, every open class is a union of *countably* many basic open classes. Although these are classes they are often called *open sets*, viewing the objects as reals.

(ii)   A set $A \subseteq 2^{<\omega}$ is an *open representation* of the open class

(8.2) $$\llbracket A \rrbracket \;=\; \bigcup_{\sigma \in A} \llbracket \sigma \rrbracket.$$

(We may assume A is closed *upwards*, i.e., $\sigma \in A$ and $\sigma \prec \tau$ implies $\tau \in A$.)[3]

(iii)   A class $\mathcal{A}$ is *effectively open (computably open)* if $\mathcal{A} = \llbracket A \rrbracket$ for a computable set $A \subseteq \omega$. (See Theorem 8.1.2 (i).)

(iv)   A class $\mathcal{A}$ is *(lightface)* Σ_1^0 (abbreviated *(lightface)* Σ_1) if there is a computable R such that

(8.3) $$\mathcal{A} = \{\, f \;:\; (\exists x)\, R(\, f \upharpoonright x\,) \,\}.$$

(v)   A class $\mathcal{A}$ is *(boldface)* Σ_1^0 if (8.3) holds with R replaced by R^X computable in some $X \subseteq \omega$. In this case we say $\mathcal{A}$ is $\Sigma_1^{0,X}$ or simply Σ_1^X.

Theorem 8.1.2 (Effectively Open Classes).   *Let* $\mathcal{A} \subseteq 2^\omega$.

(i)   If $\mathcal{A} = \llbracket A \rrbracket$ *with A c.e., then* $\mathcal{A} = \llbracket B \rrbracket$ *for some* computable *set* $B \subseteq \omega$.

(ii) $\mathcal{A}$ is effectively open iff $\mathcal{A}$ is (lightface) Σ_1^0.

(iii) $\mathcal{A}$ is open iff $\mathcal{A}$ is (boldface) Σ_1^0.

Proof. (i)   Let $\mathcal{A} = \llbracket A \rrbracket$ with A c.e. and upward closed. Let $A = \cup_s A_s$ for a computable enumeration $\{A_s\}_{s \in \omega}$. Define a computable set B with $\mathcal{A} = \llbracket B \rrbracket$ as follows. At stage s, for every σ with $|\sigma| = s$ put σ into B if $(\exists \rho \preceq \sigma)[\rho \in A_s]$, and put σ into $\overline{B}$ otherwise. If $\sigma \in A$, then $\sigma \in A_s$ for some s, and every $\tau \succeq \sigma$ with $|\tau| \geq s$ is put into B. Hence $\llbracket \sigma \rrbracket \subseteq \llbracket B \rrbracket$. Therefore, $\llbracket A \rrbracket \subseteq \llbracket B \rrbracket$. Clearly, $B \subseteq A$ since A is upward closed. Therefore, $\llbracket B \rrbracket \subseteq \llbracket A \rrbracket$.

(ii)   Let $\mathcal{A}$ be effectively open. Then $\mathcal{A} = \llbracket B \rrbracket$ for some B computable. Define $R(\sigma)$ iff $\sigma \in B$. Now $f \in \mathcal{A}$ iff $(\exists x)\, R(f \upharpoonright x)$. Hence, $\mathcal{A}$ is Σ_1^0. Conversely, assume $\mathcal{A}$ is Σ_1^0 via a computable R satisfying (8.3). Define $A = \{\sigma : R(\sigma)\}$. Then $\mathcal{A} = \llbracket A \rrbracket$.

(iii)   Relativize the proof of (ii) to a set $X \subseteq \omega$. $\square$

8.2   Closed Classes in Cantor Space

Recall the tree notation defined in §3.7.

---

[3]If $\sigma \in A$ and $\sigma \prec \tau$ but $\tau \notin A$ we may add τ to A without changing $\llbracket A \rrbracket$.

Definition 8.2.1. (i) A *tree* $T \subseteq 2^{<\omega}$ is a set closed under initial segments, i.e., $\sigma \in T$ and $\tau \prec \sigma$ imply $\tau \in T$. (By our canonical coding of strings $\sigma \in 2^{<\omega}$ we may think of T as a subset of ω.) The set of *infinite paths* through T is

$$(8.4) \qquad\qquad [T] = \{\, f \;:\; (\forall n) \lfloor f \restriction n \in T \rfloor \,\}.$$

(ii) A class $\mathcal{C} \subseteq 2^\omega$ is *(lightface)* Π^0_1 if there is a computable relation $R(x)$ such that

$$(8.5) \qquad\qquad \mathcal{C} = \{\, f \;:\; (\forall x)\, R(\, f \restriction x\,) \,\}.$$

A class $\mathcal{C}$ is *(boldface)* $\mathbf{\Pi}^0_1$ if (8.5) holds for R^X computable in some $X \subseteq \omega$. This is also written $\Pi^{0,X}_1$ and is abbreviated Π^X_1.

(iii) A class $\mathcal{C} \subseteq 2^\omega$ is *effectively closed* (*computably closed*) if its complement is effectively open. A set $\mathcal{C} \subseteq 2^\omega$ is *closed* if its complement is open.

Theorem 8.2.2 (Effectively Closed Sets and Computable Trees). *Fix* $\mathcal{C} \subseteq 2^\omega$. *TFAE:*

(i) $\mathcal{C} = [T]$ *for some computable tree* T;

(ii) $\mathcal{C}$ *is effectively closed;*

(iii) $\mathcal{C}$ *is a* Π^0_1 *class.*

Corollary 8.2.3 (Closed Sets and Trees). *Fix* $\mathcal{C} \subseteq 2^\omega$. *The following are equivalent (TFAE):*

(i) $\mathcal{C} = [T]$ *for some tree* T;

(ii) $\mathcal{C}$ *is closed;*

(iii) $\mathcal{C}$ *is a (boldface)* $\mathbf{\Pi}^0_1$ *class.*

Proof. Relativize the proof of Theorem 8.2.2 to $X \subseteq \omega$.   □

Remark 8.2.4. (Representing Closed Classes). The most convenient way of representing open and closed classes is with trees. If $\mathcal{C}$ is closed we choose a tree T such that $\mathcal{C} = [T]$. Define $A = \omega - T$. Then T is downward closed, A is upward closed, as sets of strings, and A defines the open set $[\![A]\!] = 2^\omega - [T] = \overline{\mathcal{C}}$. Note that the representations A and T are complementary in ω and the open and closed classes $[\![A]\!]$ and $[T]$ are complementary in 2^ω. The only difference between the effective case and general case is whether the tree T is computable or only computable in some set X.

We may imagine a path $f \in 2^\omega$ trying to climb the tree T without passing through a node $\sigma \in A$. If f succeeds, then $f \in \mathcal{C} = [T]$. However, if $f \succ \sigma$ for even *one* node $\sigma \in A$, then f falls off the tree forever and $f \notin \mathcal{C}$.

8.3 The Compactness Theorem

Particularly useful features of Cantor space are the well-known Compactness Theorem and the Effective Compactness Theorem 8.5.1, both of which lead to the study of one of our main topics, Π_1^0 classes.

Theorem 8.3.1 (Compactness Theorem). *The following easy and well-known properties hold for Cantor Space 2^ω. The term "compactness" refers to any of them, but particularly to (iv).*

(i) *(Weak König's Lemma, WKL).   If $T \subseteq 2^{<\omega}$ is an infinite tree, then $[T] \neq \emptyset$.*

(ii) *If $T_0 \supseteq T_1 \ldots$ is a decreasing sequence of trees with $[T_n] \neq \emptyset$ for every n, and intersection $T_\omega = \cap_{n \in \omega} T_n$, then $[T_\omega] \neq \emptyset$.*

(iii) *If $\{\mathcal{C}_i\}_{i \in \omega}$ is a countable family of closed sets such that $\cap_{i \in F} \mathcal{C}_i \neq \emptyset$ for every* finite *set $F \subseteq \omega$, then $\cap_{i \in \omega} \mathcal{C}_i \neq \emptyset$ also.*

(iv) *(Finite subcover).   Any open cover $[\![A]\!] = 2^\omega$ has a finite open subcover $F \subseteq A$ such that $[\![F]\!] = 2^\omega$.*

Proof. (i) Let T be infinite. We construct a sequence of nodes $\sigma_0 \prec \sigma_1 \ldots$ such that $f = \cup_{n \in \omega} \sigma_n$ and $f \in [T]$. Define a node σ to be *large* if there are infinitely many $\tau \succ \sigma$ such that $\tau \in T$. Define $\sigma_0 = \emptyset$, which is large. Given σ_n large, one of $\sigma_n{}^\frown 0$ and $\sigma_n{}^\frown 1$ must be large by the pigeon-hole principle. (This fails for Baire space ω^ω, where there may be infinitely many possible successors none of which is large.) Let $\sigma_{n+1} = \sigma_n{}^\frown 0$ if it is large and $\sigma_{n+1} = \sigma_n{}^\frown 1$ otherwise.

(ii)   Build a new tree S by putting σ of length n into S if $\sigma \in \cap_{i \leq n} T_i$ (which is also a tree). Note that S is infinite because $[T_n] \neq \emptyset$ for every n. By König's Lemma (i) there exists $f \in [S]$, but $[S] = [T_\omega]$.

(iii)   Define $\widehat{\mathcal{C}}_i = \cap_{j \leq i} \mathcal{C}_j$. Hence, $\widehat{\mathcal{C}}_0 \supseteq \widehat{\mathcal{C}}_1 \ldots$ is a decreasing sequence of nonempty closed sets. Choose a decreasing sequence of computable trees $T_0 \supseteq T_1 \ldots$ such that $[T_i] = \widehat{\mathcal{C}}_i$ and apply (ii).

(iv)   Suppose $[\![A]\!]$ is an open cover of 2^ω but $[\![F]\!] \not\supseteq 2^\omega$ for any finite subset $F \subset A$. Hence, the closed set $[T_F] = 2^\omega - [\![F]\!]$ is nonempty for all $F \subseteq A$. Therefore, $\mathcal{C} = \cap_{F \subset A} [T_F] \neq \emptyset$ by (iii), but $\mathcal{C} = 2^\omega - [\![A]\!] \neq \emptyset$. Hence, $[\![A]\!] \not\supseteq 2^\omega$.   □

8.4 Notation for Trees

Recall the notation in §3.7 for a tree $T \subseteq 2^{<\omega}$:

$$T_\sigma = \{ \tau \in T : \sigma \preceq \tau \quad \text{or} \quad \tau \prec \sigma \};$$

$$T^{\text{ext}} = \{ \sigma \in T : (\exists f \succ \sigma)[f \in [T]] \}.$$

A path $f \in [T]$ is *isolated* if $(\exists \sigma)[[T_\sigma] = \{ f \}]$. We say that σ *isolates* f because $[\![\sigma]\!] \cap [T] = \{f\}$ and we call σ an *atom*. If f is isolated we say it has Cantor-Bendixson rank 0. If f is not isolated, then f is a *limit point* and has rank ≥ 1. (See Definition 8.7.5 and surrounding exercises for Cantor-Bendixson rank.)

8.5 Effective Compactness Theorem

For a *computable* tree $T \subseteq 2^{<\omega}$ we can establish the following effective analogues of the Compactness Theorem 8.3.1.

Theorem 8.5.1 (Effective Compactness Theorem). *Let $T \subseteq 2^{<\omega}$ be a computable tree.*

(i) T^{ext} *is a Π^0_1 set. Hence,* $\overline{T^{\text{ext}}}$ *is Σ^0_1,* $\overline{T^{\text{ext}}} \leq_{\text{m}} \emptyset'$, *and* $T^{\text{ext}} \leq_{\text{T}} \emptyset'$.

(ii) *(Kreisel Basis Theorem)* $[T] \neq \emptyset \implies (\exists f \leq_{\text{T}} \emptyset')[f \in [T]]$. *(This was generalized in the* Low Basis Theorem 3.7.2.)

(iii) *If $f \in [T]$ is the lexicographically least member, then f has c.e. degree.*

(iv) *If $f \in [T]$ is isolated, then f is computable. If $[T]$ is finite, then all paths are isolated and therefore computable.*

(v) *Given an open cover $[\![A]\!] = 2^\omega$ with A c.e. there is finite subset $F \subseteq A$ such that $[\![F]\!] = 2^\omega$ and a canonical index for F can be found uniformly in a c.e. index for A.*

Proof. (i) The formal definition of T^{ext} in (3.22) has one function quantifier, and it is in Σ^1_1 form. Indeed is this the best we can do for Baire space ω^ω. However, for Cantor space 2^ω we can use the Compactness Theorem 8.3.1 (i) to reduce it to one arithmetical quantifier.

$$(8.6) \quad \sigma \in \overline{T^{\text{ext}}} \iff T_\sigma \text{ is finite} \iff (\exists n)(\forall \tau \succ \sigma)_{|\tau|=n} [\tau \notin T].$$

This is a Σ^0_1 condition because the second quantifier on τ is bounded by $|\tau| = n$ and acts like a finite disjunction. (See Theorem 4.1.4 (vi).)

(ii) Now use a $\emptyset'$ oracle to choose $f \in [T]$ such that $f = \cup_n \sigma_n$. Given $\sigma_n \in T^{\text{ext}}$, let $\sigma_{n+1} = \sigma_n{^\frown}0$ if $\sigma_n{^\frown}0 \in T^{\text{ext}}$, and $\sigma_n{^\frown}1$ otherwise.

(iii) (This gives a stronger conclusion than (ii).) Let f be the lexicographically least member of $[T]$, i.e., in the dictionary ordering $<_L$ on the alphabet $\{0,1\}$. (Think of the tree T as growing downwards and $\sigma <_L f$ as

denoting that σ is to the left of f lexicographically.) Define the following
c.e. set of nodes $M \subseteq \overline{T^{\text{ext}}}$ such that $M \equiv_{\text{T}} f$:

$$M = \{\sigma : (\forall \tau)_{|\tau| = |\sigma|}\, [\, [\, \tau \in T \; \& \; \tau \leq_L \sigma\,] \quad \Longrightarrow \quad \tau \in \overline{T^{\text{ext}}}\,]\, \}$$

(We just wait until σ and all its predecessors of length $|\sigma|$ have appeared
nonextendible. Then we put σ into M. In this way we enumerate all nodes
$\tau <_L f$. Therefore, f determines a *left c.e* set, one where when σ is
enumerated, all later strings τ enumerated satisfy $\sigma \leq_L \tau$.)

(iv)   Choose $\sigma \in T$ with $[T_\sigma] = \{f\}$. To compute f assume we have
computed $\tau = f \restriction n$. Exactly one of $\tau \hat{\ } 0$ and $\tau \hat{\ } 1$ is extendible. Enumerate
$\overline{T^{\text{ext}}}$ until one of these nodes appears and take the other one.

(v)   Assume $[\![A]\!] = 2^\omega$ with A c.e. Enumerate A until a finite set $F \subseteq A$
is found with $[\![F]\!] = 2^\omega$ by the Compactness Theorem (iv). We can search
until we find it. $\square$

Remark 8.5.2. Note that the *conclusions* in the Effective Compactness
Theorem 8.5.1 have various levels of effectiveness even though the *hypotheses* are all effective. In (v) if $[\![A]\!]$ covers 2^ω then the passage from A to
F is *computable* because we simply enumerate A until F appears (as with
any Σ_1 process). However, if $[\![A]\!]$ *fails* to cover 2^ω then the complementary
closed class $[T] = 2^\omega - [\![A]\!]$ is nonempty. Then (ii) gives a path $f \in [T]$
with $f \leq_{\text{T}} \emptyset'$ and (iii) even produces a path of c.e. degree, but neither
produces a *computable* path f because, given an extendible string $\sigma \prec f$,
the process for the proof of König's Lemma in Theorem 8.3.1 (i) does not
computably determine whether to extend to $\sigma \hat{\ } 0$ or $\sigma \hat{\ } 1$. In Theorem 9.3.2
we shall construct a computable tree with paths but no computable paths.

8.6   Dense Open Subsets of Cantor Space

The following important notion of *dense sets* will be developed more later.

Definition 8.6.1. Let $\mathcal{S}$ be Cantor space 2^ω.

(i)   A set $\mathcal{A} \subseteq \mathcal{S}$ is *dense* if $(\forall \sigma)\,(\exists f \succ \sigma)\,[\,f \in \mathcal{A}\,]$.

(ii)   An open set $\mathcal{A} \subseteq \mathcal{S}$ is *dense open* if

(8.7) $(\forall \tau)(\exists \sigma \succeq \tau)(\forall f \succ \sigma)\,[\,f \in \mathcal{A}\,]$.

(iii)   A class $\mathcal{B} \subseteq \mathcal{S}$ is G_δ, i.e., boldface $\mathbf{\Pi_2^0}$, if $\mathcal{B} = \cap_i \mathcal{A}_i$, a countable
intersection of open sets $\mathcal{A}_i$.

To be *dense* $\mathcal{A}$ must contain a point f in every basic open set $[\![\sigma]\!]$. To be *dense open* $\mathcal{A}$ must contain an *entire basic open set* $[\![\tau]\!] \subseteq [\![\sigma]\!]$ for every basic open set $[\![\sigma]\!]$. Notice that a set is dense open iff it is both dense and open.

After open and closed sets, much attention has been paid in point set topology to G_δ sets. If the sets $\mathcal{A}_i$ are *dense open* sets, then they have special significance. In §14.1.3 we shall explore Banach-Mazur games for finding a point $f \in \cap_i \mathcal{A}_i$ where the $\mathcal{A}_i$ are *dense open* sets. This is the paradigm for the *finite extension* constructions in Chapter 6, where we used the method to construct sets and degrees meeting an infinite sequence of "requirements." Meeting a requirement R_i amounts to meeting the corresponding dense open set $\mathcal{A}_i$.

8.7 Exercises

Exercise 8.7.1. We use the notation and definitions of §8.1, including the open representation A of $[\![A]\!]$ and the closed representation $T = \overline{A}$ of the closed set $[T] = 2^\omega - [\![A]\!]$, and we use the tree notation of §3.7 on the Low Basis Theorem.

(i) Define the open representation A to be the set of strings σ containing at least *two* 0's, and let $T = \overline{A}$. Describe the paths $f \in [T]$. Which are the limit points and which are the isolated ones?

(ii) Next define the open representation A to be the set of strings σ containing at least *three* 0's, and let $T = \overline{A}$. Describe the paths $f \in [T]$. (See Exercise 8.7.9 for the Cantor-Bendixson rank of these points, which gives much deeper insight into the structure of $[T]$ when three 0's are replaced by n 0's.)

Exercise 8.7.2. Prove that if T is computable and $[T]$ has exactly *one* limit point f, then $f \leq_T \emptyset''$.

Exercise 8.7.3. Prove that there is a computable tree $T \subset 2^{<\omega}$ such that $[T]$ contains a unique limit point $f \equiv_T \emptyset''$.

Exercise 8.7.4. (i) Define

$$(8.8) \qquad \Gamma(T) = \{\, \sigma \, : \, \mathrm{card}\,([T_\sigma]) < \infty \,\},$$

i.e., the nodes σ with only finitely many paths $f \in [T]$ with $f \succ \sigma$.

(ii) If $T = T^{\mathrm{ext}}$, define the set of *splitting nodes*,

$$(8.9) \quad \mathcal{S}(T) = \{\, \sigma : (\exists \rho \in T)(\exists \tau \in T)\,[\,\sigma \prec \rho \,\&\, \sigma \prec \tau \,\&\, \rho \,|\, \tau\,]\,\},$$

the nodes σ which *split* in T in the sense that some ρ and τ split σ, where $\rho \mid \tau$ denotes that $(\exists x) [\, \rho(x) \! \downarrow \, \neq \tau(x) \! \downarrow \,]$.

(i) Prove that if $\sigma \in \Gamma(T)$ and $f \in [T_\sigma]$ then f is isolated.

(ii) Prove that if $f \in [T]$ is not isolated then every $\sigma \prec f$ lies in $S(T)$.

(iii) Prove that $S(T)$ is Σ_1 in T and hence $S(T) \leq_T T'$.

Definition 8.7.5. (Cantor-Bendixson Derivative for tree T). Fix a tree T. For $\sigma \in T$ define the *Cantor-Bendixson rank* $r(\sigma)$ of σ relative to T.

$$D^0(T) = T.$$

$$D^{\alpha+1}(T) = D^\alpha(T) - \Gamma(D^\alpha(T)) \text{ for } \Gamma \text{ defined in } (8.8).$$

$$D^\lambda(T) = \bigcap \{D^\alpha(T) : \alpha < \lambda\} \text{ for } \lambda \text{ a limit ordinal.}$$

$$r(\sigma) = (\mu\alpha)[\, \sigma \in D^\alpha(T) - D^{\alpha+1}(T) \,].$$

If there is no such α, define $r(\sigma) = \infty$.

Definition 8.7.6. (Cantor-Bendixson derivative for closed set $\mathcal{A}$).   If $\mathcal{A}$ is a closed set, choose a tree T such that $[T] = \mathcal{C}$ and let $r(\sigma)$ be the rank above for tree T. If $r(\sigma) = \alpha$ and σ isolates f in $D^\alpha(T)$, then define $r(f) = \alpha$. If there is no such α then define $r(f) = \infty$.

The derivative of a closed set $\mathcal{C}$ is the set of all points which are not isolated points of $\mathcal{C}$, and we are iterating this derivative. Note that derivative of a closed set is closed.

Exercise 8.7.7. Prove that Definition 8.7.6 for the Cantor-Bendixson derivative of a closed set does not depend on the choice of the tree such that $[T] = \mathcal{C}$. Take any two trees T_1 and T_2 such that $[T_1] = [T_2] = \mathcal{C}$ and prove that the tree derivative of Definition 8.7.5 gives the same rank in both trees for any $f \in \mathcal{C}$. *Hint.* Keep applying the fact that $T_1^{\text{ext}} = T_2^{\text{ext}}$.

Exercise 8.7.8. $^\diamond$ (i) Prove that $D^\alpha(T)$ is a tree and hence $[D^\alpha(T)]$ is closed subset of $\mathcal{A} = [T]$.

(ii) Show that there is an ordinal β such that $D^\beta(T) = D^\alpha(T)$ for all $\alpha > \beta$. Define $D^\infty(T) = D^\beta(T)$. Prove that there is an $\alpha < \omega_1$ such that $D^\alpha(T) = D^\infty(T)$. We call $D^\infty(T)$ and $[D^\infty(T)]$ the *perfect kernel*.

(iii) Prove that either $D^\infty(T) = \emptyset$ or else $D^\infty(T)$ is a perfect tree, namely every $\sigma \in D^\infty(T)$ splits as defined above. In this case $D^\infty(T)$ has $2^{\aleph_0}$ many infinite paths.

(iv) Let β be as in (ii). Prove that $[D^\alpha(T)] - [D^{\alpha+1}(T)]$ is countable for every $\alpha < \beta$. Therefore, $\cup_{\alpha<\beta}[D^\alpha(T)]$ is countable, namely $[T] - [D^\infty(T)]$ is countable.

Exercise 8.7.9. Define the open representation A as in Exercise 8.7.1 and define $T = \overline{A}$.

(i) Analyze the Cantor-Bendixson rank of all points $f \in [T]$.

(ii) How does the rank change if we define A to be all strings having at least n 0's?

(iii) Define a computable tree T such that $[T]$ has a point of rank ω.

9
Basis Theorems

9.1 Bases and Nonbases for Π_1^0-Classes

The main theme of this chapter is this: Given a nonempty Π_1^0 class $\mathcal{C}$ what are the Turing degrees of members $f \in \mathcal{C}$?

Definition 9.1.1. A nonempty Π_1^0 class $\mathcal{C}$ is *special* if it contains no computable member.

It follows that if $T \subseteq 2^{<\omega}$ is a computable tree such that $[T]$ is special, then T^{ext} must be a *perfect* tree, meaning that every $\sigma \in T^{\text{ext}}$ admits incompatible extensions in T^{ext} because any isolated path would be computable. Therefore, every special Π_1^0 class has $2^{\aleph_0}$ members.

Definition 9.1.2. (i) Let $\mathcal{D} \subseteq 2^\omega$ be a class of sets. We call $\mathcal{D}$ a *basis for* Π_1^0 *classes* if every nonempty Π_1^0 class has a member $f \in \mathcal{D}$.

(ii) Let $\mathbf{D}$ be the corresponding class of Turing degrees of sets $X \in \mathcal{D}$. Then $\mathbf{D}$ is a *basis for* Π_1^0 *classes* if $\mathcal{D}$ is. Otherwise, we call $\mathbf{D}$ a *nonbasis*.

(iii) We call $\mathbf{D}$ an *antibasis* if whenever a Π_1^0 class contains a member of every degree in $\mathbf{d} \in \mathbf{D}$, it contains a member of every degree $\mathbf{d}$.

9.2   Previous Basis Theorems for Π_1^0-Classes

In §3.7 the Low Basis Theorem and exercises included some of the following basis theorems which we now list again. By the Kreisel Basis Theorem 8.5.1 (ii) we can always find $f \leq_T \emptyset'$. In 1960 Shoenfield improved the Kreisel Basis Theorem to f *strictly* below $\emptyset'$, namely $f <_T \emptyset'$.

Theorem 9.2.1 (Kreisel-Shoenfield Basis Theorem). *Every nonempty Π_1^0 class C has a member $f <_T \emptyset'$.*

Proof. Given a Π_1^0 class C, Shoenfield considered the Π_1^0 class $\mathcal{D}$ of all $\langle f, g \rangle$ such that $f \in C$ and

$$(\forall e)[\ \Phi_e^f(e)\!\downarrow \quad \Longrightarrow \quad \Phi_e^f(e) \neq g(e)\ ].$$

He then applied Kreisel's result to $\mathcal{D}$. $\qquad\qquad\qquad\qquad\qquad\square$

The previous Low Basis Theorem 3.7.2 substantially generalized these results by Kreisel and Shoenfield and will itself be generalized below.

Theorem 9.2.2 (Low Basis Theorem). *The low sets form a basis for Π_1^0.*

Theorem 9.2.3. *The sets of c.e. degree form a basis for Π_1^0.*

We proved this in the Effective Compactness Theorem 8.5.1 (iii). We shall see that it is false for the sets of *incomplete* c.e. degree.

9.3   Nonbasis Theorems for Π_1^0-Classes

Definition 9.3.1. If A and B are disjoint sets, then S is a *separating set* if $A \subseteq S$ and $B \cap S = \emptyset$.

Theorem 9.3.2. *(i) If W_e and W_i are disjoint c.e. sets, then the class of separating sets is a Π_1^0-class.*

(ii) There is a nonempty Π_1^0-class with no computable members.

Proof. (i) Define a computable tree T with $[T]$ the class of separating sets of W_e and W_i. For σ with $|\sigma| = s$, put σ in T if $\forall x < |\sigma|$

$$x \in W_{e,s} \implies \sigma(x) = 1 \quad . \& . \quad x \in W_{i,s} \implies \sigma(x) = 0.$$

Hence, $f \in [T]$ iff

$$(\forall x)[x \in W_e \implies f(x) = 1 \quad . \& . \quad x \in W_i \implies f(x) = 0].$$

(ii) Let W_e and W_i be disjoint c.e. sets which are computably inseparable as defined in Exercise 1.6.26. $\qquad\qquad\qquad\qquad\qquad\square$

Corollary 9.3.3. *The class of computable sets is not a basis for* Π_1^0 *classes* *(i.e.,* $\{0\}$ *is a nonbasis).*

We can generalize the preceding corollary as follows.

Theorem 9.3.4 (Jockusch and Soare, 1972a, Theorem 4). *The class of* *sets of incomplete c.e. degree is not a basis for* Π_1^0 *classes (i.e., the class of* *c.e. degrees* $\mathbf{d} < \mathbf{0}'$ *is a nonbasis).*

Proof. Let A be the Post simple set of Theorem 5.2.3. Then $\overline{A}$ and every infinite subset $S \subseteq \overline{A}$ is effectively immune via $f(x) = 2x + 1$, and therefore is not of incomplete c.e. degree by Exercise 5.4.6. Furthermore, $\overline{A}$ is computably bounded by $f(x) = 2x$ and therefore $\overline{A}$ is not hyperimmune by Theorem 5.3.3. Let $\{F_x\}_{x \in \omega}$ be a disjoint strong array witnessing that $\overline{A}$ is not hyperimmune. Define the Π_1^0 class

$$\mathcal{C} = \{\, S : S \cap A = \emptyset \ \& \ (\forall x)[\, F_x \cap S \neq \emptyset \,] \}.$$

This produces a nonempty Π_1^0 class $\mathcal{C}$ containing only infinite subsets of $\overline{A}$ and therefore having no members of incomplete c.e. degree. $\square$

Note that $\mathcal{C}$ has no c.e. members and no members of incomplete c.e. degree.

9.4 The Super Low Basis Theorem (SLBT)

The proof of the Low Basis Theorem 3.7.2 gives even more information about the jump f' than was explicitly claimed, but explaining it requires some definitions.

Definition 9.4.1. A set $A \leq_T \emptyset'$ is *super low* if $A' \leq_{tt} \emptyset'$ or equivalently if A' is ω-c.e. by Theorem 3.8.8.

Theorem 9.4.2 (Super Low Basis Theorem (SLBT)). *Every nonempty* Π_1^0 *class* $\mathcal{C} \subseteq 2^\omega$ *has a member* A *which is super low and indeed* A' *is* 2^{e+1}*-c.e.*

We now give what was historically the first proof of the SLBT from c. 1969, by Jockusch and Soare. This unpublished result was subsequently obtained independently by others.

Proof. We construct a computable a sequence of strings $\{\sigma_s\}_{s \in \omega}$ such that $A := \lim_s \sigma_s$ is super low. Fix a computable tree T with $[T] = \mathcal{C}$. Define the computable tree,

$$(9.1) \qquad\qquad U_{e,s} \;=\; \{\, \sigma : \Phi_{e,s}^\sigma (e)\!\uparrow \,\}$$

Let $T_{0,s} = T$ for all s. For every s given $T_{e,s}$: (1) define $T_{e+1,s} = T_{e,s} \cap U_{e,s}$, the e-black strings, if the latter contains a string σ of length s; and (2) define $T_{e+1,s} = T_{e,s}$, the e-white strings, otherwise.

To visualize this e-strategy, fix e and the previous tree $T_{e,s}$. Begin by playing the e-black strategy of choosing σ_s to be e-black if possible until for some n all nodes of length n are e-white. In other words, try to outrun letting $\Phi_e^\sigma(e)\downarrow$ as long as possible. This may involve many changes in σ_s but no change in the e-black strategy. During this phase nest the i-strategies within the e-strategy for all $i > e$.

If ever there is a stage when there is an n such that all strings of length n are e-white, then make *one* change of e-strategy from e-black to e-white. Thereafter, the e-strategy exerts no influence on the i-strategies for $i > e$. To prove that this construction succeeds define the following computable function.

$$\widehat{g}(e,s) := \begin{cases} 1 & \text{if} \quad \Phi_{e,s}^{\sigma_s}(e)\downarrow; \\ 0 & \text{otherwise.} \end{cases}$$

Clearly, $\widehat{g}(e,s)$ is computable. Fix e and assume by induction that $g(j) = \lim_s \widehat{g}(j,s)$ for all $j < e$ and that $g(j) = A'(j)$. Now the e-strategy begins in the e-black case and $\sigma_s \neq \sigma_{s+1}$ only if σ_s becomes e-white. If this happens finitely often, then the final σ_s is e-black and $\lim_s \widehat{g}(e,s) = 0 = A'(e)$. If it happens infinitely often, then the e-white nodes cover T_e. By compactness there is a finite subcover and therefore an n when all strings of length n are e-white. At this point we change once from the e-black to the e-white strategy. Thereafter, σ_s never changes, $\widehat{g}(e,s) = g(e) = A'(e)$.

Furthermore, assume by induction that for $e - 1$ there are at most 2^e stages when $\widehat{g}(e-1,s) \neq \widehat{g}(e-1,s+1)$. The e-strategy adds one more to each so that there are at most 2^{e+1} stages when $\widehat{g}(e,s) \neq \widehat{g}(e,s+1)$. (This is the same injury pattern as for the Friedberg-Muchnik finite injury construction.) $\qquad\square$

9.5   The Computably Dominated Basis Theorem

The key idea in the next theorem is to use a $\emptyset''$ oracle to build a member f of a given Π_1^0 class with the property that we can decide whether Φ_e^f is total or not at a definite stage of the construction. This differs from the proof of the Low Basis Theorem, where we needed only a $\emptyset'$ oracle to similarly decide whether $\Phi_e^f(e)$ converges or not. In both cases, however, we use the same technique (known as *forcing with Π_1^0 classes*) of continually pruning an infinite computable tree while preserving certain desired properties.

Recall that a function f is *computably dominated* (*hyperimmune-free*) if every function $h \equiv_T f$ is dominated by some computable function g. (See also Definition 5.6.1.)

Theorem 9.5.1 (Computably Dominated Basis Theorem, Jockusch and Soare, 1972b). *Every nonempty Π_1^0 class has a member f which is low$_2$ and computably dominated.*

Proof. Fix a nonempty Π_1^0 class $\mathcal{C}$ and a computable tree $T \subseteq 2^{<\omega}$ such that $\mathcal{C} = [T]$. We build a sequence of infinite computable trees

$$T = T_0 \supseteq T_1 \supseteq \cdots$$

as follows. Given T_e, define for each $x \in \omega$ the set

$$U_{e,x} = \{\sigma \in T_e : \Phi_{e,|\sigma|}^\sigma(x) \uparrow\},$$

noting that this is a computable subtree of T_e whose index as such can be found effectively from e, x, and an index for T_e. Now $\emptyset''$ can determine whether any of these subtrees is infinite, since this amounts to answering the following Σ_2^0 question:

$$(\exists x)(\forall n)(\exists \sigma)_{|\sigma|=n}\, [\, \sigma \in U_{e,x}\,]?$$

If so, let $T_{e+1} = U_{e,x}$ for the least x such that $U_{e,x}$ is infinite, and otherwise let $T_{e+1} = T_e$. In the former case, $\Phi_e^f(x)\uparrow$ for all $f \in [T_{e+1}]$, so Φ_e^f is not total, and in the latter, $\Phi_e^f(y)\downarrow$ for all y and all $f \in [T_{e+1}]$, so Φ_e^f is total.

As usual, take $f \in \cap_{e\in\omega} [T_e]$. Then $\emptyset''$ can compute the set Tot^f of all $e \in \omega$ such that Φ_e^f is total, and hence also $f'' \equiv_T \mathrm{Tot}^f$, because the above construction was $\emptyset''$-effective. Therefore, whether or not $e \in \mathrm{Tot}^f$ was decided during the construction at a finite stage. Hence, f is low$_2$. To show that f is computably dominated, let h be an f-computable function and fix e such that $h = \Phi_e^f$. In particular, Φ_e^f is total, so during the construction it must have been that $U_{e,x}$ was finite for all x. Hence, for every x, there must exist an n such that $\Phi_{e,|\sigma|}^\sigma(x) \downarrow$ for all $\sigma \in T_e$ of length n; let n_x be the least such n for a given x. Since T_e is computable, we can effectively find n_x for every x, meaning that the function

$$g(x) = \max\{\Phi_{e,|\sigma|}^\sigma(x) : |\sigma| = n_x \wedge \sigma \in T_e\}$$

is computable. Note that g bounds h. $\square$

Note that if $\mathcal{C}$ is a *special* Π_1^0 class, i.e., one with no computable members, then the above theorem yields a low$_2$ nonlow$_1$ member $f \in \mathcal{C}$, because no noncomputable, computably dominated f can be computable in $\emptyset'$, let alone be low, as we saw in Theorem 5.6.7.

9.6 Low Antibasis Theorem

For the purposes of the following theorem, we will say that a set $S \subseteq 2^{<\omega}$ is *isomorphic to* $2^{<\omega}$ provided there is a bijection $g : 2^{<\omega} \to S$ such that for all $\sigma, \tau \in 2^{<\omega}$, $\sigma \preceq \tau$ if and only if $g(\sigma) \preceq g(\tau)$. Notice that if a tree T has a subset isomorphic to $2^{<\omega}$ via a computable such bijection, then $[T]$ has a member of every degree. Indeed, for every real X, we have $Y = \cup_n g(X \restriction n) \in [T]$. Clearly, $Y \leq_T X$, while to compute $X(n)$ from Y for a given n we search for a $\sigma \in 2^{<\omega}$ until we find one of length greater than n with $g(\sigma) \subset Y$, and then $\sigma(n) = X(n)$.

Theorem 9.6.1 (Low Antibasis Theorem, Kent and Lewis, 2009). *Every Π_1^0 class that has a member of every nonzero low degree has one of every degree.*

Proof. [1] Fix a nonempty Π_1^0 class $\mathcal{C}$ not containing a member of every degree and let $T \subseteq 2^{<\omega}$ be a computable tree such that $\mathcal{C} = [T]$. We define a noncomputable low set A such that for all $e \in \omega$,

$$(9.2) \qquad \Phi_e^A = h \in 2^\omega \qquad \Longrightarrow \qquad [\, h \leq_T \emptyset \;\; \vee \;\; h \notin [T] \,].$$

In particular, $[T]$ has no member $h \equiv_T A$. We obtain A as $\sqcup_s \sigma_s$ where $\sigma_0 \preceq \sigma_1 \preceq \cdots$ are built in a $\emptyset'$-construction. Write $\Phi_e^\rho = \tau$ if

$$(\forall x < |\tau|)[\, \Phi_e^\rho(x)\downarrow = \tau(x) \,].$$

Let $\sigma_0 = \emptyset$. At stage s+1 we are given σ_s.

Stage s+1 = 3e. Let $n = |\sigma|$. Using $\emptyset'$, define $\sigma_{s+1} \succ \sigma_s$ such that $\sigma_{s+1}(n) \neq \varphi_e(n)$.

Stage s+1 = 3e+1. Ask $\emptyset'$ whether there exists $\rho \succ \sigma_s$ such that $\Phi_e^\rho(e)$ converges. If so, define σ_{s+1} to be the least such ρ, and define $\sigma_{s+1} = \sigma_s$ otherwise.

Stage s+1 = 3e+2. There are two cases.

Case 1. There exist strings $\alpha \succ \sigma_s$ and τ such that $\Phi_e^\alpha = \tau$ and $\tau \notin T$. In this case let σ_{s+1} be the least such α.

Case 2. Otherwise. In this case it follows that if $\Phi_e^A = h$ total, then $h \in [T]$. We proceed as follows. For a given σ define the c.e. set

$$\begin{aligned}
V_\sigma \;=\; & \{\langle \alpha, \beta \rangle : [\sigma \prec \alpha, \beta] \\
& \&\; (\exists \rho)(\exists \tau)[\, \Phi_e^\alpha = \rho \quad \&\quad \Phi_e^\beta = \tau \,] \\
& \&\; (\exists x < \min\{|\rho|, |\tau|\})[\, \rho(x)\downarrow \neq \tau(x)\downarrow \,]\}.
\end{aligned}$$

(We say that $\langle \alpha, \beta \rangle$ form an *e-splitting of* σ.) Using $\emptyset'$ we search for a $\sigma \succ \sigma_s$ such that $V_\sigma = \emptyset$. We claim that this search must succeed, and we define $\sigma_{s+1} = \sigma$ for the least such σ found.

---

[1] This proof is due to Dzhafarov and Soare with comments by Jockusch.

Suppose the claim is false. We shall contradict the assumption that $[T]$ does not have a member of every degree. Define a map $h : 2^{<\omega} \mapsto 2^{<\omega}$ as follows. Let $h(\emptyset) = \sigma_{s+1}$. Having defined $h(\sigma)$ for some σ, search computably for the least member $\langle \alpha, \beta \rangle$ of the nonempty c.e. set V_σ. Then define $h(\sigma \hat{\,} 0) = \alpha$ and $h(\sigma \hat{\,} 1) = \beta$. Now define $g : 2^{<\omega} \mapsto T$ by letting $g(\sigma) = \Phi_e^{h(\sigma)}$ for all σ. Since Case 1 does not hold, it is clear that $g(\sigma) \in T$. Therefore, g defines an isomorphic copy of $2^{<\omega}$ in T, contrary to hypothesis.

The first two types of stages guarantee that $A = \cup_s \sigma_s$ is a low noncomputable set. It remains to prove the following lemma.

Lemma 9.6.2. *If $\Phi_e^A = h$ is total, then h is computable or $h \notin [T]$.*

Proof. If Case 1 held at Stage $s+1 = 3e+2$, then h would not be in $[T]$. So suppose Case 2 held. By construction, $\sigma_{s+1} \preceq A$ was such that $V_{\sigma_{s+1}} = \emptyset$. In other words, there are no e-splittings above σ_{s+1}. Thus, to compute $h(n)$ find the first $\alpha \succeq \sigma_{s+1}$ such that $\Phi_e^\alpha(n) \downarrow$. Now $\Phi_e^\alpha(n) = \Phi_e^A(n) = h(n)$, else there would have been an e-splitting above σ_s. □

□

Corollary 9.6.3. *If C is a nonempty Π_1^0 class which does not have a member of every degree, then there are infinitely many low degrees with no members in C.*

Proof. Combine the proof of this theorem with Exercise 6.3.7, where we avoided the cone above a nonzero low degree and repeat for infinitely many low degrees uniformly below $\mathbf{0}'$. □

There are two notable features of the proof of the Low Antibasis Theorem 9.6.1. As in Exercise 6.3.7 we do not try to force the functional to be undefined. We merely look for e-splittings, which is a Σ_1 process, and then apply Lemma 9.6.2 if we cannot find them. Second, we do not actually build the computable bijection g but we *threaten* to. This is analogous to constructing a simple set A below a noncomputable c.e. set C where we threatened to build a computable characteristic function $g = C$. We did not build all of g but enough of g to force C to permit elements to enter A.

9.7 Proper Low_n Basis Theorem

The following generalization of the Low Basis Theorem says that, up to degree, the restriction of the jump operator to any special Π_1^0 class is surjective. The trick used for pushing the jump of the member up to the desired set is like the one used in the standard proof of the Friedberg Completeness Criterion.

The following theorem was stated with proof by Jockush and Soare in 1972 after Theorem 2.1 and later by Cenzer in 1999.

Theorem 9.7.1. *For every set $A \geq_T \emptyset'$, every special Π^0_1 class has a member f satisfying $f \oplus \emptyset' \equiv_T f' \equiv_T A$.*

Proof. Fix a nonempty Π^0_1 class $\mathcal{C}$ and a computable tree $T \subseteq 2^{<\omega}$ such that $\mathcal{C} = [T]$. We build a sequence of infinite computable trees $T = T_0 \supseteq T_1 \supseteq \cdots$ as follows. Let T_e be given. If e is even, define T_{e+1} from T_e as in the proof of the Low Basis Theorem. If e is odd, say $e = 2i + 1$, note that T_e^{ext} must be perfect since $\mathcal{C}$ is special, so $\emptyset'$ can find the smallest extendible nodes $\sigma, \tau \in T_e$ such that $\sigma(x) = 0$ and $\tau(x) = 1$ for some x. Let T_{e+1} consist of all the nodes in T_e comparable with σ or τ, depending on whether $A(i) = 0$ or $A(i) = 1$, respectively.

Take $f \in \cap_{e \in \omega} [T_e]$. If e is even, T_{e+1} can be obtained from T_e computably in $\emptyset'$, and hence both $f \oplus \emptyset'$-effectively and A-effectively because $A \geq_T \emptyset'$. If e is odd, say $e = 2i + 1$, then to obtain T_{e+1} from T_e we need an oracle for $\emptyset'$ to find the extendible nodes σ and τ and the position x on which they disagree, and then an oracle for A since we need to know $A(i)$. But in this case, $i \in A$ iff $f(x) = 1$, so an oracle for f suffices to determine whether to let T_{e+1} consist of the nodes comparable with σ or the nodes comparable with τ. Since f' is decided during the construction, we consequently have that $f \oplus \emptyset' \leq_T f' \leq_T A \leq_T f \oplus \emptyset'$, as desired. $\square$

10
Peano Arithmetic and Π_1^0-Classes

10.1 Logical Background

One of the earliest purposes of computability theory was the study of logical systems and theories. We consider theories in a computable language: one which is countable, and whose function, relation, and constant symbols and their arities are effectively given. We also assume that languages come equipped with an effective coding for formulas and sentences in the languages, i.e., a *Gödel numbering*, and identify sets of formulas with the corresponding set of Gödel numbers. We can then speak of the Turing degree of a theory in a computable language. Here we will examine the language $\mathcal{L} = \{+, \cdot, <, 0, 1\}$ of arithmetic, and theories extending PA, the theory of Peano arithmetic.

Definition 10.1.1. Let $\mathcal{D}_{\mathrm{PA}}$ be the set of (Turing) degrees of complete consistent extensions of Peano arithmetic; such a degree is called a *PA degree*.

The following is surely the best known theorem in mathematical logic.

Theorem 10.1.2 (Gödel, 1931; Rosser, 1936).

1. *The theory of Peano arithmetic is incomplete.*

2. *Furthermore, any consistent computably axiomatizable extension of PA is also incomplete.*

Corollary 10.1.3. $\mathbf{0} \notin \mathcal{D}_{PA}$.

Thus, there is no complete consistent extension of PA which is computable. However, there are many ways to extend PA to a complete theory, and we can think of them as paths on a computable tree. We identify a completion of Peano Arithmetic with the set of Gödel numbers of its sentences.

10.2 Π_1^0 Classes and Completions of Theories

Theorem 10.2.1. *There exists a Π_1^0 class whose members are precisely the completions of Peano Arithmetic. Thus, $\mathcal{D}_{PA}$ is the degree spectrum of a Π_1^0 class.*

Proof. (Sketch).   Fix a bijective Gödel numbering $G : \omega \to \text{Sent}_{\mathcal{L}}$ for sentences of arithmetic. Given $\sigma \in 2^{<\omega}$, we identify σ with the sentence

$$\theta(\sigma) = \bigwedge_{\sigma(i)=1} G(i) \quad \& \quad \bigwedge_{\sigma(j)=0} \neg G(j).$$

We say that a sentence θ "appears to be consistent at stage t" if there is no derivation of $\neg\theta$ from the first t axioms of PA in fewer than t lines. Since there are finitely many such derivations, the relation $R(\sigma, t) = $ "$\theta(\sigma)$ appears to be consistent at stage t" is computable. Therefore, the class

$$\mathcal{C} \quad = \quad \{\, f \in 2^\omega : (\forall n)(\forall t < n)R(f \upharpoonright t, n) \,\}$$

is a Π_1^0 class. Some f is an element of this class if and only if the corresponding set of sentences $G(\{n : f(n) = 1\})$ is a complete consistent extension of PA. $\square$

Remark 10.2.2. *This theorem follows from an analysis of Lindenbaum's Lemma. Note that no special properties of PA were used, beyond the fact that it is a computably axiomatizable theory in a computable language. Therefore, the same theorem applies to all such theories.*

Lindenbaum's Lemma says that a consistent theory T has a complete consistent extension. This follows by the Compactness Theorem.

We defined a PA degree as a degree of a completion of Peano Arithmetic. From this definition, it may be surprising that the class of degrees is closed upwards. This is true, however, and to demonstrate it we need an important fact arising from Gödel's incompleteness theorem: the proof actually constructs a "Gödel sentence" which is independent of the axioms.

Theorem 10.2.3 (Gödel's Incompleteness Theorem, effective version).

From a description of a consistent, computably axiomatizable theory T extending PA, we can effectively find a sentence, called the Gödel-Rosser sentence of T, which is independent of T.

10.3   Equivalent Properties of PA Degrees

The PA degrees arise naturally in a variety of contexts, especially those relating to trees and weak König's lemma. This is because the PA degrees are exactly those degrees which can achieve weak König's lemma by finding paths through trees. For this reason, there are several equivalent properties which all serve to define the PA degrees. We shall highlight a few of these properties.

Definition 10.3.1. A function $f : \omega \to \omega$ is *diagonally noncomputable (d.n.c.)* if, for all e, if $\varphi_e(e)\downarrow$, then $f(e) \neq \varphi_e(e)$.

Recall that up to Turing degree this is equivalent to f being fixed point free by Exercise 5.4.5.

Definition 10.3.2. A function is *n-valued* if $f(e) < n$ for each $e \in \omega$.

The term "diagonally noncomputable" derives from the particular way that d.n.c. functions are noncomputable. We see that if f is d.n.c., f cannot be computable, because then f would be φ_e for some e, but f and φ_e differ on argument e; thus d.n.c. functions diagonalize against the list of all (partial) computable functions. We will be primarily interested in 2-valued d.n.c. functions.

Theorem 10.3.3 (Scott, 1962; Jockusch and Soare, 1972b; Solovay, unpublished). [1]

For a Turing degree $\mathbf{d}$, the following are equivalent:

(i) $\mathbf{d}$ is the degree of a complete consistent extension of Peano arithmetic.

(ii) $\mathbf{d}$ computes a complete consistent extension of Peano arithmetic.

(iii) $\mathbf{d}$ computes a 2-valued d.n.c. function.

(iv)   Every partial computable 2-valued function has a total $\mathbf{d}$-computable 2-valued extension.

(v)   Every nonempty Π_1^0 class has a member of degree at most $\mathbf{d}$.

(vi)   Every computably inseparable pair has a separating set of degree at most $\mathbf{d}$.

Proof. (i) $\implies$ (ii). This implication is trivial.

(ii) $\implies$ (iii). Let $\mathbf{d}$ compute a complete consistent extension T of PA, and let f be the (partial computable) diagonal function $f(e) = \varphi_e(e)$. By results of Gödel and Kleene, there is a formula ψ representing f, in the

---

[1]In 1962 Scott proved the equivalence of conditions (i) and (v). In 1972b Jockusch and Soare proved the equivalence of conditions (ii) and (vi); the equivalence with (iii) and (iv) is also implicit in their work. Jockusch and Soare left the equivalence of (i) and (ii) as an open question, which was answered by Solovay (unpublished).

sense that

$$f(x){\downarrow} = y \qquad \Longleftrightarrow \qquad PA \vdash \psi(x,y), \text{ and}$$

$$f(x){\downarrow} \neq y \qquad \Longleftrightarrow \qquad PA \vdash \neg\psi(x,y).$$

Since $PA \vdash \psi(x,y)$ implies that $\psi(x,y) \in T$, and T is complete and **d**-computable, the function

$$\widehat{f}(e) = \begin{cases} 1 & \psi(e,0) \in T \\ 0 & \neg\psi(e,0) \in T \end{cases}$$

is a **d**-computable 2-valued d.n.c. function.

(iii) $\Longrightarrow$ (iv). Suppose g is a 2-valued d.n.c. function, and let f be a partial computable 2-valued function. There is a computable function $\widehat{f}$ such that $f(x) = \varphi_{\widehat{f}(x)}(\widehat{f}(x))$ for all x. Then $1 - (g \circ \widehat{f})$ is a total **d**-computable 2-valued function extending f.

(iv) $\Longrightarrow$ (v). Let $\mathcal{P}$ be a nonempty Π_1^0 class, and T a computable tree with $\mathcal{P} = [T]$. Fix a computable bijection $h : \omega \to 2^{<\omega}$. Let f be the function

$$f(e) = \begin{cases} 0 & \begin{array}{l} h(e) \in T \text{ and there is a level } l \text{ such that } h(e)\widehat{}0 \\ \text{has a descendent at level } l \text{ in } T, \text{ but } h(e)\widehat{}1 \text{ does not} \end{array} \\[2em] 1 & \begin{array}{l} h(e) \in T \text{ and there is a level } l \text{ such that } h(e)\widehat{}1 \\ \text{has a descendent at level } l \text{ in } T, \text{ but } h(e)\widehat{}0 \text{ does not.} \end{array} \end{cases}$$

This function f is partial computable, since to compute $f(e)$ one simply searches for a level l such that one case or the other holds. If $h(e) \in T$ is extendible, then either both $h(e)\widehat{}0$ and $h(e)\widehat{}1$ are extendible, in which case $f(e){\uparrow}$, or only one is, so $f(e){\downarrow}$, and $h(e)\widehat{}f(e)$ is extendible. Let $\widehat{f}$ be a 2-valued **d**-computable extension of f. Then using $\widehat{f}$, we can find an element of $[T]$ as follows: starting with any string $\sigma \in T^{\text{ext}}$, apply $\widehat{f} \circ h^{-1}$ to get either 0 or 1, which we can append to σ to get a longer string still in T^{ext}. Starting with the empty string, we can iterate this process to get an infinite **d**-computable path through $[T]$, i.e., an element of $\mathcal{P}$.

(v) $\Longrightarrow$ (vi). If A, B is a computably inseparable pair, the class of separating sets is a Π_1^0 class by Theorem 9.3.2. If property (v) holds, this has a **d**-computable member.

(vi) $\Longrightarrow$ (i). Fix some order of $\mathcal{L}$-sentences, and some order for generating proofs. Let A be the set of pairs (F, ψ), where F is a finite set of $\mathcal{L}$-sentences and ψ is an $\mathcal{L}$-sentence, such that a proof of a contradiction is found from $PA \cup F \cup \{\psi\}$ before (if ever) finding a proof of a contradiction from $PA \cup F \cup \{\neg\psi\}$. Similarly, let B be the set of pairs (F, ψ), such that a proof of contradiction is found from $PA \cup F \cup \{\neg\psi\}$ before (if ever) finding one from $PA \cup F \cup \{\psi\}$. Clearly A and B are disjoint c.e. sets. Suppose the pair A, B has a **d**-computable separating set C. Let $D \in \mathbf{d}$. We shall construct a completion T of PA, of degree **d**, in stages, along with a bijective function $g : \omega \to \text{Sent}_{\mathcal{L}}$, also defined in stages. At stage n we shall determine $g(n)$, and decide whether $g(n) \in T$. Define the set of sentences,

$$F_n \quad = \quad (T \cap g[0 \ldots n-1]) \ \cup \ \{\neg\psi : \psi \in g[0 \ldots n-1] \setminus T\}.$$

In other words, F_n keeps track of every sentence we decided by the beginning of stage n. It contains those sentences we have declared to be in T, together with the negations of those sentences we have declared not to be in T. At stage n, do the following:

1. If n is even, let $g(n)$ be the Gödel sentence of $PA \cup F_n$. If n is odd, let $g(n)$ be the first $\mathcal{L}$-sentence not yet in the range of g.

2. If $n = 2s$ is even, consider whether s is an element of D. If $s \in D$, then $g(n) \in T$; otherwise, $g(n) \notin T$.

3. If n is odd, consider the pair $(F_n, g(n))$. If this pair is in C, then $g(n) \notin T$; otherwise, $g(n) \in T$.

We shall show that T is a complete consistent extension of PA, of degree **d**. Assume (for the sake of induction) that F_n is consistent with PA. (Since $F_0 = \emptyset$, it is consistent with PA.) Note that F_{n+1} is either $F_n \cup \{g(n)\}$ or else $F_n \cup \{\neg g(n)\}$. Since F_n is consistent with PA, at least one of $F_n \cup \{g(n)\}$ and $F_n \cup \{\neg g(n)\}$ must be consistent with PA. Furthermore, if n is even, both are consistent since $g(n)$ is the Gödel sentence for $PA \cup F_n$. If both are consistent with PA, then clearly F_{n+1} is as well. Suppose instead only one of the two is consistent (so we know n is odd). If only $F_n \cup \{g(n)\}$ is consistent with PA, then a proof of contradiction will be found from $PA \cup F_n \cup \{\neg g(n)\}$ before finding one from $PA \cup F_n \cup \{g(n)\}$, so $(F_n, g(n)) \in B$. Thus $(F_n, g(n)) \notin C$; by the construction, $g(n) \in T$, and F_{n+1} is consistent with PA. Similarly, if only $F_n \wedge \neg g(n)$ is consistent with PA, then the construction goes the opposite way and again F_{n+1} is consistent with PA. By induction, F_n is consistent with PA for all n, so $T = \bigcup_n F_n$ is consistent with PA. Since F_n decides $g(0) \ldots g(n-1)$, T is complete. Therefore, T is a complete consistent extension of PA.

In order to show that T has degree **d**, we first show that $g \leq_T T$. To see this, note that $g(n)$ is either the first $\mathcal{L}$-sentence which is not one of $g(0) \ldots g(n-1)$, if n is odd, or else $g(n)$ is the Gödel sentence of $PA \cup F_n$, where F_n is determined entirely by T and the values $g(0) \ldots g(n-1)$. Thus $g(n)$ can be computed from n, $g(0) \ldots g(n-1)$, and T, so $g \leq_T T$. From the construction, we see that $s \in D$ if and only if $g(2s) \in T$, so we have $D \leq_T g \oplus T \leq_T T$. However, the entire construction was **d**-computable, so $T \in \mathbf{d}$. $\qquad\qquad\square$

11
Randomness and Π_1^0-Classes

11.1 Martin-Löf Randomness

In this chapter, we explore some of the relationships between Π_1^0 classes, algorithmic randomness, and computably dominated degrees.

Let μ be the Lebesgue measure on Cantor space, with which we assume the reader is familiar. For completeness, we define the measure of an open class $\mathcal{A} \subseteq 2^\omega$. Let $A \subset 2^{<\omega}$ be any set with $\mathcal{A} = [\![A]\!]$ which is prefix-free (i.e., if $\sigma \in A$ and $\tau \prec \sigma$ then $\tau \notin A$). Alternatively, let A could be the class of strings σ such that $[\![\sigma]\!] \subseteq \mathcal{A}$ and σ is minimal with respect to this property. Such an A can be seen to exist for example as follows. Since $\mathcal{A}$ is open, its complement is closed and hence is equal to $[T]$ for some tree $T \subseteq 2^{<\omega}$ (which is not necessarily computable). Then A can be taken to consist of all elements of $\overline{T}$ whose predecessors all belong to T. Now the measure of $\mathcal{A}$ is defined as

$$\mu(\mathcal{A}) = \sum_{\sigma \in A} 2^{-|\sigma|}.$$

the Lebesgue measure on Cantor space has all the same properties we are familiar with from the Lebesgue measure on the real line. Recall that a sequence of c.e. sets $A_0, A_1, \ldots$ is *uniformly c.e.* (abbreviated u.c.e.) if there exists a computable function f such that $A_n = W_{f(n)}$ for all n.

Definition 11.1.1.

1. A sequence $\mathcal{A}_0, \mathcal{A}_1, \ldots$ of subclasses of 2^ω is *uniformly (lightface)* Σ_1^0 if there exists a u.c.e. sequence $A_0, A_1, \ldots$ of subsets of $2^{<\omega}$ such that $\mathcal{A}_n = [\![A_n]\!]$ for all n.

2. A *Martin-Löf (ML) test* is a uniformly Σ_1^0 sequence $\mathcal{A}_0, \mathcal{A}_1, \ldots$ of subclasses of 2^ω such that $\mu(\mathcal{A}_n) \le 2^{-n}$ for all n.

3. A set $X \in 2^\omega$ *fails* a Martin-Löf test $\mathcal{A}_0, \mathcal{A}_1, \ldots$ if $X \in \bigcap_{n\in\omega} \mathcal{A}_n$. Otherwise, X *passes* the test.

4. A set $X \in 2^\omega$ is *Martin-Löf random (ML-random)* if it passes every Martin-Löf test.

The key point here is that the ML test must be effective in two ways. The sequence $\{\mathcal{A}_n\}_{n\in\omega}$ must be uniformly c.e., and it must converge computably fast in measure to 0. The intuition is that a non-ML-random set X is "caught" by an infinite sequence $\{\mathcal{A}_n\}_{n\in\omega}$ which reveals some of its information even though the measure of $\bigcap_n \{\mathcal{A}_n\}$ is effectively 0. For example, if the set X is computable then it is non-ML-random because it fails the ML test in which $\mathcal{A}_n = [\![X \upharpoonright n]\!]$. Schnorr proved that a set is ML-random iff it is 1-random, a closely related concept, so one may use the terms interchangeably.

11.2   A Π_1^0 Class of ML-Randoms

A Martin-Löf test $\mathcal{A}_0, \mathcal{A}_1, \ldots$ is called *universal* if $\bigcap_{n\in\omega} \mathcal{A}_n \supseteq \bigcap_{n\in\omega} \mathcal{B}_n$ for every other Martin-Löf test $\mathcal{B}_0, \mathcal{B}_1, \ldots$. Thus, if X passes a universal test, it must pass every test, and hence

$$\bigcap_{n\in\omega} \mathcal{A}_n = \{ X \in 2^\omega : X \text{ is not ML-random} \}.$$

This is a (lightface) Π_1^0 class and therefore an effective analogue of the (boldface) $\mathbf{\Pi}_2^0$ classes (i.e., G_δ classes) such as those we studied in Chapter 8, and which we shall study in the Banach-Mazur theorem in Chapter 14.

The following theorem is thus useful when trying to show that a given set is not ML-random.

Theorem 11.2.1 (Martin-Löf, 1966). *There exists a universal Martin-Löf test.*

Proof. Let $\{V_n^0\}_{n\in\omega}, \{V_n^1\}_{n\in\omega}, \ldots$ be an effective listing of all uniformly c.e. subsets of $2^{<\omega}$. Let $\mathcal{B}_n^e = [\![V_n^e]\!]$ where we stop enumerating if the measure exceeds 2^{-n}. Then $\{\mathcal{B}_n^e\}_{n\in\omega}$ for $e \in \omega$ lists all ML tests. Define $\mathcal{A}_n = \mathcal{B}_{e+n+1}^e$. Then the $\{\mathcal{A}_n\}$ are uniformly c.e. and $\mu(\mathcal{A}_n) \le 2^{-n}$.

$$\mu(\mathcal{A}_n) = \Sigma_e \, \mu(\mathcal{B}_{n+e+1}^e) \le \Sigma_e \, 2^{-n+e+1} = 2^{-n}.$$

Therefore, $\{A_n\}_{n \in \omega}$ is a universal ML test. □

Notice that this implies that the class of ML-randoms has measure 1. Indeed, each member of a universal Martin-Löf test $U_0, U_1, \ldots$ is an open set covering $\{X \in 2^\omega : X$ is not ML-random$\}$, implying that

$$\mu(\{X \in 2^\omega : X \text{ is not ML-random}\}) \leq \mu(U_n) \leq 2^{-n}$$

for all n. Essentially the same argument, in reverse, yields the following:

Corollary 11.2.2. *(F. Stephan) There is a nonempty Π_1^0 class all of whose elements are ML-random.*

Proof. Let $U_0, U_1, \ldots$ be a universal Martin-Löf test. For every $n > 0$, U_n is a proper Σ_1^0 subclass of 2^ω, implying that $\overline{U_n}$ is a nonempty Π_1^0 class. By the definition of a universal Martin-Löf test,

$$\overline{U_n} \subseteq \bigcup_{n \in \omega} \overline{U_n} = \overline{\bigcap_{n \in \omega} U_n} = \{X \in 2^\omega : X \text{ is ML-random}\},$$

as desired. □

From this and the various basis theorems in Chapter 9, we can conclude that there are ML-random sets which are of c.e. degree, hyperimmune-free (computably dominated), low, even superlow, and of PA degree. However, any set which is ML-random and of PA degree must be of degree $\geq \mathbf{0}'$.

11.3 Π_1^0 Classes and Measure

Given the measure-theoretic definition of ML-randomness, it is natural to ask about the measure of Π_1^0 classes containing ML-randoms. The following theorem gives a full answer to this question.

Theorem 11.3.1. *Let $\mathcal{C}$ be a Π_1^0 class. If $\mu(\mathcal{C}) = 0$, then $\mathcal{C}$ contains no ML-random sets.*

Proof. Suppose $\mathcal{C}$ has measure 0. Let $T \subseteq 2^{<\omega}$ be a tree such that $\mathcal{C} = [T]$, and for each $n \in \omega$, let $A_n = [\![\{\sigma \in T : |\sigma| = n\}]\!]$. Then $A_0, A_1, \ldots$ is a nested sequence of open classes whose intersection is the measure 0 class $\mathcal{C}$, so it must be that $\lim_n \mu(A_n) = 0$. As the sequence $\{A_n\}_{n \in \omega}$ is given by a strong array of finite sets of strings, the map $n \mapsto \mu(A_n) \in \mathbb{Q}$, the rationals, is computable. Therefore, we can find a computable function p such that $\mu(A_{p(n)}) \leq 2^{-n}$ for all n. Now since $A_0, A_1, \ldots$ is uniformly Σ_1^0, $A_{p(0)}, A_{p(1)}, \ldots$ is a Martin-Löf test. But for all $f \in \mathcal{C}$, $f \in \bigcap_{n \in \omega} A_{p(n)}$, so f is not ML-random. □

Note that we can view this as a generalization of the remark earlier that any computable set is not ML-random beginning with a similar sequence defined by strings of length n.

Theorem 11.3.2 (Kucera). *Let $\mathcal{C}$ be a Π_1^0 class. If $\mu(\mathcal{C}) > 0$, then every ML-random set computes a member of $\mathcal{C}$.*

Proof. Suppose $\mathcal{C}$ has positive measure and let X be a ML-random set. Let V_0 be a prefix-free c.e. subset of $2^{<\omega}$ such that $\overline{\mathcal{C}} = [\![V_0]\!]$. For each $n \in \omega$, let $V_{n+1} = [\![\{\sigma^\frown\tau : \sigma \in V_n \ \& \ \tau \in V_0\}$, and let $\mathcal{A}_n = [\![V_n]\!]$. Notice that for all n, V_n is prefix-free since V_0 is, so we have

$$
\begin{aligned}
\mu(\mathcal{A}_{n+1}) &= \sum_{\sigma \in V_{n+1}} 2^{-|\sigma|} \\
&= \sum_{\sigma \in V_n} \sum_{\tau \in V_0} 2^{-|\sigma\tau|} \\
&= \sum_{\sigma \in V_n} 2^{-|\sigma|} \sum_{\tau \in V_0} 2^{-|\tau|} \\
&= \mu(\mathcal{A}_n)\mu(\mathcal{A}_0).
\end{aligned}
$$

It follows that $\mu(\mathcal{A}_n) = \mu(\mathcal{A}_0)^{n+1} = \mu(\overline{\mathcal{C}})^{n+1}$, and hence that $\lim_n \mu(\mathcal{A}_n) = 0$ because $\mu(\overline{\mathcal{C}}) = 1 - \mu(\mathcal{C}) < 1$. Since $\mathcal{A}_0, \mathcal{A}_1, \ldots$ is uniformly Σ_1^0, and the measures $\mu(\mathcal{A}_n)$ converge to zero faster than the (computable) function $p(n) = q^n$, where $q > \mu(\mathcal{A}_0)$ is rational, there is some subsequence of the sequence $\{\mathcal{A}_n\}$ which is a Martin-Löf test. Since X is ML-random, it is not in the intersection of this test, so $X \notin \mathcal{A}_n$ for some least n. If $n = 0$, then $X \notin \overline{\mathcal{C}}$ and hence $X \in \mathcal{C}$. If $n > 0$, since $X \in \mathcal{A}_{n-1}$, we can choose $\sigma \in V_{n-1}$ such that $\sigma \prec X$. Since no $\tau \in V_0$ can satisfy $\sigma^\frown\tau \prec X$, it follows that $Y = \{x - |\sigma| : x \in X \ \& \ x \geq |\sigma|\} \notin \mathcal{A}_0$ as $X = \sigma^\frown Y$. Thus, $Y \in \mathcal{C}$, which, since $Y \equiv_T X$, completes the proof. $\square$

We saw in Chapter 9 that the PA degrees are precisely those which, for every nonempty Π_1^0 class, bound the degree of a member of that class. The preceding theorem can be seen as saying that the degrees of ML-random sets are precisely the analogues of PA degrees with respect to Π_1^0 classes of positive measure. This is a surprising fact because, in most other settings, the PA degrees and degrees of ML-random sets behave very differently. It is fact that if a set X is both ML-random and of PA degree, then $X \geq_T \emptyset'$ although we do not prove it.

11.4   Randomness and Computable Domination

We conclude by looking at applications of some of the ideas from computable domination to two other notions studied in the area of algorithmic randomness. We begin with the following.

Definition 11.4.1. [Terwijn and Zambella] (i) A set X is *computably traceable* if there is a computable function p such that, for each $f \leq_T X$, there is a computable function h with $|D_{h(n)}| \leq p(n)$ and $f(n) \in D_{h(n)}$ for all n.

(ii) A set X is *c.e. traceable* if there is a computable function p such that, for each $f \leq_T X$, there is a computable function h with $|W_{h(n)}| \leq p(n)$ and $f(n) \in W_{h(n)}$ for all n.

The idea of computably traceable is that there is for any function $f \leq_T X$ a strong array of "boxes" $D_{h(n)}$ such that the value $f(n)$ lies in box $D_{h(n)}$. In addition, there is a single computable function $p(n)$ which uniformly bounds the size of the boxes over all such f. The idea of c.e. traceable is the same except with a weak array $W_{h(n)}$ in place of a strong array. This is the analogous change in weakening h-simple to hh-simple by replacing a strong array by a weak one.

Clearly, every computably traceable set is c.e. traceable, and it can be shown that this implication is strict (see Downey and Hirschfeldt [2010]). On the other hand, the following theorem shows that the reverse implication is true if we restrict ourselves to sets of computably dominated degree.

Theorem 11.4.2 (Kjos-Hanssen, Nies, and Stephan, 2005). *If X is a set of computably dominated degree, then X is c.e. traceable if and only if it is computably traceable.*

Proof. Let X be a c.e. traceable set of computably dominated degree, and let p be a bound as in Definition 11.4.1 (ii). Given $f \leq_T X$, let h_0 be a computable function with $|W_{h_0(n)}| \leq p(n)$ and $f(n) \in W_{h_0(n)}$ for all n. Define a function g by

$$g(n) = (\mu s)[\, f(n) \in W_{h_0(n),s}\,],$$

so that g is total and X-computable. By Theorem 5.6.2 (ii), there exists a computable function h_1 with $h_1(n) \geq g(n)$ for all n. If we define h by letting $h(n)$ be the canonical index of the finite set $W_{h_0(n),h_1(n)}$, we have

$$|D_{h(n)}| = |W_{h_0(n),h_1(n)}| \leq |W_{h_0(n)}| \leq p(n)$$

and $f(n) \in W_{h_0(n),h_1(n)} = D_{h(n)}$. Hence, X is computably traceable. $\quad\square$

We obtain a similar result by looking at the following notion of randomness due to Kurtz. In view of Theorem 11.3.1 (i), it is implied by ML-randomness, and, as above, it can be shown that this implication is strict.

Definition 11.4.3. A *Kurtz test* is an effective sequence of clopen classes $\{\mathcal{A}_n\}_{n \in \omega}$ such that

$$(\forall n)[\, \mu(\mathcal{A}_n) < 2^{-n}\,].$$

A set X is *Kurtz random* or *weakly 1-random* if it passes every Kurtz test.

Kurtz tests are equivalent to Π_1^0 classes of measure 0 in a uniform way. Therefore, a set X is weakly 1-random iff X avoids all Π_1^0 classes of measure 0 iff X is contained in every Σ_1^0 class of measure 1.

Theorem 11.4.4 (Nies, Stephan, and Terwijn, 2005). *If X is a set of computably dominated degree, then X is ML-random if and only if it is weakly 1-random.*

Proof. Let X be a set of computably dominated degree which is not 1-random. Let $\mathcal{A}_0, \mathcal{A}_1, \ldots$ be a Martin-Löf test which X does not pass, and let f be a computable function such that $\mathcal{A}_n = [\![\, W_{f(n)}\,]\!]$ for all n. Define a function g by

$$g(n) = (\mu s)(\exists \sigma \prec X)[\, \sigma \in W_{f(e),s}\,],$$

noting that since $X \in [\![\, W_{f(n)}\,]\!]$ for all n, g is total and X-computable. By Theorem 5.6.2 (ii), there exists a computable function h with $h(n) \geq g(n)$ for all n. Define

$$\mathcal{C} = \bigcap_{n \in \omega} W_{f(n),\, h(n)}.$$

Therefore, $\mathcal{C}$ is a Π_1^0 class with $X \in \mathcal{C}$ and

$$\mu(\mathcal{C}) \leq \mu([\![W_{f(n),h(n)}]\!]) \leq \mu(S_n) = 2^{-n}$$

for all n. Hence, $\overline{\mathcal{C}}$ is a Σ_1^0 class of measure 1 not containing X, so X is not weakly 1-random. $\square$

It follows by a result of Kurtz (see [Downey and Hirschfeldt 2010]), that every hyperimmune degree contains a set which is weakly 1-random but not 1-random. Thus, the degrees separating these two randomness notions are *precisely* the hyperimmune degrees.

Part III

Minimal Degrees

12
Minimal Degrees Below $\emptyset''$

12.1 Function Trees and e-Splitting Strings

This chapter presents the important method of forcing with trees to contruct minimal degrees. Variations on this method have produced many results on degrees and their initial segments as presented in Lerman [1983]. Theorems using the minimal degree construction can be found in the bibliographies of Epstein [1975] and [1979].

Definition 12.1.1. A degree $\mathbf{a}$ is *minimal* if $\mathbf{a} > \mathbf{0}$ and there is no degree $\mathbf{b}$ such that $\mathbf{0} < \mathbf{b} < \mathbf{a}$.

Spector [1956] proved the existence of minimal degrees below $\mathbf{0}''$ and Sacks [1963a] proved their existence below $\mathbf{0}'$. Our method is a revision of that of Shoenfield [1966], which is a simplification of the Sacks method. We begin with some terminology and lemmas which will be useful in both proofs.

Definition 12.1.2. (i) Let $\alpha, \beta, \gamma \in 2^{<\omega}$ be strings. We say β and γ *split* α if $\alpha \prec \beta$, $\alpha \prec \gamma$, and β and γ are incompatible in the sense of Definition 6.5.8.

(ii) A *function tree* (abbreviated f-*tree*) is a partial computable function

$$T : 2^{<\omega} \to 2^{<\omega}$$

such that if one of $T(\alpha\hat{\ }0)$, and $T(\alpha\hat{\ }1)$ is defined, then all of $T(\alpha), T(\alpha\hat{\ }0)$, and $T(\alpha\hat{\ }1)$ are defined and $T(\alpha\hat{\ }0)$ and $T(\alpha\hat{\ }1)$ split $T(\alpha)$. (For example, the identity function $\mathrm{Id}(\sigma) = \sigma$ is an f-tree.)

Remark 12.1.3. Recall the definition of a tree $T \subseteq 2^{<\omega}$ and its associated trees in Definition 3.7.1. In this chapter we need a stronger notion which gives more information about splittings. We use the term *"f-tree"* (function tree) to distinguish this notion from the previous notion of tree. Note that if T is a total f-tree, and $\widehat{T}$ is the downward closure of $\mathrm{rng}(T)$ under initial segments, then $\widehat{T}$ is an ordinary computable tree as defined in Definition 3.7.1. This is used in the e-white Lemma 12.2.3. We use the concept of f-tree in this Part III only, and elsewhere "tree" will mean an ordinary tree.

Definition 12.1.4. (Subtrees).

(i) We say that a string σ is *on* an f-tree T if $\sigma \in \mathrm{rng}(T)$.

(ii) A *set* A is *on* T if $\sigma \prec A$ for infinitely many σ on T.

(iii) An f-tree T_1 is a *sub-f-tree* of T (written $T_1 \subseteq T$) if σ on T_1 implies σ on T (and hence A on T_1 implies A on T).

(iv) If T is an f-tree and $\nu = T(\alpha)$ is a node on T then we define the sub-f-tree T_ν which contains exactly those nodes ρ in T such that $\rho \succeq \nu$. Define for all $\beta \in 2^{<\omega}$,

$$(12.1) \qquad\qquad T_\nu\,(\beta)\ =\ T(\alpha{}^\frown\beta).$$

(This closely resembles the Definition 3.7.1 (i) of the restricted subtree T_ν of an ordinary tree T, which consisted of nodes $\rho \in T$ such that either $\rho \preceq \nu$ or $\nu \prec \rho$.)

Notation 12.1.5. *We use α, β, and γ for strings in the domain of a function tree T and ρ, σ, and τ for strings in the range.*

Definition 12.1.6. (e-Splitting a Node).

(i) Strings ρ and τ *e-split* ν on f-*tree* T, and we say ν *e-splits on* T if ρ and τ are on T, $\nu \prec \rho$, $\nu \prec \tau$, and Φ_e^ρ and Φ_e^τ are incompatible, that is,

$$(12.2) \quad (\exists x)\,(\exists y)\,(\exists z)\,(\exists t)\,[\ \Phi_{e,t}^\rho(x){\downarrow} = y\ \ \&\ \ \Phi_{e,t}^\tau(x){\downarrow} = z\ \ \&\ \ y \neq z\ ].$$

(If (12.2) holds, then ρ and τ necessarily split ν as in Definition 12.1.2 (i).)

(ii) An f-tree T is *e-splitting* if whenever $T(\alpha{}^\frown 0)$ and $T(\alpha{}^\frown 1)$ are defined, they e-split $T(\alpha)$.

Definition 12.1.7. (e-Black or e-White)

(i) A node ν on f-tree T is *e-black* if there are no e-splittings on T above ν. (This is a Π_1 question, as shown in (12.4).)

(ii) A node ν on f-tree T is *e-white* if the *immediate* extensions of ν on T form an e-splitting above ν, that is

$$T(\alpha) = \nu\ \ \&\ \ T(\alpha{}^\frown 0) = \rho\ \ \&\ \ T(\alpha{}^\frown 1) = \tau\ \ \&\ \ \rho, \tau \text{ is an } e\text{-splitting of } \nu.$$

(iii) A total f-tree T is *e-white* if every node on T is e-white, is *e-black* if every node on T is e-black, and is *e-gray* otherwise,

12.2 The e-Splitting Lemmas

We shall show that to construct a set A of minimal degree it suffices to meet for all e the following requirements, as proved in the next two lemmas below.

(12.3) R_e : A lies on a total f-tree T which is e-white or e-black.

We could add requirements to make A noncomputable, $S_e : A \neq \varphi_e$, but the following simple lemma makes this unnecessary.

Lemma 12.2.1 (Noncomputability Lemma, Posner-Epstein). *If A meets all the minimality requirements $\{R_e\}_{e\in\omega}$, then A is not computable.*

Proof. Assume toward a contradiction that A is computable. Define the computable functional Φ_e^X as follows:

$$\Phi_e^\sigma(x) = \begin{cases} \sigma(x) & \text{if } \sigma \text{ is incompatible with } A; \\ \text{undefined} & \text{if } \sigma \prec A. \end{cases}$$

If A satisfies the minimality requirement R_e then A must lie on the tree T for Φ_e and T must be either e-black or e-white. First, note that T is a total f-tree and Φ_e is the identity of the initial segments of A. Therefore, T cannot be e-black because every node ν on T has an e-splitting on T. But T cannot be e-white because no $\nu \prec A$ is half of an e-splitting. $\square$

The intuition for the following procedure to meet R_e is as follows. We start with an f-tree T which probably has many e-gray nodes and we attempt to find a sub-f-tree consisting of either all e-black nodes or all e-white nodes. The next two lemmas say that this suffices for meeting requirement R_e.

Lemma 12.2.2 (e-Black Lemma). *If A is on T, a total f-tree, ν is on T, $\nu \prec A$, T_ν is e-black, and $\Phi_e^A = g$ is total, then g is computable.*

Proof. Suppose $\Phi_e^A = g$ is total. To compute $g(x)$ find *any* τ on T_ν such that $\Phi_e^\tau(x)$ converges, and let $y = \Phi_e^\tau(x)$. We claim $g(x) = y$. Such a τ exists because $\Phi_e^A(x)$ converges, and $\nu \prec A$, so $\Phi_e^\rho(x)$ converges for some ρ on T_ν with $\rho \prec A$. Furthermore, $\Phi_e^\rho(x) = \Phi_e^\tau(x)$ because otherwise ρ and τ form an e-splitting of ν on T_ν. Therefore, $\Phi_e^\tau(x) = \Phi_e^A(x)$. Finally, g is computable because T is computable, and therefore $\text{rng}(T)$ is c.e. and can be enumerated until τ is found. $\square$

Lemma 12.2.3 (e-White Lemma). *If A is on T, a total f-tree, ν is on T, $\nu \prec A$, T_ν is e-white, and $\Phi_e^A = g$ is total, then $A \leq_T g$.*

Proof. Fix g as an oracle. We shall g-computably define a sequence of strings $\{\sigma^s\}_{s \in \omega}$ on T such that $A = \cup_s \sigma^s$. Define $\sigma^0 = \nu$. Now suppose we are given $\sigma^s = T(\alpha)$ for some α such that $\sigma^s \prec A$. Compute $\rho = T(\alpha^\frown 0)$ and $\tau = T(\alpha^\frown 1)$. These exist and e-split σ^s because T_ν is e-white. Therefore, (12.2) holds for ρ and τ. Exactly one value y or z of (12.2) agrees with $g(x)$ because $\Phi_e^A(x) = g(x)$ and $\Phi_e^A(x) = y$ or $\Phi_e^A(x) = z$. Enumerate the quadruples $\langle x, y, z, t \rangle$ until the first is found satisfying (12.2). Define $\sigma^{s+1} = \rho$ if $g(x) = y$ and define $\sigma^{s+1} = \tau$ if $g(x) = z$. Now $\sigma^{s+1} \prec A$ because A extends exactly one of ρ and τ, and $\Phi_e^A(x) = g(x)$. $\square$

12.3   The Splitting Procedure

The *splitting procedure* presented next will prune an e-gray f-tree T to obtain an e-white sub-f-tree $\widehat{T} \subseteq T$. It succeeds provided that T contains no e-black nodes. Otherwise, the procedure stalls on any e-black node ν and produces only a *partial* f-tree $\widehat{T}$ because it never finds an e-splitting on T above ν.

Definition 12.3.1. Given an f-tree T, a string σ on T, and $e \in \omega$, define $\widehat{T} = Sp(T, \sigma, e)$, the *$e$-splitting sub-$f$-tree of T above σ*, by induction on $|\alpha|$ as follows. Set $\widehat{T}(\emptyset) = \sigma$. If $\widehat{T}(\alpha) = \nu$ is defined, enumerate all tuples $\langle \rho, \tau, t, x, y, z \rangle$ such that $\nu \prec \rho$, $\nu \prec \tau$, and ν, ρ, and τ are on T, until (if ever) the first such tuple is found satisfying the e-splitting in the matrix of (12.2). Define $\widehat{T}(\alpha^\frown 0) = \rho$ and $\widehat{T}(\alpha^\frown 1) = \tau$, so that ν becomes e-*white*. If they do not exist, then $\widehat{T}(\alpha^\frown i)$ is undefined for $i = 0, 1$ and ν remains e-*black* forever. (This enumeration of tuples is done using the canonical indices of strings presented in the Notation section.)

Remark 12.3.2. The Splitting Automaton $Sp(T, \sigma, e)$. In this chapter most constructions use some oracle. However, it is crucial that this splitting procedure used as a submodule in those constructions be entirely *effective* with no oracle. The function $\widehat{T} = Sp(T, \sigma, e)$ is a partial *computable* function with inputs T, σ and e. From a fixed node ν it waits for a splitting to appear and adds it to the e-white tree it is building. If there is no such splitting, then the procedure stalls at ν and $Sp(T, \sigma, e)$ produces only a *partial* f-tree. This is how partial trees arise in the $\emptyset'$ but not the $\emptyset''$ case. We can think of $Sp(T, \sigma, e)$ as a kind of *automaton* which chugs along adding e-splittings whenever they appear, but gets stuck forever if it stumbles across a node ρ which is e-black, although the automaton cannot recognize this. If not, and if T_σ is a *total* f-tree, then $\widehat{T} = Sp(T, \sigma, e)$ is a total e-white f-tree.

12.4   The Basic Module for Minimality

We begin with the f-tree $T_{-1} = Id$, the identity. Now suppose by induction we have thinned the tree to obtain a total sub-f-tree T which has satisfied all the minimal requirements R_i for $i < e$. By Lemmas 12.2.2 and 12.2.3, to guarantee the minimality condition for e it suffices to find A and a total sub-f-tree $\widehat{T} \subseteq T$ with A on $\widehat{T}$ such that $\widehat{T}$ is either e-black or e-white. Whether a *given* node ν on T is e-black is a Π_1-question, as follows:

(12.4) ν is e-black on T $\iff$ $\neg(\exists \rho \succ \nu)(\exists \tau \succ \nu)[\,\rho$ and τ e-split $\nu\,]$?

where all strings range over T. Therefore, the following question is Σ_2 :

(12.5) $(\exists \nu$ on $T)[\,\nu$ is e-black $]$?

The basic module is as follows. Ask the Σ_2 question (12.5) above. (This assumes one has a $0''$-oracle as in Spector's theorem that $\mathbf{a} < 0''$. If not, we need to make various approximations to the Σ_2 question.)

Case 1. (12.5) holds. Choose the first e-black node ν on T. Let $\widehat{T} = T_\nu$, which is an e-black tree.

Case 2. (12.5) fails. Define $\widehat{T} = \mathcal{S}p(T, \sigma, e)$ of Definition 12.3.1, which is a total e-white tree because every node on $\widehat{T}$ e-splits on T and hence on $\widehat{T}$.

12.5   A Minimal Degree Below a $\emptyset''$-Oracle

Theorem 12.5.1 (Spector, 1956).   *There is a minimal degree* $\mathbf{a} < 0''$.

Proof. Let $T^{-1} = Id$, the identity tree, and $\sigma^{-1} = \emptyset$, the empty node.

Stage $e \geq 0$. Assume by induction on e that we are given a total f-tree T^{e-1} and string σ^{e-1} on T^{e-1} such that for $\sigma = \sigma^{e-1}$ and $T = T_\sigma^{e-1}$, and for all i, $0 \leq i < e$, we have that T is either i-black or i-white. Use the $\emptyset''$-oracle to decide the Σ_2-question of whether (12.5) holds for e and tree $T = T_\sigma^{e-1}$.

Case 1. (12.5) holds for e and T. Use a $\emptyset''$-oracle to choose the first e-black node $\nu \succ \sigma$ on T. Let $T^e = T_\nu^{e-1}$, which is an e-black tree, and let $\sigma^e = \nu$.

Case 2. (12.5) fails for e and T. Define $\widehat{T} = \mathcal{S}p(T, \sigma, e)$ using the splitting procedure in Definition 12.3.1. This produces an e-white sub-f-tree $\widehat{T} \subseteq T$. Find an extension $\widehat{\sigma} \succ \sigma$ such that $\widehat{\sigma}$ is on $\widehat{T}$. Define $T^e = \widehat{T}_{\widehat{\sigma}}$ and $\sigma^e = \widehat{\sigma}$.

This completes stage e. Let $A = \cup_e \sigma^e$. Note that the sequence of trees $\{T_{\sigma^e}^e\}_{e \in \omega}$ is a decreasing sequence of closed sets and $\sigma^e \prec \sigma^{e+1} \prec f$. Hence, by the Compactness Theorem 8.3.1 (iii) the intersection is nonempty and A lies on all these trees.

Furthermore, A has minimal degree by Lemmas 12.2.2 and 12.2.3 because for every e we have shown that A lies on an f-tree $T^e_{\sigma_e}$ which is either e-black or e-white. Furthermore, $A \leq_T \emptyset''$ because the above construction uses a only a $\emptyset''$-oracle. $\qquad\qquad\square$

12.6   Exercises

Exercise 12.6.1. (Sacks). Show that any countable ascending sequence of degrees $\mathbf{b}_0 < \mathbf{b}_1 < \cdots$ has a *minimal* upper bound $\mathbf{a}$. (Namely, $\mathbf{a}$ is an upper bound but there is no upper bound $\mathbf{d} < \mathbf{a}$.) *Hint.* Choose a set $B_n \in \mathbf{b}_n$ for each n. Use Theorem 12.5.1 to find a *total* computable f-tree T_1 satisfying R_0. Define a total B_0-computable f-tree $\widetilde{T}_1 \subseteq T_1$ which still satisfies R_0 and such that $B_0 \leq_T A$ for every A on $\widetilde{T}_1$. Define $\widetilde{T}_s(\alpha)$ by induction on $|\alpha|$ as follows. Let $\widetilde{T}_1(\emptyset) = T_1(\emptyset)$. Assume $\widetilde{T}_1(\alpha)$ has been defined and equals $T_1(\rho)$ for some ρ. If $|\alpha| = n$, define $\widetilde{T}_1(\alpha\hat{\ }i) = T_1(\rho\hat{\ }B_0(n)\hat{\ }i)$ for $i = 0, 1$. Choose an appropriate σ_1 on $\widetilde{T}_1$, $|\sigma_1| \geq n$. In general, apply the above procedure to $\widetilde{T}_n$ to obtain a B_n-computable total f-tree $\widetilde{T}_{n+1} \subseteq \widetilde{T}_n$ such that $\widetilde{T}_{n+1}$ satisfies R_n and $B_n \leq_T A$ for every A on $\widetilde{T}_{n+1}$. Next, choose $\sigma_{n+1} \in \widetilde{T}_{n+1}$ such that $\sigma_n \prec \sigma_{n+1}$. Let $A = \cup_n \sigma_n$.

Exercise 12.6.2. Show that there exist $2^{\aleph_0}$ different minimal degrees. *Hint.* Use the method of Theorem 12.5.1 with that of Exercise 6.1.6 to build a tree of f-trees.

13
Minimal Degrees Below $\emptyset'$

13.1 The Sacks Minimal Degree $\mathbf{a} < \mathbf{0'}$

The Spector proof of a minimal degree used a $\emptyset''$ oracle to prune one tree T^{e-1} to obtain the next tree T^e. Sacks used a variation of the finite injury method. He saw that the Spector method could be approximated to find a minimal degree below a $\emptyset'$ oracle.

Theorem 13.1.1 (Sacks). *There is a minimal degree $\mathbf{a} < \mathbf{0'}$.*

Proof. We begin with the *basic module* to satisfy a single requirement R_e and then give the full construction to satisfy all the requirements $\{R_e\}_{e \in \omega}$.

13.2 The Basic Module for One Requirement R_e

We use a $\emptyset'$-oracle and present the basic module to construct a tree T^e to meet a single requirement R_e as defined in (12.3). For simplicity let us fix e, the preceding tree T^{e-1}, and a string $\sigma \in T^{e-1}$ which we are trying to extend. We must define a sub-f-tree $T^e \subseteq T_\sigma^{e-1}$, which is either e-white or e-black, and a string $\sigma' \succ \sigma$ on it. As a first example, let us consider $e = 0$ so $T^{e-1} = Id$, the identity tree. Other trees $T^{e,s}$ for $e \geq 0$ will change as s changes, but $T^{-1,s} = Id$ for all s. This simplifies the description of T^0.

In the basic module in §12.4 for the minimal degree $\mathbf{a} < \mathbf{0''}$ we asked the Σ_2 question (12.4) of whether there is a node $\nu \succ \sigma$ with $\nu \in T^{e-1}$ such that ν is e-black on T^{e-1}. If so, we built $T^e = T_\nu^{e-1}$, an e-black f-tree. If not, we launched the e-splitting automaton and let $T^e = Sp(T, \sigma, e)$. We can no longer ask the oracle whether there exists an e-black node ν on T^{e-1}. Therefore, we must begin with the e-white strategy. We start with the e-white f-tree

(13.1) $$T^{e,s} = Sp(T_\sigma^{e-1}, \sigma, e).$$

Of course, this strategy may stall if it encounters a node which does not e-split. To recover from this obstacle we use the $\emptyset'$-oracle to test whether

(13.2) $(\exists \nu)_{|\nu| \leq s} [\, \nu \text{ on } T^{e,s} \quad \& \quad \nu \text{ is } e\text{-black on } T_\sigma^{e-1,s} \,].$

If so, we define $T^{e,s+1} = T_\nu^{e-1}$, an e-black f-tree, and $\sigma' = \nu$.

Notice that this is the same strategy as in the $\emptyset''$-case except that we no longer have a $\emptyset''$-oracle to determine whether there *exists* an e-black node $\nu \in T^{e-1}$. We have only a $\emptyset'$-oracle which can recognize whether a *given* node ν is e-black if we stumble across it. (The final conjuct in (13.2) is Π_1 because T_σ^{e-1} is total and computable, and hence the bounded quantifier makes this a $\emptyset'$-question.)

13.3 Putting the Strategies Together

So long as $T^{e-1,s}$ remains fixed this strategy for $T^{e,s}$ has at most one *reverse* of strategy when the e-strategy is forced to change from the e-white to the e-black stategy. If this occurs then we have a definite e-black node and as in the $\emptyset''$ case there is no reason to ever change back to the e-white strategy again.

However, suppose $e = 1$ and we are constructing a tree T^1 to meet requirement R_1. Now $T^{-1} = Id$ never changes, and the 0-tree $T^{0,s}$ changes at most once from the white to the black strategy. Next $e = 1$ is given the previous tree $T^{0,s}$. While $T^{0,s}$ is 0-white the strategy for $e = 1$ plays as if this is the final tree T^0 that we have at the end of the construction. However, if $T^{0,s}$ suddenly switches to the 0-black tree, then T^1 is injured and must restart the 1-strategy as defined in (13.3) with $e = 1$.

When the tree $T^{e-1,s} = T^{e-1}$ finally becomes fixed, the e-strategy reverses from e-white to e-black at most once and succeeds in building a tree T^e which satisfies requirement R_e. However, for a general e the tree T^{e-1} may change 2^e many times because for each $j > 0$ the j-strategy may be reset whenever the preceding tree T^{j-1} changes. Therefore, whenever tree T^{e-1} changes we must *reset* the e-strategy by redefining

(13.3) $$T^{e,s+1} = Sp(T^{e-1,s}, \sigma, e).$$

This causes the e-strategy to begin anew on the current tree T^{e-1} with the e-white strategy.

13.4   A Subtle Point

If $T^{e-1,s}$ is currently in the $(e-1)$-white strategy, then we have currently recognized only finitely many nodes as $(e-1)$-white. What does it mean that the e-strategy uses a $\emptyset'$-oracle to recognize that a node $\nu \in T^{e-1}$ is e-black? Note that e-black means no futher e-splittings on the previous tree $T^{e-1,s}$, but we have so far examined only finitely many nodes $\sigma \in T^{e-1,s}$, perhaps only σ of length less than s. How can we determine whether there are longer ones which do e-split?

The answer is that at stage s we have an index j for a (potentially) total tree $T_j = T^{e-1,s}$. In the first case, $T^{e-1,s}$ is in the $(e-1)$-black mode, in which case $T^{e-1,s} = T^{e-2,s}_\nu$ and is total if $T^{e-2,s}$ is total and is not later reset. In the second case, $T^{e-1,s}$ is in the $(e-1)$-white mode and of the form $Sp(T^{e-2,s}, \sigma, e-1)$. In this case it is also a total tree provided that the $(e-1)$-strategy never changes from white to black. If it does, then the $(e-1)$-strategy is reset. In either case we may assume by induction that we have an index for $T^{e-2,s}$ as a total computable f-tree. From this we have an index for T_j regardless of which case holds currently for $T^{e-1,s}$. Therefore, we can behave as if T_j is a total tree and the final tree T^{e-1}, because either this is true or the $(e-1)$-strategy is reset and we begin again.

We ask the $\emptyset'$-oracle the same Π_1 question as in (12.4),

$$(13.4) \quad \nu \text{ is } e\text{-black on } T_j \iff \neg(\exists \rho \succ \nu)(\exists \tau \succ \nu)[\rho \text{ and } \tau \ e\text{-split } \nu]?$$

where all strings range over T_j.

Either $T^{e,t} = T_j = T^e$ for all $t > s$, in which case T_j is total and the question gives the correct answer, or else $T^{e,s} \neq T^{e,t}$ at some future stage $t > s$, in which case we reset the strategy at stage t and it does not matter what action we took at stage s. The main point is that if we are in the $(e-1)$-white mode at stage s, we cannot let T_j be only the finite set of nodes enumerated in the white tree *so far*. Finding no e-splitting in this small tree is not sufficient reason to change from the e-white to the e-black strategy.

13.5   Constructing A to Meet Requirements $\{R_e\}_{e \in \omega}$

Define the initial string $\sigma^{-1} = \emptyset$ and the initial tree $T^{-1,s} = Id$ for all s.

Stage $s \geq 0$. Given σ^s and trees $T^{e,s}$ with $T^{e,s} \supseteq T^{e+1,s}$ for all $-1 \leq e < s$.

Case 1. Using the $\emptyset'$-oracle find the least $e \leq s$ (if it exists) such that $T^{e,s}$ is in the e-white mode but

$$(\exists \nu)_{|\nu| \leq s} [\, \nu \in T^{e-1,s} \;\; \& \;\; \nu \succ \sigma^s \;\; \& \;\; \nu \text{ is } e\text{-black on } T^{e-1,s} \,].$$

(This question is computable in $\emptyset'$ because the quantifier is bounded, the first two clauses in the matrix are computable, and the third is Π_1 by (12.4).) Choose the first such ν in the canonical listing of the nodes of T. Define

$$\sigma^{s+1} = \nu \;\; \& \;\; T^{e,s+1} = T_\nu^{e-1,s+1},$$

which is the e-black sub-f-tree of $T^{e-1,s}$ above ν. For all i, $e \leq i \leq s$, define

$$T^{i+1,s+1} \;=\; Sp(T^{i,s+1}, \sigma^{s+1}, i).$$

For all $i < e$ define $T^{i,s+1} = T^{i,s}$.

Case 2. Case 1 fails. Namely, there is no such e. For all $e \leq s$ define $T^{e,s+1} = T^{e,s}$ and $\sigma^{s+1} = \sigma^s$. For $e = s+1$ let $T^{s+1,s+1} = Sp(T^{s,s}, \sigma^s, s)$.

At the end of stage s go to stage $s+1$. Define $A = \cup_s \sigma^s$. Therefore, $A \leq_T \emptyset'$ because the sequence of strings $\{\sigma^s\}_{s \in \omega}$ is computable in $\emptyset'$.

Lemma 13.5.1. *For every e the strategy for e is reversed at most finitely often, requirement R_e is satisfied, $T^e = \lim_s T^{e,s}$ exists and A is on T^e.*

Proof. Fix e and assume these hypotheses true for every $i < e$. Choose the last stage s at which Case 1 applied to some $i < e$ if it exists, and let $s = 0$ if there is no such stage. Then $T^{e,s} = Sp(T^{e-1}, \sigma^s, e)$, the e-white sub-f-tree. The e-strategy reverses at most once after stage s, say at some stage $t > s$. Hence, $T^e = T^{e,v}$ for all $v > t$. Note that the sequence of trees $\{T^e\}_{e \in \omega}$ is a decreasing sequence of closed sets and $\sigma^s \preceq \sigma^{s+1} \prec f$. Hence, by the Compactness Theorem 8.3.1 (iii), the intersection is nonempty and A lies on all these trees. Therefore, A meets every requirement R_e and A has minimal degree. $\qquad\qquad\qquad\qquad\qquad\qquad\qquad\qquad\qquad\qquad\qquad\square$

13.6   A Limit Computable Minimal Degree

In the preceding section we built the set of minimal degree $A = \cup_s \sigma^s$ for a sequence of strings $\{\sigma^s\}_{s \in \omega}$ determined by an oracle such as $\mathbf{0}'$. In this section we give a *limit computable* construction of a set A of minimal degree. This construction produces a *computable* sequence of strings $\{\sigma^s\}_{s \in \omega}$ such that $A(x) = \lim_s \sigma^s(x)$. By the Limit Lemma 3.6.2 we have $A \leq_T \emptyset'$. Therefore, this gives another proof of Theorem 13.1.1. This limit computable construction[1] is necessary in §13.7 on a minimal degree below a nonzero c.e. degree, and is very flexible for other applications.

13.6.1 Meeting a Single Requirement R_e

We give a very brief sketch to meet a single requirement. Fix e and assume that T^{e-1} is a fixed tree. Begin with the e-splitting c.e. tree $T^{e,s}$ of (13.1) and color all these nodes e-white. The nodes extending these which have not yet e-split we color e-black and fill in the identity tree of T^{e-1} above them. If any nodes later split, they are colored e-white and the e-white boundary is extended. An e-white node never changes back to e-black so the tree $\lim_s T^{e,s}$ exists.

Choose a node α_s as follows. First, choose the longest e-white node β with the least index with this property, and then choose a γ through the black nodes so that $\alpha_s = \beta^\frown\gamma$ has length at least s. Later, if there is an e splitting ρ, τ of β, we move γ_{s+1} to one of these, say ρ, to stay as long as possible within the e-white nodes. Such a change occurs for γ at most once, so $\lim_s \alpha_s$ exists. Therefore, the infinite path A is Δ_2. It either lies on an e-white tree if the e-white boundary advances to infinity along the path A, or else there is a last e-black node on A, in which case A lies on an e-black tree. Whenever we have a choice of nodes, we choose the one with the least index satisfying the given condition.

To put the requirements together we play two versions of the strategy for $e + 1$, one within the e-white nodes and one outside the e-white boundary within the e-black nodes. The second strategy for $e + 1$ guesses that the e-white boundary never advances further along the present approximation to A, and this strategy is reset whenever that occurs. This leads to injury of the strategy. If the e-white boundary advances to infinity along A, then the strategy for $e + 1$ plays exactly as formerly, but among the e-white nodes. If the e-white boundary advances only finitely often along the path A, then the strategy for $e + 1$ is reset at most finitely often and eventually behaves exactly like the former strategy, but played now on e-black nodes.

13.7 A Minimal Degree Below a Nonzero C.E. Degree

Previously, we used an oracle such as $\emptyset''$ or $\emptyset'$ to determine for a given node ν whether it ever e-split or remained e-black forever. Now, we have no oracle, only a *computable* construction, and therefore in this version of the minimal degree method we must begin with the assumption that every

---

[1]This was previously called a *full approximation construction*. Now it is called a *limit computable construction* as in Definition 3.5.4 and in the Limit Lemma 3.6.2.

node ν is e-black. If e-splittings appear during the construction, we can begin building the e-splitting white tree as in §12.3, where we started to construct the e-white tree $\widehat{T} = \mathcal{S}p(T, \sigma, e)$.

We now give only a very small sketch to convey the main idea for a reader familiar with the limit computable minimal degree construction above and with the permitting construction of §5.2. For more details on this theorem see Lerman [1983, Chapter XI], or Epstein [1979, Chapter XI], or Epstein [1975, Chapter II].

Theorem 13.7.1. *If C is a noncomputable c.e. set, then there is a set $A \leq_T C$ of minimal degree.*

Proof. (Very brief sketch). We give a very brief sketch for meeting a single requirement. Fix e and assume that T^{e-1} is a fixed tree. Begin with the e-splitting c.e. tree $T^{e,s}$ of (13.1) and color all these nodes e-white as in §13.6.1. Begin by defining $\alpha_s = \beta_s\widehat{\ }\gamma_s$, as before. However, if we later see an e-splitting ρ, τ of β_s, we cannot immediately move γ_s to it, but can only later at stage $t > s$ if there is a $C_t \upharpoonright n$ change where $n = |\beta|$, which permits this move. We start to define a computable function $g(t) = n$ to encourage C to make at least one such change. We continue defining $g(t_i) = n_i$ as more such points appear.

There are three cases. If infinitely many points appear and none is permitted, then C is computable. If one is permitted, we make the change and begin the strategy again until A lies on an e-white tree. If there are at most finitely many such requests for a change to an e-white splitting, then A lies on an e-black tree. $\square$

Corollary 13.7.2. *There is a low minimal degree.*

Part IV

Games in Computability Theory

14

Banach-Mazur Games

14.1 Banach-Mazur Games and Baire Category

14.1.1 Meager and Comeager Sets

We use the definitions and notation on the topology of Cantor space and Baire space from §8.1 with basic open sets $[\![\sigma]\!]$ defined in (8.1). We extend Definition 8.6.1, which introduced dense sets and dense open sets.

Definition 14.1.1. Let $\mathcal{S}$ be Cantor space 2^ω. (Baire space ω^ω is similar.)

(i) For a string $\tau \in 2^{<\omega}$ a set $\mathcal{A}$ is *dense in* the basic open set $[\![\tau]\!]$ if

$$(14.1) \qquad (\forall \sigma \succeq \tau)(\exists f \succ \sigma)[\, f \in \mathcal{A}\,].$$

(ii) A set $\mathcal{A} \subseteq \mathcal{S}$ is *dense* if $\mathcal{A}$ is dense in $[\![\nu]\!]$ for ν the null string, i.e.,

$$(14.2) \qquad (\forall \sigma)(\exists f \succ \sigma)[\, f \in \mathcal{A}\,].$$

(iii) A set $\mathcal{A} \subseteq \mathcal{S}$ is a *dense open set* if

$$(14.3) \qquad (\forall \tau)(\exists \sigma \succ \tau)(\forall f \succ \sigma)[\, f \in \mathcal{A}\,].$$

Therefore, a set $\mathcal{A}$ is *dense open* if it is both dense and open. Note that $\mathcal{A}$ contains the dense open set $\mathcal{D} := \cup_\sigma [\![\sigma]\!]$ for all σ in the second quantifier of (14.3).

(iv) $\mathcal{A}$ is *comeager* if it contains the intersection of a countable family of dense open sets, namely $\mathcal{A} \supseteq \cap_i [\![A_i]\!]$ for $[\![A_i]\!]$ dense open.

(v) $\mathcal{A}$ is *meager* if its complement is comeager. (Also $\mathcal{A}$ is meager if it is of first category, See Exercise 14.1.6 for a discussion of first and second category and their relationship to comeager.)

The intuition is that comeager sets are *large*. They form a filter, are dense, uncountable, and are closed under countable intersections. Meager sets are *small*. They form an ideal, and countable sets are meager.

14.1.2 The Baire Category Theorem

In many theorems we are given a sequence $\{\mathcal{A}_n\}_{n \in \omega}$ of subsets of 2^ω, and we want to construct a point $f \in 2^\omega$ which meets every one. If the sets $\mathcal{A}_n$ were merely *dense* this might not be possible. (See Exercise 14.1.4.) However, if every $\mathcal{A}_n$ is dense and *open*, then the dense open property (14.3) ensures that we can do exactly that. Hence, with a finite extension at stage $s+1$ from σ_s to $\sigma_{s+1} = \rho$ we can satisfy once and for all the requirement $R_n : f \in \mathcal{A}_n$ and move on to other requirements on the list. This is the essence of the following theorem by Baire.

Theorem 14.1.2 (Baire Category Theorem, 1899). *Let $\mathcal{A}$ be comeager.*

(i) $\mathcal{A}$ is not empty.

(ii) Indeed, $(\forall \rho \in 2^\omega)\,[\,\mathcal{A} \cap [\![\rho]\!] \neq \emptyset\,]$. Therefore, $\mathcal{A}$ is dense.

Proof. If $\mathcal{A}$ is comeager, then $\mathcal{A} \supseteq \cap_i [\![A_i]\!]$ for A_i dense open. Given any $\tau \succeq \rho$ and i there exists $\sigma_i \succ \tau$ such that $\sigma_i \in A_i$. Construct $f = \cup_i \sigma_i$ where $\sigma_i \prec \sigma_{i+1}$ and $\sigma_i \in A_i$ for all i. Hence, $f \in (\cap_i [\![A_i]\!]) \cap [\![\rho]\!]$.   □

14.1.3 Banach-Mazur Games

In 1928 the Polish mathematician S. Mazur invented the following game, called a *Banach-Mazur game*. We fix ahead of time a set $\mathcal{A} \subseteq 2^\omega$. Player I chooses string $\sigma_0 = \emptyset$. Player II chooses $\sigma_1 \succ \sigma_0$. For $i \geq 1$, given σ_{2i-1}, Player I first chooses $\sigma_{2i} \succ \sigma_{2i-1}$ and then Player II chooses $\sigma_{2i+1} \succ \sigma_{2i}$. The *play* of the game is the infinite sequence $f = \cup_n \sigma_n$ so constructed. If $f \in \mathcal{A}$, then Player I wins, and otherwise Player II wins. (Note that f is a unique point because $\sigma_i \prec \sigma_{i+1}$. This strict extension guarantees that $\cap_i [\![\sigma_i]\!] = f$.)

This is similar to the Gale-Stewart game in Chapter 15 except that in the latter each σ_{i+1} is exactly a *one* point extension of σ_i. As before, a *winning strategy* for one of the players is a function from finite positions telling him which string to play next. It follows from the proof of the Baire Category Theorem 14.1.2 that if $\mathcal{A}$ is comeager, then Player I has a winning strategy. Mazur conjectured the converse, which Banach proved.

Theorem 14.1.3 (Banach-Mazur). *Player I has a winning strategy in the Banach-Mazur game for $\mathcal{A}$ iff $\mathcal{A}$ is comeager. (Equivalently, Player II has a winning strategy iff $\mathcal{A}$ is meager.)*

Proof. ($\Longleftarrow$). Let Player I apply the strategy of Theorem 14.1.2 (ii) with $\rho = \sigma_{2i-1}$ to find $\sigma_{2i} \succ \rho$ such that $\sigma_{2i} \in \mathcal{A}_i$.

($\Longrightarrow$). (The reverse is more complicated and will be omitted.)   □

14.1.4   Exercises

Exercise 14.1.4. Construct dense sets $\mathcal{A}_0, \mathcal{A}_1 \subset 2^\omega$ such that $\mathcal{A}_0 \cap \mathcal{A}_1 = \emptyset$.

Exercise 14.1.5. For any set $\mathcal{A} \subseteq 2^\omega$ define the *closure* of $\mathcal{A}$, denoted by $\mathcal{A}^{cl}$, to be the smallest closed set $\mathcal{C} \supseteq \mathcal{A}$. Given $\mathcal{A}$ define the tree

$$T_\mathcal{A} = \{\, \sigma \;:\; (\exists f \succ \sigma)[\, f \in \mathcal{A}\,]\,\}.$$

(i) Show that $T_\mathcal{A}$ is an extendible tree. Prove that $\mathcal{A}^{cl} = [\, T_\mathcal{A}\,]$.

(ii) Consider the points $f \in \mathcal{A}^{cl} - \mathcal{A}$. Are they limit points or isolated points?

(iii) Let $\mathcal{A} = \{\, 1^n 0^\infty \,\}_{n \geq 1}$, where $1^n 0^\infty$ denotes a string of n 1's followed by an infinite string of 0's. Describe $\mathcal{A}^{cl}$ and $\mathcal{A}^{cl} - \mathcal{A}$. What is the Cantor-Bendixson rank of $f \in \mathcal{A}^{cl} - \mathcal{A}$? (See Definition 8.7.5 and surrounding exercises for the Cantor-Bendixson rank.)

(iv) Construct a set $\mathcal{A}$ such that there exists $f \in \mathcal{A}^{cl} - \mathcal{A}$ of Cantor-Bendixson rank 2.

Exercise 14.1.6. A set $\mathcal{A}$ is *nowhere dense* if there is no τ such that $\mathcal{A}$ is dense in $[\![\, \tau\,]\!]$ in the sense of (14.1).

(i) Prove that $\mathcal{A}$ is nowhere dense iff its complement $\overline{\mathcal{A}} = 2^\omega - \mathcal{A}$ contains a dense open set.

(ii) Prove that if $\mathcal{C} = [T]$ is a nowhere dense closed set then

$$(\forall \sigma \in T)(\exists \tau \succ \sigma)[\, \tau \notin T\,].$$

Hence, the tree T has lots of "holes."

(iii) R. Baire defined a set $\mathcal{A}$ to be *of first category* if it is a countable union of nowhere dense sets. Prove that $\mathcal{A}$ is of first category iff $\mathcal{A}$ is meager as in Definition 14.1.1 (vi).

(iv) Baire defined an $\mathcal{A}$ to be *of second category* if it is not of first category. Prove that if $\mathcal{A}$ is comeager then it is of second category. Assume that $\mathcal{A}$ is both comeager and of first category (meager) and derive a contradiction. (It is false that a set $\mathcal{A}$ of second category is necessarily comeager.)

Exercise 14.1.7. Prove the following. (i) The comeager sets form a filter, i.e., are closed under supersets and intersections.

(ii) A comeager set is uncountable.

(iii) The comeager sets are closed under countable intersections.

Exercise 14.1.8. Fix a set $X \leq_T \emptyset'$. Let $\mathcal{A}_i$, $i \in \omega$, be a (countable) uniformly Δ_2^0 sequence of Δ_2^0 sets which are dense open.

(i) Prove that if $X \equiv_T \emptyset'$, then there is an $f \in \cap_i \mathcal{A}_i$ such that $f \leq_T X$.

(ii) Now assume that the $\mathcal{A}_i$ are not necessarily all dense. What oracle X suffices to construct $f \leq_T X$ such that $f \in \mathcal{A}_i$ for every dense $\mathcal{A}_i$?

14.2   The Finite Extension Paradigm

The following theorem expresses the essence of all the finite extension oracle constructions seen in §6.1–§6.4, although they were not exactly stated this way. All these and many more can be derived immediately from this one paradigm theorem by appropriately defining the sets $\{V_e\}_{e\in\omega}$.

Theorem 14.2.1 (Finite Extension Paradigm). *As in Definition 2.6.1, given a u.c.e. sequence $\mathbb{V} = \{V_e\}_{e\in\omega}$ of c.e. sets, there exists $f \leq_T \emptyset'$ which is $\mathbb{V}$-generic, i.e., f forces every V_e in the sense of (6.11).*

Proof. We construct a $\emptyset'$-computable sequence of strings $\{\sigma_s\}_{s\in\omega}$.

Stage $s = 0$. Let $\sigma_0 = \epsilon$ the null string.

Stage $s + 1$. Given σ_s let $e = s$ and ask the $\emptyset'$-oracle whether:

(14.4) $(\exists \rho \succ \sigma_s)\,[\,\rho \in V_e\,]$ (denoted $\rho \Vdash f \in V_e$).

If so, find the first such ρ and define $\sigma_{s+1} = \rho$. If not, define σ_{s+1} to be the first $\rho \succ \sigma_s$. Define $f = \cup \sigma_s$. Clearly, $f \leq_T \emptyset'$. Note that if (14.4) fails, then automatically

(14.5) $(\forall \rho \succ \sigma_s)\,[\,\rho \notin V_e\,]$.   (denoted $\sigma_s \Vdash f \notin V_e$).

If (14.4) holds we pronounce $\rho \Vdash f \in [\![V_e]\!]$ as asserting that ρ *forces* $f \in [\![V_e]\!] = V_e$ and if (14.5) holds we pronounce $\rho \Vdash f \notin [\![V_e]\!]$ as asserting that ρ *forces* $f \notin [\![V_e]\!]$. In either case, $\sigma_{s+1} \Vdash f \in V_e$ or $\sigma_{s+1} \Vdash f \notin V_e$ so f *forces (decides)* V_e via string $\sigma = \sigma_{s+1}$.   □

Theorem 14.2.1 can be used to derive most of the finite extension results such as those in §6.1–§ 6.4.

For the Friedberg Jump Theorem 6.4.1 we also had to code a set B. It is easy to add this feature to the paradigm Theorem 14.2.1 whether or not $B \geq_T \emptyset'$. Given a sequence of strings, $S = \{\sigma_s\}_{n\in\omega}$, define the function which codes the sequence to be $g_s(n) = \sigma_n$.

Theorem 14.2.2 (Finite Extension Coding Paradigm). *Fix a uniformly c.e. sequence* $\mathbb{V} = \{V_e\}_{e\in\omega}$ *of c.e. sets and any set* $B \subseteq \omega$.

(i)   There exists a sequence $S = \{\sigma_s\}_{s\in\omega}$ *such that* $f = \cup_s \sigma_s$ *is* $\mathbb{V}$*-generic (forces every* V_e*) and*

(14.6) $$g_s \leq_\mathrm{T} B \oplus \emptyset' \quad \& \quad g_s \leq_\mathrm{T} A \oplus \emptyset'.$$

(ii)   If $\mathbb{V} = \{W_e\}_{e\in\omega}$ *then* f *is 1-generic and* $A' \equiv_\mathrm{T} A \oplus \emptyset' \equiv_\mathrm{T} B \oplus \emptyset'$.

Proof. Do the same as in Theorem 14.2.1 except that at stage $s + 1$ after finding $\rho \succ \sigma_s$ which decides V_s, let $\sigma_{s+1} = \rho^\frown B(s)$. $\square$

14.2.1   Finite Extension Games

It is useful to think of the method in the Finite Extension Paradigm Theorem 14.2.1 as a modification of the Banach-Mazur game in §14.1.3 with the following changes.

1. Player I plays the sequence $\{\sigma_s\}_{s\in\omega}$ of strings $\sigma_s \prec \sigma_{s+1}$, and defines $f = \cup_s \sigma_s$. (Player II plays *none* of the strings σ_s. The sequence $\{\sigma_s\}_{s\in\omega}$ is usually constructed computably in some oracle X, often $X = \emptyset'$ or some related set such as $X = D'$ for some set $D \subseteq \omega$.)

2. Player II plays a u.c.e. sequence $\mathbb{V} = \{V_e\}_{e\in\omega}$ of c.e. sets $V_e \subseteq \omega$.

3. Player I *wins* if f forces V_e for every e.

4. A *winning strategy* for Player I is a function from finite positions describing which string σ_{s+1} to play next. (For example, the proof of the Finite Extension Paradigm Theorem 14.2.1 yields a winning strategy for Player I computable in $\emptyset'$.)

Definition 14.2.3. Constructions of the type in Finite Extension Paradigm Theorem 14.2.1 and in the finite extension games are called *finite extension oracle* constructions or sometimes *Kleene-Post* constructions because of Kleene-Post [1954].

14.2.2   Exercises

Exercise 14.2.4. Show that every c.e. set $A >_\mathrm{T} \emptyset$ bounds a 1-generic set B. *Hint.* Fix a computable enumeration $\{A_s\}_{s\in\omega}$ of A. Build a Δ_2 set $B = \lim_s B_s$ by a standard permitting argument so that $B \leq_\mathrm{ibT} A$ as in Theorem 5.2.7. Construct B to meet, as in (6.11), every e requirement,

$$R_e: \quad (\exists \sigma \prec A)\, [\, \sigma \in V_e \quad \vee \quad (\forall \tau \succ \sigma)\, [\, \tau \notin V_e\,]\,]$$

as follows. Let σ_1 be the first string (if any) which appears in V_e, say at stage s_1, and let $x_1 = (\mu y)\,[\sigma_1(y) \neq B_{s_1}(y)]$. Given σ_i, x_i, and s_i for $i < j$, $x_j > x_{j-1}$, $s_j > s_{j-1}$ and $\sigma_j \in V_{e,s_j}$ such that $\sigma_j \restriction x_j \prec A_{s_j}$, define $f_e(y) = A_{s_j}(y)$ for all $y \leq x_j$. If $A_t \restriction x_j \neq A_{s_j} \restriction x_j$ for some j and $t > s_j$ then A permits us to change B_t so that $B_{t+1} \succ \sigma_j$. If not, then A is computable via f_e providing V_e is dense along B.

15
Gale-Stewart Games

15.1 Gale-Stewart Games and Open Games

Gale-Stewart games illustrate the applications of Π_1^0-classes. In a Gale-Stewart game there are two players who alternately choose elements $a_i \in \{0,1\}$. Player I chooses a_0, then player II chooses a_1, and so on. The infinite sequence f chosen, namely $f(n) = a_n$, is the particular *play* of the game. We fix ahead of time a set $\mathcal{A} \subseteq 2^\omega$. In game $\mathcal{G}(\mathcal{A})$ player I wins if the play $f \in \mathcal{A}$ and II wins otherwise. A *winning strategy* for player I is a function g on finite positions in the game, namely strings $\sigma \in 2^{<\omega}$ of even length (nodes at which I is to play), such that $g(\sigma) \in \{0,1\}$ and if I follows strategy g then he wins the game. Likewise, a winning strategy for player II is defined on strings σ of odd length (where Player II is to play), and guarantees a win for player II. The game $\mathcal{G}(\mathcal{A})$ is *determined* if one player or the other has a winning strategy. The first easy theorem about these games is that $\mathcal{G}(\mathcal{A})$ is determined if $\mathcal{A}$ is open, namely boldface $\mathbf{\Sigma}_1$. This means that $\mathcal{A} = [\![A]\!]$ for some set $A \subseteq \omega$ as defined in (8.2). We can play as if this were an *effectively* open set by fixing the parameter A as an oracle. We analyze the computable content of this game and the winning strategies.

Theorem 15.1.1 (Gale-Stewart, 1953). *If $\mathcal{A} \subseteq 2^\omega$ is open, then the game $\mathcal{G}(\mathcal{A})$ is determined.*

Proof. Let $\mathcal{A} = [\![A]\!]$ as in (8.2). Define an open set $[\![B]\!] \supseteq [\![A]\!]$, namely a certain A-c.e. set $B \subseteq 2^{<\omega}$, $B \supseteq A$, by induction as follows,

$$(15.1) \qquad\qquad\qquad \sigma \in A \quad\Longrightarrow\quad \sigma \in B$$

$$(15.2) \qquad |\sigma| \text{ even} \quad \& \quad (\exists i)\,[\sigma^\frown i \in B\,] \quad . \quad\Longrightarrow\quad \sigma \in B$$

$$(15.3) \qquad |\sigma| \text{ odd} \quad \& \quad (\forall i)\,[\sigma^\frown i \in B\,] \quad . \quad\Longrightarrow\quad \sigma \in B.$$

The set B represents the nodes σ from which Player I has a winning strategy to eventually get into the open set $\mathcal{A}$. In (15.1) if $\sigma \in A$ then $\sigma \in B$ because Player I has already ensured that $f \in [\![A]\!]$. Now $|\sigma|$ even means that Player I is to play next. Therefore, if (15.2) holds, then there is an immediate extension $\tau = \sigma^\frown i \in B$ which Player I can play. Hence, by moving from σ to $\tau \in B$ Player I can ensure inductively that he has a winning strategy from position σ. The case $|\sigma|$ odd, namely Player II to play, is similar but *every* extension $\tau = \sigma^\frown i$ must be in B or else Player II can move to avoid nodes in B.

Note that B is an A-computably enumerable set. (This uses the compactness of 2^ω because in ω^ω for (15.3) we would have to examine infinitely many i before putting σ into B.) Choose an A-computable tree T such that $[T] = 2^\omega - [\![B]\!]$. If T is infinite then Player II has a winning strategy g which consists of always choosing nodes $\sigma \notin B$. This strategy g amounts to choosing a path on $[T]$. This strategy is not necessarily A-computable because although the tree T is A-computable, the tree of extendible nodes T^{ext} is only computable in A' by the Effective Compactness Theorem 8.5.1. If the tree T is finite, then player I has a winning strategy $h \leq_T A$. $\square$

15.1.1 Exercises

Exercise 15.1.2. Now assume that A is computable with $\chi_A = \varphi_k$.

(i) Prove that if Player I has a winning strategy h, then Player I has a computable winning strategy.

(ii) Prove that if Player II has a winning strategy, then he has a winning strategy $g \leq_T \emptyset'$.

(iii) Prove that if Player II has a winning strategy g then he has a low winning strategy h, namely such that $h' \equiv_T \emptyset'$. (You must consider the Π^0_1 class of *strategies* for player II, not just the class of plays.)

(iv) (Slaman) Prove that (i) is not uniform. Use the Recursion Theorem to prove there is no total computable function ψ such that for all k, if $\chi_A = \varphi_k$, then $\psi(k)$ converges, and if Player I has a winning strategy then $\varphi_{\psi(k)}$ is an effective winning strategy for Player I.

15.1.2   Remarks on the Axiom of Determinacy

D.A. Martin [1975] proved determinacy for all Borel sets. The Axiom of Determinacy (AD) asserts that all games are determined. The assumption that definable sets are determined plays an important role in set theory. Full AD contradicts the Axiom of Choice (AC) but nevertheless is an important tool. A *cone* of degrees is a set of degrees of the form $\{d : d \geq a\}$ for some degree a.

Theorem 15.1.3 (Martin, 1968). *If AD holds, then every set of degrees either contains a cone or is disjoint from a cone.*

Proof. Given a set $\mathcal{A}$ of degrees, let $\mathcal{A}^*$ be the class of sets whose degrees are in $\mathcal{A}$. Suppose Player I has a winning strategy for $G(\mathcal{A}^*)$. That strategy has a degree a and every degree $b \geq a$ must be in $\mathcal{A}$. Choose any function $f \in b$, let Player II play according to f and Player I according to the winning strategy. The final outcome will be a function of degree b. The outcome must be in $\mathcal{A}^*$ because Player I was following a winning strategy. Hence, $b \in \mathcal{A}$. Similarly, if Player II has a winning strategy of degree d then all degrees $c \geq d$ lie in $\overline{\mathcal{A}}$. $\qquad\square$

16
More Lachlan Games

16.1 Increasingly Complicated Constructions

With the Kleene-Post [1954] paper on oracle constructions presented in Chapter 6, and the finite injury computable approximations to them in Friedberg [1957] and Muchnik [1956] presented in Chapter 7, constructions in computability entered a new and much more complicated phase in the 1960s. Shoenfield [1961] and independently Sacks [1963a, 1963c, and 1964a] invented the *infinite injury* method for constructing c.e. sets and degrees. These infinite injury constructions were very difficult to read, even more so because they were presented in the Kleene T predicate notation of Kleene's papers fashionable at the time. Lachlan [1966b] and independently Yates [1966a] invented a still more difficult infinitary method, called the *minimal pair method*, to construct a minimal pair of c.e. degrees.

By the middle of the 1960s computability theory was in danger of being crushed by proofs which were so complicated that it was difficult for a reader to even verify them, much less extend them to new theorems. By the late 1970s, Lachlan had invented an intuitive game theory model for constructing c.e. sets which clearly revealed the intuition. This remains one of the most important tools in the subject for understanding the material, presenting theorems, and solving new problems.

16.2   Lachlan Games in Computability Theory

Several kinds of games have played a role in computability theory. In the Banach-Mazur game of Chapter 14, the players I and II alternately choose finite sequences $\sigma \in 2^{<\omega}$ to construct a set $A = \cup_s \sigma_s$. Player I has a winning stategy for forcing A into a target set $\mathcal{A} \subseteq 2^\omega$ iff $\mathcal{A}$ is comeager. The Banach-Mazur game is particularly well suited to the finite extension constructions of a set A in Chapter 6 computable in some oracle X because the sequence of strings $\{\sigma_s\}_{s\in\omega}$ is X-computable. Therefore, we can X-computably determine $A(x)$ for every x. The Gale-Stewart game in Chapter 15 is similar except that the strings must be of length 1.

These games are not suitable for a construction of a *computable enumerable* set A because the latter requires a computable construction with no oracle X. After studying Martin's advances [1970] on Gale-Stewart games to study measurable cardinals and analytic games, Lachlan proposed a new kind of game to analyze problems and to construct *computably enumerable* sets. Lachlan [1970] observed that many theorems in computability theory can be viewed as a game between two players, Player I (RED) and Player II (BLUE). We refer the reader to the definition in §2.5 of a Lachlan game.

16.2.1   Playing Turing Reductions

This definition may seem too restrictive because in many theorems in computability we may allow one or both players to construct other objects such as partial computable (p.c.) functions, or Turing reductions $\Phi_e^A(x)$, e.g. in the Friedberg-Muchnik theorem, to meet a requirement such as $R_e : B \neq \Phi_e^A$. However, no generality is lost because these objects can be constructed as c.e. sets. For example, to define φ_e we enumerate the c.e. set

$$\text{graph}\,(\varphi_e) \; := \; \{\, \langle x, y \rangle \; : \; \varphi_e(x) = y \,\}$$

which we defined in Definition 2.1.7. Likewise, we can identify the Turing functional Φ_e with the c.e. set which is the oracle graph G_e from Definition 3.3.7:

$$(16.1) \qquad G_e \; := \; \{\, \langle \sigma, x, y \rangle \; : \; \Phi_e^\sigma(x) \; = \; y \,\}.$$

To play a Turing functional we simply enumerate axioms of this type in the oracle graph on moves during the game. Therefore, allowing the players to each build infinitely many c.e. sets is about as general as any theorem we prove using computable constructions, and oracle constructions are simply computable constructions relativized to the oracle. Therefore, these Lachlan games are as general as we need to prove most theorems about c.e. sets.

16.3 Some Easy Examples of Lachlan Games

We begin by analyzing some known theorems from preceding chapters in terms of Lachlan games. Since BLUE has a winning strategy for these games we can let RED play $V_n = W_n$ without loss of generality. We give the reference to the earlier mentioned result and sketch the winning strategy for BLUE as a game.

16.3.1 Theorem 5.2.3: Post's Simple Set

To prove Theorem 5.2.3 as a game we allow RED to play $U_n = W_n$ and allow BLUE to play a single set $A = V_0$ which he must guarantee is coinfinite and satisfies for all e the requirement (5.1),

$$P_e: \quad |W_e| = \infty \quad \Longrightarrow \quad W_e \cap A \neq \emptyset.$$

The method in Theorem 5.2.3 gives a winning strategy for BLUE to satisfy P_e by waiting roughly until $W_{e,s} \cap A_s = \emptyset$ and there is some $x \in W_{e,s}$, $x > 2e$. Then BLUE enumerates x into A. Theorem 5.2.5 gives a second winning strategy for BLUE .

16.3.2 Theorem 5.2.7: Permitting a Simple Set $A \leq_T C$

In Theorem 5.2.7 RED plays a noncomputable c.e. set C and $\{W_n\}_{n \in \omega}$ while BLUE plays a simple set A and a Turing reduction $\Psi^C = A$ by permitting. To ensure that C permits often enough for BLUE to satisfy every P_e requirement, BLUE also plays for every e a p.c. function g_e. Suppose that some requirement P_e fails because W_e is infinite but $W_e \cap A = \emptyset$. Then, as in Lemma 5.2.8, with each new element x appearing in W_e, BLUE extends $g_e(y)$ on arguments $y \leq x$ and $C \upharpoonright x$ will not later change. Hence, C is computable.

   (Note that the blue function g_e is rarely explicitly mentioned in proofs of this theorem, but it is implicit in any such proof because it shows how BLUE can force a given W_e to eventually permit. There are often such implicit elements in proofs which the Lachlan game makes explicit and provides a method to resolve the conflict.)

16.3.3 Theorem 7.4.1: A Simple Set $A \not\geq_T C$

In this theorem RED plays a noncomputable c.e. set C and a Turing reduction Φ_e while BLUE plays a noncomputable (say simple) c.e. set A such that $\Phi_e^A \neq C$. To assist in this, BLUE again plays a computable function g_e and attempts to preserve agreements between Φ_e^A and C while g_e records these agreements. If RED allows these agreements to go to infinity, then BLUE achieves $g_e = C$, refuting the hypothesis that C is noncomputable.

16.3.4   Friedberg-Muchnik Theorem 7.3.1

RED plays Turing functional Φ_e and BLUE plays c.e. sets A and B. For e even, BLUE wants to satisfy the requirement $R_e : A \neq \Phi_e^B$, and vice versa for e odd. To aid in meeting requirement R_e, BLUE builds a wall of restraint $r(e,s)$, not allowing any element $x < r(e,s)$ to enter B for the sake of any lower priority requirement R_j, $j > e$. At some stage t, BLUE chooses a fresh element $x > t$, and therefore $x > r(i,s)$ for all higher priority $i < e$. BLUE waits until, if ever, $\Phi_e(x)[s]\!\downarrow\,= 0$ for some $s > t$. He puts x into A and defines $r(e,s) = s$ to preserve the computation.

This restraint is not an official element of the Lachlan game, but rather an internal notation to aid BLUE. This strategy ensures that BLUE's action will be injured at most finitely often and he will eventually succeed in meeting every requirement.

Part V

History of Computability

17
History of Computability

17.1 Hilbert's Programs

Around 1880, Georg Cantor, a German mathematician, invented naive set theory. A small fraction of this is sometimes taught to elementary school children. It was soon discovered that this naive set theory was inconsistent because it allowed unbounded set formation, such as the set of all sets. David Hilbert, the world's foremost mathematician from 1900 to 1930, defended Cantor's set theory but suggested a formal axiomatic approach to eliminate the inconsistencies. He proposed two programs. *First,* Hilbert wanted an axiomatization for mathematics, beginning with arithmetic, and a finitary consistency proof of that system. *Second,* Hilbert suggested that the statements about mathematics be regarded as formal sequences of symbols, and his famous *Entscheidungsproblem* (decision problem) was to find an algorithm to decide whether a statement was valid or not. Hilbert characterized this as the fundamental problem of mathematical logic.

Hilbert retired and gave a special address in 1930 in Königsberg, the city of his birth. Hilbert spoke on the importance of mathematics in science and the importance of logic in mathematics. He asserted that there are no unsolvable problems and stressed, "We must know. We will know." At a mathematical conference preceding Hilbert's address, a quiet, obscure young man, Kurt Gödel, only a year beyond his Ph.D. in 1931, refuted Hilbert's consistency program with his famous incompleteness theorem and changed forever the foundations of mathematics. Gödel soon joined other leading figures, Albert Einstein and John von Neumann, at the Institute for Advanced Study in Princeton.

17.2   Gödel, Church, and Recursive Functions

The refutation of Hilbert's first program on consistency gave hope for refuting his second program on the *Entscheidungsproblem*. However, this was no ordinary problem in number theory or analysis. To prove the unsolvability of a certain problem, such as Hilbert's famous Tenth Problem on Diophantine equations of 1900, one must: (1) find a precise mathematical definition for the intuitive idea of algorithm; (2) demonstrate beyond doubt that every algorithm has been captured; (3) prove that no algorithm on the list can be the solution of the Diophantine equation problem.

Work began independently at Princeton and Cambridge. Alonzo Church completed an A.B. degree at Princeton in 1924 and his Ph.D. degree there under Oswald Veblen in 1927. Church joined the Department of Mathematics at Princeton from 1929 until his retirement in 1967, when he moved to UCLA. Church worked from 1931 through 1934 with his graduate student, Stephen Cole Kleene, on the formal system of λ-definable functions. They had such success that in 1934 Church proposed privately to Gödel that a function is effectively calculable (intuitively computable) if and only if it is λ-definable. Gödel rejected this first version of Church's Thesis. In addition, Kleene reported "chilly receptions from audiences around 1933–35 to disquisitions on λ-definability."

However, in the spring of 1934 Gödel lectured on recursive functions. By 1935 Church and Kleene had moved enthusiastically to the formalism of Gödel's recursive functions as a vehicle to capture the intuitive idea of effectively calculable. In 1931 Gödel had used the *primitive* recursive functions, those where one computes a value $f(n)$ by using previously computed values $f(m)$, for $m < n$, such as the factorial function $f(0) = 1$ and $f(n+1) = (n+1) \cdot f(n)$.

In his 1934 lectures at Princeton, Gödel extended this to the (Herbrand-Gödel) (*general*) recursive functions. Church eagerly embraced them and formulated his famous *Church's Thesis* in [Church 1935] and [Church 1936] that the effectively calculable functions coincide with the recursive functions. Again, Gödel failed to accept this thesis even though he was the author of the recursive functions. Gödel noted that recursive functions are clearly effectively calculable, but the converse "cannot be proved, since the notion of finite computation is not defined, but it serves as a heuristic principle."

17.2.1 The Concept of Recursion

The term *recursion* refers to a function f defined by induction. We first define $f(0)$ and then define $f(x+1)$ in terms of previously defined functions using as inputs x and $f(x)$. For example, the factorial function $f(x) = x!$ is defined by the recursion schemes

(17.1) $f(0) = 1$ and $f(x+1) = (x+1) \cdot f(x)$,

where we assume that multiplication has been previously defined. Dedekind in [Dedekind 1888] showed that certain functions could be uniquely defined by recursion. The concept of recursion gradually developed during the early 1900s particularly in the work of [Skolem 1923], [Hilbert 1926] and especially [Gödel 1931]. Many logic papers may be found in the source book [van Heijenoort 1967].

17.2.2 The Primitive Recursive Functions

Up until the early 1930s, the term "recursive function" meant what we now call a *primitive recursive function* to distinguish it from the *Herbrand-Gödel general recursive function* defined in §17.2.4. In 1931 Gödel used primitive recursive functions in the proof of his famous incompleteness theorem and called them simply by the German term "rekursiv." The main property of recursion is the primitive recursion scheme (V) below, which yields an inductive definition of $f(n+1)$ using the preceding value $f(n)$ and previously defined functions g and h. [Kleene 1952] put the primitive recursive functions in the following succinct form which has become standard.

Definition 17.2.1. The class of *primitive recursive functions* is the least class $\mathcal{C}$ of functions closed under the following Schemes (I)–(V).

(I) The *successor function* $f(x) = (x+1)$ is in $\mathcal{C}$.

(II) The *constant functions* $f(x_1, \ldots, x_n) = m$ are in $\mathcal{C}$, $0 \le m, n$.

(III) The *identity functions* $f(x_1, x_2, \ldots x_n) = x_i$, $1 \le i \le n$, are in $\mathcal{C}$.

(IV) (*Composition*)   If $g_1, g_2, \ldots, g_m, \ h \in \mathcal{C}$, then

$$f(\overline{x}) = h(g_1(\overline{x}), \ldots, g_m(\overline{x}))$$

is in $\mathcal{C}$, where $g_1, \ldots, g_m$ are functions of n variables, $\overline{x} = (x_1, \ldots, x_n)$, and h is a function of m variables.

(V) (*Primitive Recursion*)   If $g, h \in \mathcal{C}$ and $n \ge 1$, then $f \in \mathcal{C}$ where

$$f(0, \overline{x}) = g(\overline{x})$$

$$f(x_1 + 1, \overline{x}) = h(x_1, f(x_1, \overline{x})\overline{x})$$

where $\overline{x} = (x_2, \ldots, x_n)$, the $n-1$ variables treated as *parameters*, assuming g and h are functions of $n-1$ and $n+1$ variables, respectively, and f is a function of n variables. (In case $n = 1$, a 0-ary function is a constant function which is in C by Scheme (II).)

Therefore, a function f is primitive recursive iff there is a *derivation*, namely a sequence $f_1, f_2, \ldots, f_k = f$ such that each f_i, $i \le k$, is either an initial function (i.e., is obtained by Schemes (I), (II), or (III)), or f_i is obtained from $\{f_j : j < i\}$ by an application of Scheme (IV) or (V). For example, the function $f(x_1, x_2) = x_1 + x_2$ has the following derivation.

$$
\begin{aligned}
f_1(x) &= x + 1 && \text{by (I)} \\
f_2(x) &= x && \text{by (III)} \\
f_3(x_1, x_2, x_3) &= x_2 && \text{by (III)} \\
f_4 &= f_1 \circ f_3 && \text{by (IV)} \\
f_5(0, x_2) &= f_2(x_2) && \\
f_5(x_1 + 1, x_2) &= f_4(x_1, f_5(x_1, x_2), x_2) && \text{by (V)}
\end{aligned}
$$

Similarly, [Kleene 1952] showed that all the usual functions on ω are primitive recursive, including $x \cdot y$, x^y, $x!$ and limited subtraction x *monus* y,

$$
x \dotdiv y \quad := \quad \begin{cases} x - y & \text{if } x \ge y, \\ 0 & \text{if } x < y. \end{cases}
$$

Definition 17.2.2. (Characteristic Functions). (i) Let $\chi_A(x)$ denote the *characteristic function* of A, i.e., $\chi_A(x) = 1$ if $x \in A$ and $\chi_A(x) = 0$ otherwise. For convenience, we often write $A(x)$ for $\chi_A(x)$. The *characteristic function* for a relation $R(x_1, x_2, \ldots x_k)$ is the function χ_R such that $\chi_R(x_1, x_2, \ldots x_k) = 1$ if $R(x_1, x_2, \ldots x_k)$ holds and $\chi_R(x_1, x_2, \ldots x_k) = 0$ otherwise.

(ii) A predicate (i.e., a relation) is *primitive recursive* (*computable*) if its characteristic function is primitive recursive (computable).

For example, it can be shown that the relation $R = \{x : x \text{ is prime}\}$ is primitive recursive. Let $p_0, p_1, \ldots$ be the prime numbers in increasing order. Any $x \in \omega$ has a unique representation

(17.2) $$x = p_0^{x_0} p_1^{x_1} \cdots p_n^{x_n} \cdots,$$

where finitely many $x_i \ne 0$. It can be shown that the function

(17.3) $$(x)_i = x_i$$

is a primitive recursive function of x and i. Thus, for any finite sequence of nonzero integers $\{a_0, a_1, \ldots, a_n\}$ there is a unique "code" number $a = p_0^{a_0} \cdots p_n^{a_n}$ such that each $a_i = (a)_i$ can be obtained primitively recursively from a. Gödel and later Kleene used this prime power coding to give a Gödel number to syntactical objects such as proofs or derivations of recursive functions.

17.2.3   Nonprimitive Recursive Functions

Most of the usual number-theoretic functions in ordinary mathematics are primitive recursive as is shown in [Kleene 1952]. Therefore, primitive recursive functions are a good first approximation to algorithmic functions, but they do not comprise *all* algorithmic functions. First, primitive recursive functions are *total*, and they can be effectively listed. Therefore, a diagonal argument produces an effectively calculable function which is not primitive recursive. Second, the primitive recursive functions do not even include all possible *recursions*. Scheme (V) allowed recursion on only *one* variable. The following *Ackermann generalized exponential* is defined by simultaneous induction on two variables. Hermes [1969] shows it is not primitive recursive because it dominates every primitive recursive function.

$$
\begin{aligned}
f(0,0,y) &= y, \\
f(0,x+1,y) &= f(0,x,y)+1, \\
f(1,0,y) &= 0, \\
f(z+2,0,y) &= 1, \\
f(z+1,x+1,y) &= f(z,f(z+1,x,y),y).
\end{aligned}
$$

This function is defined informally as in [Rogers 1967] as follows.

$$ f(0,x,y) = y+x $$

$$ f(1,x,y) = y \cdot x $$

$$ f(n+1,x,y) = \text{the result of applying } y \text{ to itself } x \text{ times} $$

under the nth level operation $\lambda uv[f(n,u,v)]$.

For example, multiplication $y \cdot x$ is the result of adding y to itself x times, exponentiation y^x is the result of multiplying y by itself x times, and the Ackermann function generalizes this notion through all levels $n \in \omega$. [Hermes 1969] reduced Ackermann's main idea to the following example, $h(x,y)$, where z' denotes $z+1$.

$$
\begin{aligned}
h(0,y) &= y' \\
h(x',0) &= h(x,1) \\
h(x',y') &= h(x,h(x',y)).
\end{aligned}
$$

(17.4)

This system of equations unambiguously defines an algorithmically computable function by recursion. The key difference here is the third line, which uses *simultaneous recursion* on x and y and cannot be duplicated by primitive recursion. [Hermes 1969] proved that for every primitive recursive function $g(x_1, \cdots, x_n)$ there is a number c such that

$$ (\forall x_1) \cdots (\forall x_n) \, [\, g(x_1, x_2, \cdots, x_n) \; < \; h(c, x_1 + x_2 + \cdots + x_n) \,]. $$

This is proved by induction on the number of primitive recursive schemes defining g. Hence, h cannot be primitive recursive.

Ackermann's function poses another serious problem. Clearly, the functions defined by multiple simultaneous recursions are algorithmically computable, but how do we define them? If we allow simultaneous recursion on $(n + 1)$ variables then we can produce a function not definable by recursion on only n variables. Therefore, it is not clear how to generalize Scheme (V) to capture even those functions defined by *recursion*, much less the algorithmically computable ones.

17.2.4   Herbrand-Gödel Recursive Functions

A succinct characterization of *all* functions defined by recursion, including *partial* ones, was achieved in [Gödel 1934]. He had used the primitive recursive functions in the proof of his incompleteness theorem in 1931, but he realized they did not constitute all effectively calculable functions. In the spring of 1934 Gödel gave a series of lectures at Princeton in which he introduced what he called the *general recursive functions* to distinguish them from the *primitive recursive functions* which up to that time had been called "recursive functions." Gödel refined a suggestion of Herbrand. Hence, these are called *(Herbrand-Gödel) general recursive functions* or simply *recursive functions*.

We avoid the formal definition (see [Kleene 1952]) but informally sketch the idea. Consider the equations of (17.1) which define the factorial function, $f(0) = 1$ and $f(x+1) = (x+1) \cdot f(x)$. Roughly, let the formal language L consist of nonlogical symbols: a unary function symbol S for successor, and the constant symbol $\underline{0}$ for 0. Let numeral $\underline{k}$ denote the term $S^k(\underline{0})$. In addition, L has a variety of function letters, one of which, F, is called the *principal function letter*, corresponding to the informal function f being defined. We write the system of equations $\mathbb{E}_F$ to define F:

(17.5) $F(\underline{0}) = \underline{1}$ and $F(\underline{x} + \underline{1}) = G((\underline{x} + \underline{1}), F(\underline{x}))$.

Here we assume that a previously specified system of equations $\mathbb{E}_G$ defines multiplication $g(x, y) = x \cdot y$. The *rules* for deriving new equations from those in (17.5) are the following.

R1 Substitution of a numeral for every occurrence of a particular variable in an equation.

R2 Replacement in the right-hand side of an equation of a term of the form $H(\underline{c})$ by a numeral $\underline{d}$, provided that $H(\underline{c}) = \underline{d}$ has already been derived and H is a function letter (such as F or G in our example of the factorial function).

Definition 17.2.3. [Gödel, 1934] A (partial) function f on the integers is *(Herbrand-Gödel) general recursive* (usually abbreviated *recursive*) if there is a finite system of equations $\mathbb{E}$ with principal function letter F such that $f(n) = m$ if and only if we can derive $F(\underline{n}) = \underline{m}$ from $\mathbb{E}$ using the rules R_1 and R_2.

It is easy to see how to derive any value of the factorial function $f(x) = y$ as an equation $F(\underline{x}) = \underline{y}$, and indeed to show that all the primitive recursive functions are (Herbrand-Gödel) general recursive. The calculations are natural in that they closely resemble those a mathematician would make with pencil and paper calculating the same values. The main point is that these two simple rules give a formal characterization which captures the notion of *all recursions* even for partial functions.

Herbrand had written Gödel a letter in 1931 describing systems of equations which uniquely define a (partial) function. In 1934 Gödel made two restrictions on this definition to make it *effective*, first that the left-hand sides of the functional equations be in standard form with F being the outermost symbol, and second that for each set of natural numbers $n_1, \ldots n_j$ there exists a unique m such that $F(\underline{n}_1, \ldots \underline{n}_j) = \underline{m}$ is a derived equation (see [Sieg 1994]). In 1936, 1943, and 1952 Kleene introduced variants of Gödel's two rules, which give an equivalent formulation of the Herbrand-Gödel definition.

In the initial definitions and advances for computability from 1931 to 1937 researchers considered only *total* computable functions as is common in other branches of mathematics. It was [Kleene 1938] who first proposed considering *partial* computable functions and this helped resolve some difficulties.

17.2.5   Kleene's μ-Recursive Functions

[Kleene 1936] adapted Gödel's 1931 method of arithmetization of syntax to give Scheme (VI), which, together with primitive recursive Schemes (I)–(V), gives an alternative and useful characterization of the general recursive functions.

Definition 17.2.4. (Kleene, 1936). The class $\mathcal{C}$ of μ-recursive *(partial) functions* is the least class obtained by closing under Schemes (I)–(V) for the primitive recursive functions and the following Scheme (VI).

(VI)   (*Unbounded Search*)   If $\theta(\overline{x}, y) \in \mathcal{C}$ is a partial function, and

$$(17.6) \quad \psi(\overline{x}) = (\mu y)[\, \theta(\overline{x}, y)\!\downarrow \, = \, 1 \quad \& \quad (\forall z < y)[\, \theta(\overline{x}, z)\!\downarrow \, \neq 1 \,]\,],$$

then ψ is in $\mathcal{C}$. (Here $\psi(\overline{x})$ diverges if there is no such y. Hence, ψ may be nontotal.)

To see that partial algorithmic functions are closed under Scheme (VI), fix $\overline{x}$ and a partial function $\theta(\overline{x}, y)$ for which we have an algorithm. Now compute in order $\theta(\overline{x}, y)$ for $y = 0$, $y = 1$, $\ldots$, and do not proceed to $y + 1$ until (if ever) the computation for y converges. If there is a first y with $\theta(\overline{x}, y)\!\downarrow = 1$, then output y. Otherwise, continue forever.

The μ-recursive functions give a compact, mathematically appealing definition and they are useful for proving the Kleene Normal Form Theorem.

Theorem 17.2.5 (Kleene Normal Form Theorem). *There exists a predicate $T(e, x, y)$ (called the Kleene T-predicate) and a function $U(y)$ both primitive recursive such that, for every e,*

$$\psi_e(x) \;=\; (\mu y)\, T(e, x, y),$$

where $\{\psi_e\}_{e \in \omega}$ is an effective listing of all partial recursive functions.

This proves that the μ-recursive partial functions include all partial recursive functions. However, proving that a nonprimitive recursive function (such as the Ackermann function) is μ-recursive requires first using the Gödel 1931 arithmetization method and then applying Scheme (VI). Furthermore, arithmetization is very tedious. In contrast, Turing programs can be decomposed into submodules and often give a more perspicuous demonstration that a function is computable. [Church 1936] and [Kleene 1936b] proved that the classes of general recursive functions and λ-definable functions are the same. Kleene also proved the equivalence with the μ-recursive functions.

17.2.6   Gödel Remained Unconvinced

By 1934 Kleene had shown that a large class of number-theoretic functions were λ-definable. On the strength of this evidence, Church informally proposed to Gödel around March 1934 that the notion of "effectively calculable" be identified with "λ-definable," a suggestion which Gödel rejected as "thoroughly unsatisfactory," according to Martin Davis's account. After hearing Gödel's lectures in 1934 on the general recursive functions Church changed the formal definition from "λ-definable" to "recursive," his abbreviation for Herbrand-Gödel general recursive, and Church presented on April 19, 1935 to the American Mathematical Society his famous proposition published in [Church 1936] and known since [Kleene 1952] as *Church's Thesis*, which asserts that the effectively calculable functions should be identified with the recursive functions.

Gödel, however, remained unconvinced of the validity of Church's Thesis through its publication in [Church 1936]. This is all the more significant, first, because Gödel had *originated* the formalism of the general recursive functions, the one upon which Church based his thesis, and the one which captured the notion of all recursions; and second, because much of the evidence for Church's Thesis rested on the coincidence of these formal classes (general recursive functions, μ-recursive functions, and λ-definable functions), and this was based largely on Kleene's use in Scheme (VI) of *arithmetization*, the method that Gödel *himself* had introduced so dramatically in his 1931 incompleteness theorem. Until seeing Turing's 1936 paper Gödel was "not at all persuaded."

17.3 Turing's Analysis

17.3.1 Turing's Discovery

Independently, Turing attended lectures in 1935 at Cambridge University by topologist M.H.A. (Max) Newman on Gödel's paper [1931] and Hilbert's *Entscheidungsproblem*. A year later, Turing submitted his solution to the incredulous Newman on April 15, 1936. Turing's paper [Turing 1936] was distinguished because: (1) Turing analyzed an idealized *human* computing agent—call it a *"computor"*—which brought together the intuitive conceptions of a "function produced by a mechanical procedure" that had been evolving for more than two millennia from Euclid to Leibniz to Babbage and Hilbert; (2) Turing specified a remarkably simple formal device (*Turing machine*) and demonstrated the equivalence of (1) and (2); (3) Turing proved the unsolvability of Hilbert's *Entscheidungsproblem*, which prominent mathematicians had been studying intently for some time; (4) Turing proposed a *universal* Turing machine, one which carried within it the capacity to duplicate any other, an idea which was later to have great impact on the development of high-speed digital computers and to have considerable theoretical importance. As a boy, Turing had been fascinated by his mother's typewriter. He devised his Turing machine as a kind of idealized typewriter with a reading head moving over a fixed unbounded tape or platen[1] on which the head writes. Turing's model was by far the most convincing then and now. From 1936 to 1938 Turing completed his Ph.D. at Princeton under Church. His Ph.D. thesis was on a different topic but contained a crucial idea (5), that of a local machine communicating with a database, the same mechanism we use today when a laptop communicates with the Internet.

---

[1]The platen is the cylindrical roller in a typewriter against which the paper is held. In 1930 the typing head was fixed in the center and the platen and a carriage moved back and forth under it as the keys struck the platen. By 1980 the IBM Selectric typewriter had a fixed carriage and a movable writing ball which passed back and forth across the platen. This was Turing's design in 1936. I do not know whether IBM paid royalties to Turing's estate.

17.3.2   Gödel Accepts Turing's Analysis

Gödel enthusiastically accepted Turing's analysis and always thereafter gave Turing credit for the definition of mechanical computability. For the Princeton Bicentennial he wrote [Gödel 1946], "one [Turing] has for the first time succeeded in giving an absolute definition of an interesting epistemological notion, i.e., one not depending on the formalism chosen." Gödel also wrote, "That this really is the correct definition of mechanical computability was established beyond any doubt by Turing." Church wrote that of the three notions—computability by a Turing machine, general recursiveness of Herbrand-Gödel-Kleene, and lambda-definability—"The first has the advantage of making the identification with effectiveness in the ordinary (not explicitly defined) sense evident immediately, i.e., without the necessity of proving preliminary theorems."

Later Gödel explained why he had accepted Turing's analysis so completely. For Gödel the essential point was to define what a procedure is. He believed that Turing had done this, but in 1935 he was not convinced that his own definition of recursive functions accomplished this.

In the *Nachlass* printed in Volume III of [Gödel 1995], page 166, Gödel wrote,

> When I first published my paper about undecidable propositions the result could not be pronounced in this generality, because for the notions of mechanical procedure and of formal system no mathematically satisfactory definition had been given at that time.... The essential point is to define what a procedure is.

Gödel believed that Turing in 1936 had done so, but Gödel was not convinced by Church's argument in 1936 for recursive functions. By 1937, the three definitions of computable functions had been proved mathematically equivalent so the definitions are *extensionally* equivalent but not intensionally equivalent. Turing's analysis is regarded as the most convincing and is the one on which most modern texts make the formal definition of a computable function.

17.3.3   Turing's Thesis: Definition or Theorem

Turing's claim in 1936 was that a function is intuitively computable if and only if it is computable by a Turing machine. He gave evidence for this claim in Sections 1 and 9. Simultaneously, Church claimed in 1936 that a function is effectively calculable iff it is Herbrand-Gödel recursive. Neither statement was intended as a thesis, but rather as a claim with a demonstration, albeit one relating an intuitive concept to a formal definition. Unfortunately, Kleene referred to these in his influential book [Kleene 1952] as "Church's Thesis" and "Turing's Thesis," implying that there might be some element of debate about them. Church's demonstration was somewhat less convincing than Turing's so we restrict attention here to Turing's.

The English term "thesis" comes from the Greek word θέασις, meaning "something put forth." In logic and rhetoric it refers to a "proposition laid down or stated, especially as a theme to be discussed and proved, or to be maintained against attack." It can be a hypothesis presented *without* proof, or it can be an assertion put forward with the intention of defending and debating it. The Harvard College Writing Center notes online says that a thesis is "not a topic; nor is it a fact; nor is it an opinion." A theorem such as the Gödel Completeness Theorem is not a thesis. It is a fact with a proof in a formal axiomatic system which cannot be refuted. Attaching the term "thesis" to such a proposition invites continual reexamination. It signals to the reader that the proposition may not be completely valid, but rather it should continually be examined more critically.

This is not a question of mere semantics, but about what Turing actually *achieved*. If we use the term "thesis" in connection with Turing's work, then we are continually suggesting some doubt about whether he really gave an authentic characterization of the intuitively calculable functions. The central question about Turing's work in 1936 is whether Turing demonstrated his assertion beyond any reasonable doubt, or whether it is merely a thesis, in need of continual verification. Neither Church nor Turing ever referred to their 1936 characterizations of effectively calculable functions as a "thesis." They thought of them as claims with demonstrations.

Turing's last published paper [Turing 1954] discussed puzzles. Turing wrote of his central assertion (about a function being effectively calculable iff it is computable by a Turing machine) that this assertion lies somewhere between a theorem and a definition.

> In so far as we know a priori what is a puzzle and what is not, the statement is a theorem. In so far as we do not know what puzzles are, the statement is a definition that tells us something about what they are.

In any case, most scholars agree that it is not a thesis, which in English weakens the claim to something in need of debate or continual verification. Nevertheless, we sometimes use the term "Turing's Thesis" to identify the claim because since [Kleene 1952] it has been referred to as such in the literature.

17.3.4 Turing's Demonstration of Turing's Thesis

Since the beginning of the *Entscheidungsproblem* the study of computability had been tied to formal axiomatic systems. Gödel's 1931 incompleteness theorem had been about axiomatic systems extending arithmetic. His recursive functions were in fact formal systems with a finite set of equations as axioms and two rules of inference for deriving new equations. Church followed this axiomatic approach. Church in [Church 1936] attempted to demonstrate that any effectively calculable function was derivable in a certain formal system, and that any function derivable there was Herbrand-Gödel recursive. Gödel was not completely convinced by this demonstration.

Turing took a completely different approach. Turing was a marathon runner. After a run one day he lay down in a meadow to rest and the idea of a Turing machine came to him. He reduced the mechanical process to its smallest parts as described in Chapter 1. What is remarkable is not only the definition of the Turing machine but also the demonstration in [Turing 1936], Section 9, that it captures the idea of effectively calculable functions. We have reproduced this demonstration in Chapter 1 before presenting the Turing machine definition.

In this masterful demonstration, which Robin Gandy considered as precise as most mathematical proofs, Turing analyzed the informal nature of functions computable by a finite procedure and demonstrated that they coincide with those computable by an a-machine. Gandy, in [Gandy 1988], page 82 of Herken's book [Herken 1988], observed,

> Turing's analysis does much more than provide an argument for Turing's Thesis; *it proves a theorem*.

Furthermore, Gandy continued, "Turing's analysis makes no reference whatsoever to calculating machines. Turing machines appear as a result, a codification, of his analysis of calculations by humans."

Wittgenstein remarked about Turing machines, "These machines are *humans* who calculate." Turing's achievement was to determine not what machines could compute, but what human beings could compute with enough resources of time and space for the computation.

17.4   Turing's Oracle Machine (o-Machine)

17.4.1   An Extraordinary but Almost Incidental Discovery

One of Turing's most important inventions, that of an oracle machine, appeared very briefly in Section 4 of [Turing 1939]. It was an aside and was unnecessary. Turing's oracle machine was developed by Post into Turing reducibility and other reducibilities. Turing reducibility allows us to measure the information content and complexity of structures and sets. It is crucially important in computability theory, because it subsumes the ordinary Turing machine and much more. Today the notion of a local machine interacting with a remote database or remote machine is central to practical computing.

After Turing's a-machine discovery in April, 1936, Max Newman at Cambridge suggested that Turing go to Princeton to study with Church. Turing completed his Ph.D. at Princeton under Church from 1936–1938. Many mathematicians found Gödel's Incompleteness Theorem unsettling. Turing's dissertation, published as [Turing 1939], was on ordinal logics, apparently a suggestion of Church, and was an attempt to get around Gödel's incompleteness theorem by adding new axioms. If T_1 is a consistent extension of Peano arithmetic, then the arithmetical sentence σ_1 asserting the unprovability of itself is independent of T_1 but we can form a new theory $T_2 = T_1 \cup \{\sigma_1\}$ which strictly extends T_1. One can continue this sequence through all the computable (constructive) ordinals.

17.4.2 Turing's Use of Oracle Machines

In one of the most important and most obscure parts of all of computability theory, Turing wrote in section 4 of [Turing 1939] a short statement about oracle machines.

> Let us suppose we are supplied with some unspecified means of solving number-theoretic problems; a kind of oracle as it were. ... This oracle ... cannot be a machine.
> With the help of the oracle we could form a new kind of machine (call them o-machines), having as one of its fundamental processes that of solving a given number-theoretic problem.

Turing introduced oracle machines for a very specific purpose. In the preceding Section 3, Turing had considered Π_2 predicates ($\forall\exists$-predicates over a computable matrix), and had shown that the Riemann Hypothesis and other common problems were Π_2, which Turing called "number-theoretic." For example, the question of whether a Turing a-machine computes a partial function with infinite domain is Π_2. More precisely, let φ_e be the partial computable function computed by the Turing program P_e with Gödel number e and let W_e be the domain of φ_e. Now W_e is a Σ_1-set. Define

$$\text{Inf} = \{ e : W_e \text{ is infinite } \}.$$

Now Inf is Π_2, and in fact Π_2-complete, i.e., for every Π_2 set V there is a computable function f such that $x \in V$ iff $f(x) \in \text{Inf}$.

Turing invented oracle machines to construct a set which was not Π_2. This could easily have been accomplished by a diagonal argument without oracle machines. Turing put the oracle Inf on the oracle tape. By the same diagonal argument as in [Turing 1936] for a noncomputable function he used the oracle machine to construct a non-Π_2 set. Turing wrote in [Turing 1939]

Given any one of these machines we may ask the question whether it prints an infinity of figures 0 or 1; I assert that this class of problem is not number-theoretic [i.e., Π_2].

In his analysis of [Turing 1939], [Feferman 2007] page 1204 wrote,

In Section 4 Turing introduced a new idea that was to change the face of the general theory of computation (also known as recursion theory) but the only use he made of it was curiously inessential. His aim was to produce an arithmetical problem that is not number-theoretic in his sense, i.e., not in $\forall\exists$-form. This is trivial by a diagonalization argument, since there are only countably many effective relations $R(x, y)$ of which we can say that $\forall x \exists y R(x, y)$ holds. Turing's way of dealing with this, instead, is through the new notion of computation relative to an *oracle*....

"He then showed that the problem of determining whether an *o*-machine terminates on any given input is an arithmetical problem not computable by any *o*-machine, and hence not solvable by the oracle itself. Turing did nothing further with the idea of *o*-machines, either in this paper or afterward."

For a fixed set $A \subseteq \omega$, the set of positive integers, effectively number all Turing oracle programs. Let $\Phi_e^A(x)$ denote the partial function computed by the *o*-machine with Gödel number e and with A on the oracle tape. Define the relative halting set

(17.7) $$K^A = \{ e : \Phi_e^A(e) \text{ halts } \}.$$

Theorem 17.4.1 (Turing 1939, page 173). *Given A the set K^A is not computable in A.*

Turing had shown in [Turing 1936] that there is a diagonal set not computable by a Turing *a*-machine. The same proof on *o*-machines relativized to A establishes the theorem. His specific application is that if $A = \text{Inf}$, then K^A is not Π_2. In 1939 Turing left the topic of oracle machines, never to return to it. It mostly lay dormant for five years until it was developed in a beautiful form by Post in 1944 and 1948, and by Kleene and Post in 1954 and in other papers.

17.4.3   Kleene's Definition of "General Recursive In"

Kleene in [Kleene 1943], page 44, gave a definition of "general recursive in." Kleene wrote,

A function φ which can be defined from given functions $\psi_1, \ldots, \psi_k$ by a series of applications of general recursive schemata we call *general recursive in* the given functions; and in particular, a function definable *ab initio* by these means we call *general recursive*.

Kleene was thinking of auxiliary functions used in some Herbrand-Gödel computation, where the functions are all recursive, such as multiplication being used in the definition of factorial. It is possible that Kleene thought of an infinite nonrecursive oracle A as the auxiliary function ψ, and the result a function ψ which interrogates A as an external database during the computation. If so, Kleene would surely have cited Turing [1939, §4], where this idea first appeared. There is no evidence in Kleene [1943] that Kleene thought this way. The notion of "general recursive in" or relative recursive does not appear again in [Kleene 1943].

Both Turing in 1939 and Kleene in 1943 very briefly alluded to a computation using an external set A, but neither developed the idea of relative computablity. That was done by Post in 1944. Soare in [Soare 2009] analyzes how the Turing oracle machine also gives a model for modern interactive computing where a local computer like a laptop communicates with an external database such as one on the Internet.

17.5   Emil Post's Contributions

More than a decade older than the others, Emil Post was the oldest principal reseacher in computability theory among Gödel, Church, Turing, and Kleene in the 1930s. In 1936 Post was teaching at City College in New York and had been in touch with Church and was aware of [Church 1936]. Independently of Turing, Post proposed in [Post 1936] a remarkably similar computing model. His paper "Finite Combinatory Processes, Formulation I" proposed a discrete machine almost identical to Turing's model, composed of boxes which could be blank or contain a mark. On a single step the worker could determine whether the box observed is marked or not; mark it or erase a mark; and move one box to the left or right.

From this it is often and erroneously written that Post's contribution here was "essentially the same" as Turing's, but, in fact, it was much less. Post did not attempt to prove that his formalism coincided with any other formalism, such as general recursiveness, but merely expressed the expectation that this would turn out to be true, while Turing in [Turing 1937b] proved the Turing computable functions equivalent to the λ-definable ones. Post gave no hint of a universal Turing machine. Most important, Post gave no analysis, as did Turing, of why the intuitively computable functions are computable in his formal system. Post offers only as a "working hypothesis" that his contemplated "wider and wider formulations" are "logically reducible to formulation 1." Lastly, Post, of course, did not prove the unsolvability of the *Entscheidungsproblem* because at the time Post was not aware of [Turing 1936], and Post believed that in 1936 Church had settled the *Entscheidungsproblem*. Furthermore, Post wrote in [Post 1936] that Church's identification of effective calculability and recursiveness was a working hypothesis which is in "need of continual verification." This irritated Church who criticized it in his review [Church 1937b] of [Post 1936]. Post's contributions during the 1930s were original and insightful, corresponding in spirit to Turing's more than to Church's, but they were not as influential as those of Church and Turing. It was only during the next phase, from 1940 to 1954, that Post's remarkable influence was fully felt.

17.5.1   Post Production Systems

As Turing left the subject of pure computability theory in 1939, his mantle fell on Post. Post continued the tradition of clear, intuitive proofs, of exploring the most basic objects such as computability and computably enumerable sets, and, most of all, of exploring relative computability and Turing reducibility. During the next decade and a half, from 1940 until his death in 1954, Post played an extraordinary role in shaping the subject.

Post in [Post 1941] and [Post 1943] introduced a *second* and unrelated formalism called a *production* system and (in a restricted form) a *normal* system, which he explained again in [Post 1944]. Post's (normal) canonical system is a *generational* system, rather than a *computational* system as is the case for general recursive functions or Turing computable functions, because it gives an algorithm for generating (listing) a set of integers rather than computing a function. This led Post to concentrate on *effectively enumerable sets* rather than computable functions. Post's normal system gives a formal definition of effectively enumerable sets which is equivalent to the definition of computably enumerable sets.

17.5.2   Post Considered the Complete Set K

One of Post's most influential contributions during this period was the remarkably clear and intuitive paper [Post 1944], *Recursively enumerable sets of positive integers and their decision problems*. Post defined the notion of one set being *reducible* to another, and in 1944 and 1948 Post introduced the term *degree of unsolvability* for the equivalence class of all sets mutually reducible to one another. In 1939 Turing had thought only in terms of the oracle as an external database. Post thought of it as a vehicle to compare the information content of two sets and to reduce one to the other.

Post's paper in 1944 revealed with intuition and great appeal the significance of the computably enumerable sets and the importance of Gödel's Incompleteness Theorem. Post called Gödel's diagonal set

$$(17.8) \qquad\qquad K = \{\, e \,:\, e \in W_e \,\}$$

the *complete set* because every c.e. set W_e is computable in K ($W_e \leq_{\mathrm{T}} K$). The set K has the same degree as the halting problem of whether a Turing machine with program P_e halts on a given input x. Moreover, Post felt that the creative property of K revealed the inherent creativeness of the mathematical process. Post posed his famous "Post's problem" of whether there exists a computably enumerable (c.e.) set A such that $\emptyset <_T A <_T K$.

17.5.3   Post Defined Relative Computability

Researchers in the 1930s concentrated on a formal definition of an effectively calculable function, not on a definition of a function computable in an oracle. The exception was Turing, who defined oracle machines in 1939. The idea lay dormant for five years until Post studied computably enumerable (c.e.) sets and their decision problems in 1944. Post believed that, along with decidable and undecidable problems, the relative reducibility (solvability) or nonreducibility of one problem to another is a crucial issue. Post studied not only the structure of the computably enumerable sets, those which could be effectively listed, but he initiated a series of reducibilities of one set to another culminating in the most general reducibility which he generously named "Turing" reducibility. Post's aim was to use these reducibilities to study the information content of one set relative to another. In his introduction in 1944 Post wrote,

> Related to the question of solvability or unsolvability of problems is that of the reducibility or non-reducibility of one problem to another. Thus, if problem P1 has been reduced to problem P2, a solution of P2 immediately yields a solution of P1, while if P1 is proved to be unsolvable, P2 must also be unsolvable. For unsolvable problems the concept of reducibility leads to the concept of degree of unsolvability, two unsolvable problems being of the same degree of unsolvability if each is reducible to the other, one of lower degree of unsolvability than another if it is reducible to the other, but that other is not reducible to it, of incomparable degrees of unsolvability if neither is reducible to the other.

Here Post is not merely introducing reducibility of one problem to another for the sake of demonstrating solvability or unsolvability. He takes it further by introducing for the first time in the subject the term "degree of unsolvability" to mean that two sets code the same information, i.e., have the same information content. For decades since then, researchers have classified objects in algebra, model theory, complexity, and computability theory according to their degree of unsolvability or information content. Post continued,

> A primary problem in the theory of recursively enumerable sets is the problem of determining the degrees of unsolvability of the unsolvable decision problems thereof.... Now in his paper

on ordinal logics in Section 4, Turing presents as a side issue a formulation which can immediately be restated as the general formulation of the "recursive reducibility" of one problem to another, and proves a result which immediately generalizes to the result that for any "recursively given" unsolvable problem there is another of higher degree of unsolvability. While his theorem does not help us in our search for that lower degree of unsolvability, his formulation makes our problem precise. It remains a problem at the end of this paper.

In 1944 Post wrote an intuitive Section 11 on the general case of Turing reducibility, but it was not well understood at the time. Post continued to study it for a decade and defined the concept of *degree of unsolvability*, now called "Turing degree." Before his death in 1954, Post gave his notes to Kleene, who published the joint paper [Kleene and Post 1954] which laid the foundation of all the subsequent results on Turing reducibility and Turing degree.

17.5.4   Developing the Turing Jump

In 1931 Gödel [1931] was the first to introduce the diagonal Π_1-set D. The complement $\overline{D}$ is computably isomorphic to the complete Σ_1-set K, defined in (17.8). By relativizing Post's results to the set K^A defined by Turing in (17.7) we see that K^A is c.e. in A and $A <_T K^A$, i.e., K^A is of strictly higher Turing degree than K. Post [1944] noted that Turing was the first to define K^A in this form relative to an oracle A, although Turing did not develop its properties. More properties of the jump and of Turing reducibility were developed in Kleene-Post [1954]. Let A' denote the Turing jump K^A. Let $\mathbf{0}'$ be the degree of $\emptyset'$. The jump operator is well defined on Turing degrees and gives a hierarchy in the diagram of Turing degrees.

Meanwhile, Kleene in [Kleene 1943] developed the arithmetic hierarchy described in Chapter 4 of predicates classified by a prefix of quantifiers. However, to establish that this is indeed a hierarchy, we need Post's theorem, which implies that $\emptyset^{n+1} \in \Sigma_{n+1}^0 - \Sigma_n^0$, so that the hierarchy does not collapse at any level n.

In his 1944 paper Post moved to other topics which were to have the most profound influence on computability theory. First, Post believed that computability should be done as informally as group theory, with a minimum of formalism. With Rogers' book [Rogers 1967] there was a return to Post's informal approach and it has prevailed since then. Second, Post understood that computably enumerable (effectively enumerable) sets occur in many areas of mathematics and wanted to study their content. Third, Post in 1944 and 1948 began studying various reducibilities of one set to another and started to classify the information content. These themes have dominated the subject ever since. Post's efforts to understand Turing reducibility culminated in the crucial paper [Kleene and Post 1954] which cast Turing reduciblity as an effectively continuous functional on Cantor space and led to several decades of work on degrees.

17.6 Finite Injury Priority Arguments

Post's problem was the question of finding an incomplete noncomputable c.e. set A. In 1954 Kleene and Post used a finite forcing argument to find an incomplete noncomputable set $A \leq_T \emptyset'$. This is slightly weaker, because A is merely Δ_2^0, not Σ_1^0. Nevertheless, their method was transformed into the finite injury priority method only a couple of years later by Friedberg in [Friedberg 1957] and Muchnik in [Muchnik 1956]. It is not accidental that these results followed so closely after the 1954 paper, which finally helped understand Turing reductions.

The name "priority argument" is a slight misnomer. Priority of requirements had been part of the Kleene and Post theorems, but the new elements included a computable construction in place of a $\emptyset'$-oracle construction and also *injury* to the previous action of a requirement R_e by the later action of a higher priority requirement R_i for $i < e$. The key idea of Friedberg and Muchnik was to simply restart the action for R_e whenever R_e is injured and simultaneously to protect that action against later injury by *lower* priority requirements R_j for $j > e$. This ensured that after finite injury the strategy for R_e will eventually succeed. This computable approximation to the Kleene-Post theorem was a great advance, but it was built upon a decade of work by Post from 1944 to 1954 in understanding a Turing reduction.

17.7 Computability and Recursion Terminology

Up until the 1930s the term "recursive" had meant defined by induction or recursion as in the primitive recursive functions. By 1935 Church and Kleene had switched from the λ-definable function to the Herbrand-Gödel recursive functions introduced by Gödel in 1934.

By 1936 Kleene and Church had begun thinking of the word "recursive" to mean "effectively calculable" (intuitively computable). Church had seen his first thesis rejected by Gödel and was heavily invested in the acceptance of his 1936 thesis in terms of recursive functions. Without the acceptance of this thesis Church had no unsolvable problem. Church wrote in [Church 1936], page 96, reprinted in [Davis 1965] that a *"recursively enumerable set"* is one which is the range of a recursive function. This is apparently the first appearance of the term "recursively enumerable" in the literature and the first appearance of "recursively" as an adverb meaning "effectively" or "computably."

Kleene in [Kleene 1936], page 238, mentioned *"recursive enumeration"* and noted that there is no recursive enumeration of Herbrand-Gödel recursive systems of equations which gives only the systems that define the (total) recursive functions. By a "recursive enumeration" Kleene states that he means "a recursive sequence (i.e., the successive values of a recursive function of one variable)." Post in his paper [Post 1944], under the influence of Church and Kleene, adopted this terminology of "recursive" and "recursively enumerable" over his own earlier terminology of "effectively generated set," "normal set," and "generated set." Thereafter, it was firmly established.

Martin Davis entitled his book [Davis 1958] *Computability and Unsolvability*, but adopted the prevailing Kleene terminology at the time that calculable functions should be called "recursive functions." Davis wrote on page vii of his Preface, "This book is an introduction to the theory of computability and noncomputability, usually referred to as the theory of recursive functions." Davis refers several times to "the theory of computability" as including "purely mechanical procedures for solving various problems."

This is very typical of usage from [Kleene 1936] through [Rogers 1967] and [Soare 1987] and beyond, where the term computability theory is often used for the concept, especially to a nontechnical general audience. However, a computable function always has been called a "recursive function," and a computably enumerable set always has been called a recursively eumerable (r.e.) set. This is even true for [Hopcroft and Ullman, 1979] and other computer science books. In contrast, Turing's epochal 1936 paper uses only the terminology "computable function" and calculable function, and never mentions "recursive" in the sense of calculable. Kleene thought of "recursive" with the dual meaning of inductive and effectively calculable. Kleene named the subject "recursive function theory" or simply "recursion theory."

17.7.1   Gödel Rejects Term "Recursive Function Theory"

Neither Turing nor Gödel ever used the word "recursive" to mean "computable." Gödel *never* used the term "recursive function theory" to name the subject; when others did, Gödel reacted sharply and negatively, as related privately by Martin Davis. In a discussion with Gödel at the Institute for Advanced Study in Princeton about 1952–54, Martin Davis casually used the term "recursive function theory" as it was used then. Davis related, "To my surprise, Gödel reacted sharply, saying that the term in question should be used with reference to the kind of work Rosza Peter did." (See Peter's work on recursions in [Peter 1934] and [Peter 1951].)

The meaning of "recursive" as "inductive" led to ambiguity. Kleene often wrote of calculations and algorithms dating back to the Babylonians, the

Greeks, and other early civilizations. However, Kleene in [Kleene 1981b], page 44, wrote,

> I think we can say that recursive function theory was born there ninety-two years ago with Dedekind's Theorem 126 ('Satz der Definition durch Induktion') that functions can be defined by primitive recursion.

Did he mean that recursion and inductive definition began with Dedekind or that computability and algorithms began there? The latter would contradict his several other statements, such as [Kleene 1988], page 19, where Kleene wrote, "The recognition of algorithms goes back at least to Euclid (c. 330 B.C.)." When one uses a term like "recursive" to also mean "computable" or "algorithmic" as Kleene did, then one is never sure whether a particular instance means "calculable" or "inductive," and our language has become imprecise. Returning "recursive" to its original meaning of "inductive" has made its use much clearer. We do not need another word to mean "computable." We already have one.

17.7.2   Changing "Recursive" Back to "Inductive"

By 1995 there was considerable ambiguity in the literature as to whether an instance of "recursive" meant "inductive" or "computable." In 1995 Leo Harrington said, "set theory is about sets, model theory is about models, but recursion theory is not about recursion."

After the articles [Soare 1996] and [Soare 1999] on the history and scientific reasons for why we should use "computable" and not "recursive" to mean "calculable," many authors changed terminology to have "recursive" mean only inductive and they introduced new terms such as "computably enumerable (c.e.)" to replace "recursively enumerable." This helped lead to an increased awareness of the relationship of Turing computability to other areas. There sprang up organizations like *Computability in Europe (CiE)* which developed these relationships.

17.8   Additional References

Gödel's papers can be found in the three volumes [Gödel 1986], [Gödel 1990], and [Gödel 1995]. This includes [Gödel 1931], [Gödel 1934], [Gödel 1936], [Gödel 193?], [Gödel 1946], [Gödel, 1951], and [Gödel 1964]. Computabiilty papers from the 1930s and 1940s are reprinted in [Davis 1965].

Papers on Gödel and Turing include: [Gandy 1980], [Gandy 1988] in [Herken 1988], [Sieg 1997], [Sieg 2006], [Sieg 2008], and [Sieg 2009], and also [Feferman 2007], [Wang 1981], [Wang 1987], and [van Heijenoort 1967]. Details of Turing's life can be found in his biography by [Hodges 1983].

Books on computability theory include: [Cooper 2004], [Davis 1958], [Enderton 2011], [Epstein 1975], [Epstein 1979], [Hermes 1969]. The books include [Kleene 1952], [Kleene 1967], [Lerman 1983], [Odifreddi 1989], and [Odifreddi 1999], [Rogers 1967], [Shoenfield 1971], [Shoenfield 1993], as well as [Soare 1987].

There are several books on algorithmic complexity including [Nies 2009] as well as [Downey and Hirschfeldt 2010].

Material on Hilbert and his programs may be found in [Hilbert 1904], [Hilbert 1926], [Hilbert and Ackermann 1928], [Hilbert and Bernays 1934]. Dedekind in [Dedekind 1888] proved that certain functions such as addition and multiplication were defined by recursion.

Papers by Church include his Church's Thesis abstract in [Church 1935], his paper about it in [Church 1936], corrections in [Church 1936b], and his reviews of Turing and Post in [Church 1937] and [Church 1937b]. [Church and Kleene 1936] extended the definitions to recursive ordinal numbers, and [Kleene 1938] gave a definition for notations for recursive ordinal numbers. Kleene in 1936 introduced the μ-recursive functions and [Kleene 1936b] proved the equivalence of recursive functions and λ-definable functions, giving more evidence for Church's Thesis.

Jockusch and Soare studied Π_1^0-classes in [Jockusch and Soare 1971], and also in [Jockusch and Soare 1972] and [Jockusch and Soare 1972b] following work in [Shoenfield 1960] on degrees of models and in [Scott 1962] on degrees of extensions of Peano arithmetic. Cenzer gave a summary of work on Π_1^0-classes in [Cenzer 1999]. Diamondstone, Dzhafarov, and Soare gave a more recent account in 2010 in [DiaDzhSoa 2010].

We presented in §4.8 results by [Jockusch 1972] on degrees in which the computable functions are uniformly computable and in §6.3 results from [Jockusch 1980] on generic sets. Other results on these topics can be found in these papers. The Shoenfield Limit Lemma of [Shoenfield 1959] characterized the Δ_2 sets. A similar characterization in terms of trial and error predicates was given by [Putnam 1965]. Permitting by Δ_2 sets was presented in [R. Miller 2001] and [Csima 2004].

Friedberg is best known for solving Post's problem [Friedberg 1957] done independently in [Muchnik 1956]. However, Friedberg wrote several other important papers on computably enumerable sets, [Friedberg 1957b], also [Friedberg 1957c] and [Friedberg 1958], and in [Friedberg-Rogers 1959] introduced weak truth-table (bounded Turing) reducibility. Infinite games with perfect information were studied in [Gale-Stewart 1953] which proved that open games are determined.

Articles on Turing and computability theory can be found in [Soare 1996], [Soare 1999], [Soare 2007], [Soare 2009], [Soare 2012], [Soare 2012b], also [Soare 2013], [Soare 2013b], [Soare 2013c], and [Soare 2015] as well as in the references in these papers. Applications of computability theory to differential geometry can be found in [Soare 2004].

In addition to his main papers, [Turing 1936] and [Turing 1939], Turiing gave some minor corrections in [Turing 1937] and proved the equivalence of Turing machines and λ-definable functions in [Turing 1937b]. In [Turing 1950] he considered computing machinery and intelligence, and in [Turing 1950b] he studied the word problem for certain semi-groups. In his final published paper, [Turing 1954], he considered puzzles and whether his main characterization is a definition or a theorem.

Dekker developed his theorem on hypersimple sets and deficiency sets in [Dekker 1954]. Lachlan introduced Lachlan games in [Lachlan 1970] and studied the priority method with the viewpoint of topology in [Lachlan 1973], in [Mathias and Rogers 1973]; [Hopcroft and Ullman, 1979] showed how to build subroutines for Turing machines to make it easier to program them; [Martin 1966] developed the results on high degrees presented in Chapter 4; and [Martin 1975] proved determinacy of Borel games.

After his retirement, Kleene wrote some retrospective papers on the origins of recursive function theory, including [Kleene 1981], [Kleene 1981b], [Kleene 1981c] and [Kleene 1987]. Kleene's papers and books had great influence on the subject, especially [Kleene 1936], [Kleene 1943], and his very influential book [Kleene 1952].

Myhill proved the creative set theorem of Chapter 2 in [Myhill 1955], noticed that the computably enumerable sets form a lattice in [Myhill 1956], and asked whether the induced partial ordering was dense. Spector in [Spector 1956] constructed a minimal degree below $0''$. In addition, Sacks in [Sacks 1963b] and [Sacks 1963c] constructed one below $0'$. Also, Sacks proved the jump theorem in [Sacks 1963] which generalizes the Shoenfield jump theorem in Chapter 6. Shoenfield in [Shoenfield 1960b] used the minimal degree construction to give an uncountable set of incomparable minimal degrees and in [Shoenfield 1966] gave an elegant exposition of the minimal degree method.

References

[Cenzer 1999]
    D. Cenzer, Π_1^0-*classes in computability theory*, Handbook of Computability Theory (E. Griffor, ed.), Studies in Logic and the Foundations of Mathematics, vol. 140, North-Holland, Amsterdam, 1999, pp. 37–85.

[Church 1935]
    A. Church, *An unsolvable problem of elementary number theory, Preliminary report (abstract)*, Bull. Amer. Math. Soc. **41** (1935), 332–333.

[Church 1936]
    A. Church, *An unsolvable problem of elementary number theory*, Amer. J. of Math. **58** (1936), 345–363.

[Church 1936b]
    A. Church, *A note on the Entscheidungsproblem*, J. Symbolic Logic **1** (1936), no. 3, 40–41, corrections on pp. 101–102.

[Church 1937]
    A. Church, *Review of Turing 1936*, J. Symbolic Logic **2** (1937), no. 1, 42–43.

[Church 1937b]
    A. Church, *Review of Post 1936*, J. Symbolic Logic **2** (1937), no. 1, 43.

[Church and Kleene 1936]
    A. Church and S. C. Kleene, *Formal definitions in the theory of ordinal numbers*, Fund. Math. **28** (1936), 11–21.

[Cooper 2004]
    S. B. Cooper, *Computability theory*, Chapman & Hall/CRC Mathematics, London, New York, 2004.

[Csima 2004]
B. F. Csima, *Degree spectra of prime models*, J. Symbolic Logic **69** (2004), 430–442.

[Davis 1958]
M. Davis, *Computability and unsolvability*, McGraw-Hill, New York, 1958, reprinted in 1982 by Dover Publications.

[Davis 1965]
M. Davis (ed.), *The undecidable. Basic papers on undecidable propositions, unsolvable problems, and computable functions*, Raven Press, Hewlett, New York, 1965.

[Dedekind 1888]
R. Dedekind, *Was sind und was sollen die Zahlen?*, 6th ed., Braunschweig, 1930.

[Dekker 1954]
J. C. E. Dekker, *A theorem on hypersimple sets*, Proc. Amer. Math. Soc. **5** (1954), 791–796.

[DiaDzhSoa 2010]   D. Diamondstone, D. Dzhafarov, and R. I. Soare, Π_1^0-*classes, Peano Arithmetic, randomness, and computable domination*, Notre Dame J. Form. Log. **51** (2010), 127–159, 50th Anniversary Issue.

[Downey and Hirschfeldt 2010]
R. G. Downey and D. R. Hirschfeldt, *Algorithmic randomness and complexity*, Theory and Applications of Computability, Springer-Verlag, New York, 2010.

[Enderton 2011]
H. B. Enderton, *Computability theory: An introduction to recursion theory*, Elsevier, Amsterdam, 2011.

[Epstein 1975]
R. L. Epstein, *Minimal degrees of unsolvability and the full approximation construction*, vol. 3, Memoirs of the American Mathematical Society, no. 162, American Mathematical Society, Providence, RI, 1975.

[Epstein 1979]
R. L. Epstein, *Degrees of unsolvability: Structure and theory*, Lecture Notes in Mathematics, no. 759, Springer-Verlag, Berlin, Heidelberg, New York, 1979.

[Feferman 2007]
S. Feferman, *Turing's Thesis*, Notices Amer. Math. Soc. **53** (2007), no. 10, 1200–1206.

[Friedberg 1957]
R. M. Friedberg, *Two recursively enumerable sets of incomparable degrees of unsolvability*, Proc. Natl. Acad. Sci. USA **43** (1957), 236–238.

[Friedberg 1957b]
R. M. Friedberg, *The fine structure of degrees of unsolvability of recursively enumerable sets*, Summaries of Cornell University Summer Institute for Symbolic Logic, Communications Research Division, Institute for Defense Analyses, Princeton, NJ, 1957, pp. 404–406.

[Friedberg 1957c]
    R. M. Friedberg, *A criterion for completeness of degrees of unsolvability*, J. Symbolic Logic **22** (1957), 159–160.

[Friedberg 1958]
    R. M. Friedberg, *Three theorems on recursive enumeration: I. Decomposition, II. Maximal set, III. Enumeration without duplication*, J. Symbolic Logic **23** (1958), 309—316.

[Friedberg-Rogers 1959]
    R. M. Friedberg and H. Rogers, Jr., *Reducibility and completeness for sets of integers*, Z. Math. Logik **5** (1959), 117–125.

[Gale-Stewart 1953]
    D. Gale and F.M. Stewart, *Infinite games with perfect information*, Contributions to the Theory of Games, Annals of Mathematics Studies, 28, vol. 2, Princeton University Press, Princeton, NJ, 1953, pp. 245–266.

[Gandy 1980]
    R. Gandy, *Church's thesis and principles for mechanisms*, The Kleene Symposium (J. Barwise, J. J. Keisler, and K. Kunen, eds.), North-Holland, Amsterdam, 1980, pp. 123–148.

[Gandy 1988]
    R. Gandy, *The confluence of ideas in 1936*, in [Herken 1988], pp. 55–111.

[Gödel 1931]
    K. Gödel, *Über formal unentscheidbare Sätze der Principia Mathematica und verwandter Systeme. I*, Monatsh. Math. und Phys. **38** (1931), 173–198 (English translation in [Davis 1965], pp. 4–38; in van Heijenoort, [1967], pp. 592–616; and in [Gödel 1986], pp. 145–195).

[Gödel 1934]
    K. Gödel, *On undecidable propositions of formal mathematical systems*, notes by S. C. Kleene and J. B. Rosser on lectures at the Institute for Advanced Study, Princeton, New Jersey, 1934, 30 pp., reprinted in [Davis 1965, pp. 39–74] and in [Gödel 1986, pp. 346–371].

[Gödel 1936]
    K. Gödel, *On the length of proofs*, in [Gödel 1986], pp. 397–399; reprinted in [Davis 1965], pp. 82–83, with a *Remark* added in proof [of the original German publication].

[Gödel 193?]
    K. Gödel, *Undecidable Diophantine propositions*, in [Gödel 1995], pp. 164–174.

[Gödel 1946]
    K. Gödel, *Remarks before the Princeton Bicentennial Conference of Problems in Mathematics, 1946*, reprinted in [Davis 1965], pp. 84–87, and in [Gödel 1990], pp. 150–153.

[Gödel, 1951]
    K. Gödel, *Some basic theorems on the foundations of mathematics and their implications*, [Gödel 1995], pp. 304–323. (This was the Gibbs Lecture delivered by Gödel on December 26, 1951 to the American Mathematical Society).

[Gödel 1964]
  K. Gödel, Postscriptum to [Gödel 1931], written in 1946, reprinted in
  [Davis 1965], pp. 71–73.

[Gödel 1986]
  K. Gödel, *Publications 1929–1936*, Kurt Gödel Collected works (S. Feferman et al., eds.), vol. I, Oxford University Press, Oxford, 1986.

[Gödel 1990]
  K. Gödel, *Publications 1938–1974*, Kurt Gödel Collected works (S. Feferman et al., eds.), vol. II, Oxford University Press, Oxford, 1990.

[Gödel 1995]
  K. Gödel, *Unpublished essays and lectures*, Kurt Gödel Collected works
  (S. Feferman et al., eds.), vol. III, Oxford University Press, Oxford, 1995.

[Herken 1988]
  R. Herken (ed.), *The universal Turing machine: A half-century survey*,
  Oxford University Press, 1988.

[Hermes 1969]
  H. Hermes, *Enumerability, decidability, computability: An introduction
  to the theory of recursive functions*, 2nd revised ed., Springer-Verlag,
  Berlin, Heidelberg, New York, 1969. (This is an English translation
  of Aufzählbarkeit, Entscheidbarkeit, Berechenbarkeit, Grundlehren der
  mathematischen Wissenschaften, Band 109), Springer-Verlag, 1965.)

[Hilbert 1904]
  D. Hilbert, *Über die Grundlagen der Logik und der Arithmetik*, Verhand-
  lungen des Dritten Internationalen Mathematiker-Kongresses in Heidelberg
  vom 8. bis 13. August 1904, Teubner, Leipzig, 1905, reprinted in van
  Heijenoort 1967, pp. 129–138, pp. 174–185.

[Hilbert 1926]
  D. Hilbert, *Über das Unendliche*, Math. Ann. **95** (1926), no. 1, 161–190
  (English translation in van Heijenoort 1967, pp. 367–392).

[Hilbert and Ackermann 1928]
  D. Hilbert and W. Ackermann, *Grundzüge der theoretischen Logik*,
  Springer-Verlag, Berlin, 1928 (English translation of 1938 edition, Chelsea,
  New York, 1950).

[Hilbert and Bernays 1934]
  D. Hilbert and P. Bernays, *Grundlagen der Mathematik*, Springer, Berlin,
  I (1934), II (1939); Second ed., I (1968), II (1970).

[Hodges 1983]
  A. Hodges, *Alan Turing: The enigma*, Burnett Books and Hutchinson, London, Simon and Schuster, New York, 1983, new edition, Vintage, London,
  1992.

[Hopcroft and Ullman, 1979]
  J. E. Hopcroft and J. D. Ullman, *Introduction to automata, languages and
  computation*, Addison-Wesley, 1979.

[Jockusch 1972]
  C. G. Jockusch, Jr., *Degrees in which the recursive sets are uniformly
  recursive*, Canad. J. Math. **24** (1972), 1092–1099.

[Jockusch 1980] C. G. Jockusch, Jr., *Degrees of generic sets*, Recursion theory: Its generalizations and applications (F. R. Drake and S. S. Wainer, eds.), Cambridge University Press, 1980, pp. 110–139.

[Jockusch and Soare 1971]
    C. G. Jockusch and R. I. Soare, *A minimal pair of* Π_1^0 *classes*, J. Symbolic Logic **36** (1971), 66–78.

[Jockusch and Soare 1972]
    C. G. Jockusch and R. I. Soare, *Degrees of members of* Π_1^0 *classes*, Pacific J. Math. **40** (1972), 605–616.

[Jockusch and Soare 1972b]
    C. G. Jockusch and R. I. Soare, Π_1^0 *classes and degrees of theories*, Trans. Amer. Math. Soc. **173** (1972), 33–56.

[Kleene 1936]
    S. C. Kleene, *General recursive functions of natural numbers*, Math. Ann. **112** (1936), 727–742.

[Kleene 1936b]
    S. C. Kleene, λ*-Definability and recursiveness*, Duke Math. J. **2** (1936), 340–353.

[Kleene 1938]
    S. C. Kleene, *On notation for ordinal numbers*, J. Symbolic Logic **3** (1938), 150–155.

[Kleene 1943]
    S. C. Kleene, *Recursive predicates and quantifiers*, Trans. Amer. Math. Soc. **53** (1943), 41–73.

[Kleene 1952]
    S. C. Kleene, *Introduction to metamathematics*, van Nostrand, New York, 1952, ninth reprint 1988, Wolters-Noordhoff Publishing Co., Groningen and North-Holland, Amsterdam.

[Kleene 1967]
    S. C. Kleene, *Mathematical logic*, John Wiley and Sons, Inc., New York, London, Sydney, 1967.

[Kleene 1981]
    S. C. Kleene, *Origins of recursive function theory*, Ann. Hist. Comput. **3** (1981), 52–67.

[Kleene 1981b]
    S. C. Kleene, *The theory of recursive functions, approaching its centennial*, Bull. Amer. Math. Soc. (N.S.) **5** (1981), no. 1, 43–61.

[Kleene 1981c]
    S. C. Kleene, *Algorithms in various contexts*, Proceedings of the Symposium on Algorithms in Modern Mathematics and Computer Science (dedicated to Al-Khwarizmi) (Urgench, Khorezm Region, Uzbek, SSR, 1979), Lecture Notes in Computer Science, vol. 122, Springer-Verlag, Berlin, Heidelberg and New York, 1981, pp. 355–360.

[Kleene 1987]
    S. C. Kleene, *Reflections on Church's Thesis*, Notre Dame J. Form. Log. **28** (1987), 490–498.

[Kleene 1988]
   S. C. Kleene, *Turing's analysis of computability, and major applications of it*, in [Herken 1988], pp. 17–54.
[Kleene and Post 1954]
   S. C. Kleene and E. L. Post, *The upper semi-lattice of degrees of recursive unsolvability*, Ann. of Math. **59** (1954), 379–407.
[Lachlan 1970]
   A. H. Lachlan, *On some games which are relevant to the theory of recursively enumerable sets*, Ann. of Math. (2) **91** (1970), 291–310.
[Lachlan 1973]
   A. H. Lachlan, *The priority method for the construction of recursively enumerable sets*, in Mathias and Rogers [1973], pp. 299–310.
[Lerman 1983]
   M. Lerman, *Degrees of unsolvability: Local and global theory*, Perspectives in Mathematical Logic, Springer-Verlag, Heidelberg, New York, Tokyo, 1983.
[Martin 1966]
   D. A. Martin, *Classes of recursively enumerable sets and degrees of unsolvability*, Z. Math. Logik Grundlagen Math. **12** (1966), 295–310.
[Martin 1975]
   D. A. Martin, *Borel determinacy*, Ann. of Math. (2) **102** (1975), no. 2, 363–371.
[Mathias and Rogers 1973]
   A. R. D. Mathias and H. Rogers, Jr. (eds.), *Cambridge Summer School in Mathematical Logic, held in Cambridge, England, August 1–21, 1971*, Lecture Notes in Mathematics, no. 337, Berlin, Heidelberg, New York, Springer-Verlag, 1973.
[R. Miller 2001]
   R. Miller, *The Δ_2^0-spectrum of a linear order*, J. Symbolic Logic **66** (2001), 470–486.
[Miller and Martin 1968]
   W. Miller and D. A. Martin, *The degree of hyperimmune sets*, Z. Math. Logik Grundlagen Math. **14** (1968), 159–166.
[Muchnik 1956]
   A. A. Muchnik, *On the unsolvability of the problem of reducibility in the theory of algorithms*, Dokl. Akad. Nauk **108** (1956), 194–197, (Russian).
[Myhill 1955]
   J. Myhill, *Creative sets*, Z. Math. Logik Grundlagen Math. **1** (1955), 97–108.
[Myhill 1956]
   J. Myhill, *The lattice of recursively enumerable sets*, J. Symbolic Logic **21** (1956), 220 (abstract).
[Nies 2009]
   A. Nies, *Computability and randomness*, Oxford Logic Guides 51, Oxford University Press, Oxford, UK, 2009.
[Odifreddi 1989]
   P. Odifreddi, *Classical Recursion Theory: The theory of functions and sets*

of natural numbers, Studies in Logic and the Foundations of Mathematics 125, vol. I, North-Holland, Amsterdam, 1989.

[Odifreddi 1999]
P. Odifreddi, *Classical Recursion Theory: The theory of functions and sets of natural numbers*, Studies in Logic and the Foundations of Mathematics 143, vol. II, North-Holland, Amsterdam, 1999.

[Owings 1973]
J. C. Owings, Jr., *Diagonalization and the recursion theorem*, Notre Dame J. Form. Log. **14** (1973), no. 1, 95–99.

[Peter 1934]
R. Péter, *Über den Zusammenhang der verschiedenen Begriffe der rekursiven Funktion*, Math. Ann. **110** (1934), 612–632.

[Peter 1951]
R. Péter, *Rekursive Funktionen*, Akadémiai Kiadó (Akademischer Verlag), Budapest, 1951, 206 pp. *Recursive Functions*, third revised edition, Academic Press, New York, 1967, 300 pp.

[Post 1936]
E. L. Post, *Finite combinatory processes–formulation I*, J. Symbolic Logic **1** (1936), 103–105, reprinted in [Davis 1965], pp. 288–291.

[Post 1941]
E. L. Post, *Absolutely unsolvable problems and relatively undecidable propositions: Account of an anticipation* (Submitted for publication in 1941, never published) Printed in [Davis 1965], pp. 338–433.

[Post 1943]
E. L. Post, *Formal reductions of the general combinatorial decision problem*, Amer. J. Math. **65** (1943), 197–215.

[Post 1944]
E. L. Post, *Recursively enumerable sets of positive integers and their decision problems*, Bull. Amer. Math. Soc. **50** (1944), no. 5, 284–316, reprinted in [Davis 1965], pp. 304–337.

[Post 1948]
E. L. Post, *Degrees of recursive unsolvability: preliminary report (abstract)*, Bull. Amer. Math. Soc. **54** (1948), 641–642.

[Putnam 1965]
H. Putnam, *Trial and error predicates and the solution to a problem of Mostowski*, J. Symbolic Logic **30** (1965), 49–57.

[Rice 1953]
H. G. Rice, *Classes of enumerable sets and their decision problems*, Trans. Amer. Math. Soc. **74** (1953), 358–366.

[Rice 1956]
H. G. Rice, *On completely recursively enumerable classes and their key arrays*, J. of Symbolic Logic **21** (1956), no. 3, 304–308.

[Rogers 1967]
H. Rogers, Jr., *Theory of recursive functions and effective computability*, McGraw-Hill, New York, 1967.

[Sacks 1963]
G. E. Sacks, *Recursive enumerability and the jump operator*, Trans. Amer. Math. Soc. **108** (1963), 223–239.

[Sacks 1963b]
G. E. Sacks, *On the degrees less than 0'*, Ann. of Math. (2) **77** (1963), no. 2, 211–231.

[Sacks 1963c]
G. E. Sacks, *Degrees of unsolvability*, Annals of Mathematics Studies, no. 55, Princeton University Press, Princeton, NJ, 1963 (see revised edition, 1966).

[Scott 1962]
Dana S. Scott, *Algebras of sets binumerable in complete extensions of arithmetic*, Proceedings of the Symposium on Pure and Applied Mathematics, vol. V, American Mathematical Society, Providence, RI, 1962, pp. 117–121.

[Shoenfield 1959]
J. R. Shoenfield, *On degrees of unsolvability*, Ann. of Math. (2) **69** (1959), 644–653.

[Shoenfield 1960]
J. R. Shoenfield, *Degrees of models*, J. Symbolic Logic **25** (1960), no. 3, 233–237.

[Shoenfield 1960b]
J. R. Shoenfield, *An uncountable set of incomparable degrees*, Proc. Amer. Math. Soc. **11** (1960), no. 1, 61–62.

[Shoenfield 1966]
J. R. Shoenfield, *A theorem on minimal degrees*, J. Symbolic Logic **31** (1966), no. 4, 539–544.

[Shoenfield 1971]
J. R. Shoenfield, *Degrees of unsolvability*, North-Holland Mathematics Studies, no. 2, North-Holland, Amsterdam, 1971.

[Shoenfield 1993]
J. R. Shoenfield, *Recursion theory*, Lecture Notes in Logic 1, Springer-Verlag, Heidelberg, New York, 1993.

[Sieg 1994]
W. Sieg, *Mechanical procedures and mathematical experience*, Mathematics and Mind (A. George, ed.), Oxford University Press, 1994, pp. 71–117.

[Sieg 1997]
W. Sieg, *Step by recursive step: Church's analysis of effective calculability*, Bull. Symbolic Logic **3** (1997), no. 2, 154–180.

[Sieg 2006]
W. Sieg, *Gödel on computability*, Philosophia Mathematica **14** (2006), 189–207.

[Sieg 2008]
W. Sieg, *Church without dogma-axioms for computability*, New Computational Paradigms (B. Löwe, A. Sorbi, and B. Cooper, eds.), Springer-Verlag, 2008, pp. 139–152.

[Sieg 2009]
W. Sieg, *On computability*, Handbook of the Philosophy of Mathematics (Andrew D. Irvine, ed.), Elsevier, 2009, pp. 535–630.

[Skolem 1923]
T. Skolem, *Begründung der elementaren Arithmetik durch die rekurrierende Denkweise ohne Anwendung scheinbarer Veränderlichen mit unendlichem Ausdehnungsbereich*, no. 6, Skrifter utgit av Videnskapsselskapet i Kristiania, I. Mathematisk-Naturvidenskabelig Klasse, 1923, 38 pp. (English translation in van Heijenoort, 1967, pp. 302–333.)

[Soare 1987]
R. I. Soare, *Recursively enumerable sets and degrees: A study of computable functions and computably generated sets*, Perspectives in Mathematical Logic, Springer, Heidelberg, 1987.

[Soare 1996]
R. I. Soare, *Computability and recursion*, Bull. Symbolic Logic **2** (1996), 284–321.

[Soare 1999]
R. I. Soare, *The history and concept of computability*, Handbook of Computability Theory (E. R. Griffor, ed.), North-Holland, Amsterdam, 1999, pp. 3–36.

[Soare 2004]
R. I. Soare, *Computability theory and differential geometry*, Bull. Symbolic Logic **10** (2004), no. 4, 457–486.

[Soare 2007]
R. I. Soare, *Computability and incomputability, computation and logic in the real world*, Computation and Logic in the Real World, Third Conference on Computability in Europe, Proceedings, CiE 2007, Siena, Italy, June 2007 (S. B. Cooper, B. Löwe, and A. Sorbi, eds.), Lecture Notes in Computer Science 4497, Springer, Berlin, Heidelberg, 2007.

[Soare 2009]
R. I. Soare, *Turing oracle machines, online computing, and three displacements in computability theory*, Ann. Pure Appl. Logic **160** (2009), 368–399.

[Soare 2012]
R. I. Soare, *An interview with Robert Soare: Reflections on Alan Turing*, XRDS, Crossroads, the ACM magazine for students **18** (2012), no. 3.

[Soare 2012b]
R. I. Soare, *Formalism and intuition in computability theory*, Phil. Trans. R. Soc. Lond. Ser. A **370** (2012), 3277–3304.

[Soare 2013]
R. I. Soare, *Turing and the art of classical computability*, Alan Turing—His Work and Impact (Barry Cooper and Jan van Leeuwen, eds.), Elsevier, 2013, pp. 65–70.

[Soare 2013b]
R. I. Soare, *Interactive computing and relativized computability*, Com-

putability: Gödel, Church, Turing, and Beyond (Jack Copeland, Carl Posy, and Oron Shagrir, eds.), MIT Press, 2013.

[Soare 2013c]
R. I. Soare, *Turing and the discovery of computability*, Turing's legacy: Developments from Turing's ideas in logic (Rodney Downey, ed.), Lecture Notes in Logic 42, Association for Symbolic Logic and Cambridge University Press, 2014.

[Soare 2015]
R. I. Soare, *Why Turing's Thesis is not a thesis*, Turing Centenary Volume (Thomas Strahm and Giovanni Sommaruga, eds.), Birkhäuser/Springer, Basel, to appear.

[Spector 1956]
Clifford Spector, *On degrees of recursive unsolvability*, Ann. of Math. (2) **64** (1956), no. 3, 581–592.

[Turing 1936]
A. M. Turing, *On computable numbers with an application to the Entscheidungsproblem*, Proc. London Math. Soc. (2) **42** (1936), 230–265, reprinted in [Davis 1965], pp. 115–153.

[Turing 1937]
A. M. Turing, *A correction*, Proc. London Math. Soc. (2) **43** (1937), 544–546.

[Turing 1937b]
A. M. Turing, *Computability and λ-definability*, J. Symbolic Logic **2** (1937), 153–163.

[Turing 1939]
A. M. Turing, *Systems of logic based on ordinals*, Proc. London Math. Soc. (2) **45** (1939), 161–228; reprinted in [Davis 1965], 154–222.

[Turing 1950]
A. M. Turing, *Computing machinery and intelligence*, Mind **59** (1950), 433–460.

[Turing 1950b]
A. M. Turing, *The word problem in semi-groups with cancellation*, Ann. of Math. **52** (1950), no. 2, 491–505.

[Turing 1954]
A. M. Turing, *Solvable and unsolvable problems*, Science News **31** (1954), 7–23.

[van Heijenoort 1967]
J. van Heijenoort (ed.), *From Frege to Gödel, A sourcebook in mathematical logic, 1879–1931*, Harvard University Press, Cambridge, MA, 1967.

[Wang 1981]
H. Wang, *Some facts about Kurt Gödel*, J. Symbolic Logic **46** (1981), no. 3, 653–659.

[Wang 1987]
H. Wang, *Reflections on Kurt Gödel*, MIT Press, Cambridge, MA, 1987.

Index

Printed in the United States
By Bookmasters